History of the Italian
Agricultural Landscape

. . . . .

GIOVANNI AGNELLI FOUNDATION SERIES
IN ITALIAN HISTORY

In recent decades Italian historians and social scientists have produced a broad range of profound and subtle works that merit wide reception among an English-speaking readership. The Giovanni Agnelli Foundation Series in Italian History was formed to ensure the wide distribution of these works in English translation. Together, the Giovanni Agnelli Foundation and Princeton University Press will publish a number of key works that illuminate Italian history and national development and serve as some of the best specimens of a rich historiographical and interpretive tradition. In so doing, the series will promote a more realistic, complete, and balanced understanding of Italian history, and hence modern Italian culture and society, in the English-speaking world.

Editorial Advisory Committee
Giovanni Agnelli Foundation Series in Italian History

Gabriele de Rosa, Professor of Modern History, University of Rome
Giuseppe Galasso, Professor of Medieval and Modern History,
University of Naples
Adrian Lyttleton, Professor of History, The Johns Hopkins
University, Bologna Center
Charles S. Maier, Professor of History, Harvard University
Massimo Salvadori, Professor of History, University of Turin

Ideological Profile of Twentieth-Century Italy, by Norberto Bobbio;
translated by Lydia G. Cochrane

Italian Foreign Policy: The Statecraft of the Founders, by Federico Chabod;
translated by William McCuaig

History of the Italian Agricultural Landscape, by Emilio Sereni;
translated, with an introduction, by R. Burr Litchfield

History of the Italian Agricultural Landscape

....

Emilio Sereni

TRANSLATED,
WITH AN INTRODUCTION,
BY
R. Burr Litchfield

GIOVANNI AGNELLI FOUNDATION
SERIES IN ITALIAN HISTORY
PRINCETON UNIVERSITY PRESS
PRINCETON, NEW JERSEY

Copyright © 1997 by Princeton University Press
Translated from *Storia del paesaggio agrario italiano*.
First published in 1961. This translation
is based on the Biblioteca Universale Laterza
edition (Rome-Bari: Guis. Laterza & Figli, 1989).
Published by Princeton University Press, 41 William Street,
Princeton, New Jersey 08540
In the United Kingdom: Princeton University Press,
Chichester, West Sussex

All Rights Reserved

Library of Congress Cataloging-in-Publication Data
Sereni, Emilio.
[Storia del paesaggio agrario italiano. English]
History of the Italian agricultural landscape / Emilio Sereni ;
translated with an introduction by R. Burr Litchfield.
p. cm. — (Giovanni Agnelli Foundation series in Italian history)
Includes bibliographical references and index.
ISBN 0-691-01216-4 (cloth : alk. paper). —
ISBN 0-691-01215-6 (pbk · alk. paper)
1. Agriculture—Italy—History. I. Title. II. Series.
S469.I8S413 1997
630′.945—dc21   96-47543

This book has been composed in Galliard

Published by arrangement with the
Edizioni della Fondazione Giovanni Agnelli.

Princeton University Press books are printed on
acid-free paper and meet the guidelines for permanence
and durability of the Committee on Production
Guidelines for Book Longevity of the
Council on Library Resources

Printed in the United States of America
by Princeton Academic Press

1 3 5 7 9 10 8 6 4 2

1 3 5 7 9 10 8 6 4 2
(Pbk)

... now in these things, a large part of what we call natural, is not; it is even quite artificial: that is to say, the tilled fields, the trees and other domesticated plants that are placed in order, the rivers kept within bounds and directed toward a certain course, and such, lack both the state and the appearance that they would have in nature. In this way the appearance of any land inhabited for a few generations by civilized people, not to say in cities and other places where men live together, is artificial, and much different from what it would be in nature.

—Leopardi, *Elogio degli uccelli*

CONTENTS

• • • • •

LIST OF PLATES AND FIGURES	xiii
FOREWORD TO THE SERIES BY CHARLES S. MAIER	xix
INTRODUCTION TO THE ENGLISH TRANSLATION	xxvii
PREFACE	3

I. Natural Landscape and Agricultural Landscape — 15

II. Ancient Italy — 19

1. *The Agricultural System of Fallow and the Landscape of Greek Colonization* — 21
2. *Greek Colonization and the Agricultural Landscape of the Mediterranean Garden in Sicily* — 22
3. *The Etruscan Urban Expansion, the Gallic Invasion, and the Landscape of the* Piantata *in Central and Northern Italy* — 24
4. *The Landscape Plan of the Roman Conquest* — 27
5. *Roads and Aqueducts in the Roman Agricultural Landscape* — 29
6. *The Roman Form of the Italian Agricultural Landscape* — 32
7. *The Lands of Common Pasturage, and the Agricultural Landscape of Pasturage in Ancient Rome* — 35
8. *The Rustic Villa and the Landscape of the Plantation* — 38
9. *The "Bel Paesaggio" of the* Villa Urbana — 40
10. *The Sylvan-Pastoral Landscape of the Saltus* — 42
11. *The System of Temporary Clearings, and the Deterioration of the Agricultural Landscape under the Late Empire* — 44
12. *The Barbarian Invasions and the Ruins of the Italian Agricultural Landscape* — 47

III. The Early Middle Ages and the Feudal Era — 51

13. *The Disaggregation of the Agricultural Landscape and Pictorial Landscape in Byzantine Italy* — 53

14. Castra, Curtes, Massae: *Centers of Reorganization of the Agricultural Landscape in Lombard and Byzantine Italy*   55

15. *The Landscape of the Wildwood, and Hunting in the Early Middle Ages*   58

16. *The Cultivation of Lesser Cereals, and the Medieval Agricultural Landscape of Open Fields*   60

17. *The Hilltop Town in the Pastoral-Agricultural Landscape of the Italian Middle Ages*   62

18. *The Agricultural Landscape of Closed Fields of the Italian Medieval City*   65

19. *The Medieval Agricultural Landscape of Closed Fields: The Low-Growing Vineyard*   68

20. *The Medieval Agricultural Landscape of Closed Fields: Kitchen Gardens*   70

21. *The Arab Invasions, and the Medieval Landscape of the "Mediterranean Garden"*   72

22. *The Castle in the Agricultural Landscape of Feudal Italy*   74

23. *The Revival of Plantations of Trees in the Agricultural Landscape of Feudal Italy*   78

24. *The Age of Improvement and the Great Clearings and Reorganization of the Agricultural Landscape in the Eleventh through Thirteenth Centuries*   80

25. *The Landscape of Large-Scale Pasturage in the Feudal Era*   84

IV. THE AGE OF THE COMMUNES   87

26. *Feudal Strongholds and Villas in the Landscape of the Early Communal Age*   89

27. *Individual Clearings, Plantations, and Settlements in the Agricultural Landscape of the Early Communal Period*   92

28. *Systematization in the Plain, and the Planting of Trees Festooned with Vines*   95

29. *Individual Tillage, and Extensive Systematization on the Hillsides*   98

30. *The Suburban Agricultural Landscape*   101

31. *The Landscape of the Countryside*   104

32. *The Pastoral Landscape of the Communal Period*   107

CONTENTS

33. *The Landscape of the Woods and Hunting* — 110
34. *The Revival of Cultivation of Grain, and the Landscape of Closed Fields in the Communal Period and the Renaissance* — 113

V. THE AGE OF THE RENAISSANCE — 117

35. *The Origins of the Contemporary Landscape: Enclosures, Systematization* a Rittochino *on Hillsides, and the Landscape of Irregular Fields* a Pìgola *in the Early Renaissance* — 119
36. *The Landscape of Enclosed Fields in the Plain and Systematization in* Porche — 122
37. *Toward a Redressed Balance of Forage: The Landscape of Enclosed Pastures and Meadows* — 125
38. *Improvements and Irrigation in the Renaissance Agricultural Landscape* — 129
39. *The Irrigated Meadows of Lombardy and the Po Valley in the Age of the Renaissance* — 133
40. *The Origins of the Contemporary Landscape: The* Piantata *of the Po Valley* — 136
41. *The Agricultural "Bel Paesaggio" of the Italian Renaissance* — 140
42. *The "Bel Paesaggio" in Tuscany* — 143
43. *The "Bel Paesaggio" of the Veneto* — 145
44. *The "Bel Paessagio" of the Italian-Style Villa* — 147
45. *An Agricultural Panorama of the Renaissance: Pastoral Landscapes* — 150
46. *The Landscape of Clearings in Hills and Mountains* — 154
47. *The Deterioration of the Landscape of Hills and Mountains in the Renaissance Period* — 157
48. *Systematization in the Hills and Mountains during the Italian Renaissance* — 160
49. *The Origins of the Contemporary Landscape: Systematization in Irregular Banks* (a Ciglioni) *on Hillsides in the Age of the Renaissance* — 163
50. *The Origins of the Contemporary Landscape: Systematization in the Mountains through Lunettes and Grading* — 165
51. *Systematization in the Hills in Terraces, and the "Works of Construction" of the Renaissance Period* — 169

52. *The Origins of the Contemporary Landscape: Road Building, and the Systematization of Hills Plowed "Crosswise" (a Cavalcapoggio) and "Roundabout" (a Girapoggio)*   171

53. *Plantations in the Hills in Central and Northern Italy, and the Landscape of Irregular Fields in the Late Renaissance*   174

54. *The Mediterranean Landscape of Preserves, and the "Mediterranean Garden"*   177

55. *The Era of the Great Geographical Discoveries: The Spread of Indian Corn, and the Landscape of Agricultural Systems with Continuous Rotation*   180

VI. THE AGE OF THE COUNTER-REFORMATION AND FOREIGN DOMINATION   185

56. *Marshlands and Improvement between the Renaissance and Counter-Reformation: The Landscape of Marshes, Wetlands, and Rice Fields*   187

57. *Agricultural Systems of Temporary Clearings, and the New Extension of Pastoral Landscapes between the Fifteenth and Eighteenth Centuries*   189

58. *The New Feudalism and the Landscape of the Italian Villa of the Renaissance and Counter-Reformation*   194

59. *Classic and Romantic Landscape in Italian Reality and Art of the Seventeenth Century*   199

60. *Open Fields, Farms, and Preserves in the Italian Agricultural Landscape of the Seventeenth and Eighteenth Centuries*   202

61. *The Landscape of Industrial Crops and Agricultural Systems of Continuous Rotation in the Seventeenth and Eighteenth Centuries*   206

62. *Origins of the Contemporary Landscape: The Southern Landscape of the "Mediterranean Garden"*   210

63. *The Alberata of Tuscany, Umbria, and the Marche, and Systematization of Fields with Trees in the Seventeenth and Eighteenth Centuries*   213

64. *The Piantata of the Po Valley in the Seventeenth and Eighteenth Centuries*   216

65. *Ecclesiastical Mortmain, and the Disordered Italian Landscape of the Age of Enlightenment*   221

CONTENTS

VII. THE AGE OF ENLIGHTENED DESPOTISM AND REFORMS — 225

66. *The Landscape of the Eighteenth-Century Villa, and the Italian Mode of Development of Capitalism in the Countryside* — 227

67. *The Landscape of Farms in the Po Valley, and the Crisis of Sharecropping in the Second Half of the Eighteenth Century* — 232

68. *The Age of Reforms in Italy, and the Agricultural Landscape of Closed Fields in the Second Half of the Eighteenth Century* — 236

69. *Capitalism in the Countryside: Deforestation, Clearings, and Erosion of the Mountainous Landscape in the Age of Reforms* — 241

70. *The Landscape of Landfills:* Colmate di Piano *in Tuscany during the Second Half of the Eighteenth Century* — 246

71. *The Origins of the Contemporary Landscape: Systematization in the Hills in Banks and Terraces* — 251

72. *Hillsides Plowed* a Tagliapoggio *in the Second Half of the Eighteenth Century* — 256

VIII. THE AGE OF THE RISORGIMENTO — 259

73. *The Po Valley Landscape of Irrigated Meadows, and Cultivation with Continuous Rotation in the Eighteenth and Nineteenth Centuries* — 261

74. *The Landscape of the Po Valley: From the Sharecropping Farm to the Great Capitalistic Rented Holding* — 267

75. *Landfills in the Hills, and Arrangements* a Prode *and* a Spina *in Tuscany in the Age of the Risorgimento* — 273

76. *The Overthrow of Feudalism in the South, and the Agricultural Landscape of Open Fields in the Age of the Risorgimento* — 280

IX. ITALIAN UNIFICATION — 289

77. *The Railroads in the Italian Agricultural Landscape in the Age of the Risorgimento and Italian Unification* — 291

78. *The* Piantata *in the Dryer Zones of the Po Valley in the Age of the Risorgimento and Italian Unification* — 295

79. *The Agricultural Landscape of the Irrigated Zones of the Po Valley, and Rice Fields* — 304

80. *The Alberata of Tuscany, Umbria, and the Marche in the Risorgimento and Italian Unification* — 314

81. *The Landscapes of the South in the Risorgimento and Italian Unification* — 321

82. *The Landscape of Campi a Pìgola: Irregular Fields in United Italy* — 331

83. *Improvements in the Po Valley, and the Agricultural Landscape of the Larga in United Italy* — 336

X. An Agricultural Panorama of Contemporary Italy — 347

84. *The Agricultural Landscapes of Contemporary Italy* — 349

GLOSSARY — 381

INDEX — 385

PLATES AND FIGURES

• • • • •

PLATES

Plate 1. Vines on trees in the frieze of the Casa dei Vettii at Pompei	26
Plate 2. Roman *villa* and plantations in a mosaic from the Museo del Bardo	39
Plate 3. The "bel paesaggio" of country villa in a wall painting from the Casa del Centenario at Pompei	41
Plate 4. The landscape of the *saltus* in a mosaic from the Villa Adriana	43
Plate 5. A pastoral landscape from the late imperial period from a mosaic in the central nave of Santa Maria Maggiore in Rome	45
Plate 6. The ruins of the Italian agricultural landscape: the Roman aqueduct at Acqui	48
Plate 7. The deterioration of the agricultural landscape in Byzantine Italy: landscape and decoration in the apse of Sant' Appolinare in Classe in Ravenna	52
Plate 8. The *castrum*, fortified villa, a center of reorganization of the agricultural landscape of Lombard and Byzantine Italy, in a mosaic of the Mosque of the Ommiadi, at Damascus	56
Plate 9. The landscape of the wildwood and of hunting in Italy under the barbarian invasions, from a bas-relief in the Duomo of Civitacastellana	59
Plate 10. The system of open fields in medieval Italy: a pig in a field of sorgum, from a miniature in the *Theatrum sanitatis*	61
Plate 11. The hilltop town of medieval Italy in *Saint Francis Giving Away His Cloak* by Giotto	64
Plate 12. The agricultural landscape of closed fields within urban walls, from a sixteenth-century map of Bologna	66
Plate 13. A plant recently introduced—spinach—in the *hortus conclusus* of the medieval *Theatrum sanitatis*	71
Plate 14. The Arab expansion and new plants: citrus fruits in the medieval landscape of the "Mediterranean garden," in a miniature from the *Theatrum sanitatis*	73
Plate 15. The dominant position of the castle in the landscape of feudal Italy, from the *History of the Blessed Agostino* by Simone Martini	76
Plate 16. The resumption of plantations and the persistent disaggregation of the agricultural landscape in feudal and communal Italy, in the *Garden of Olives* from the nave of San Marco	79
Plate 17. The landscape of large-scale pasturage in feudal Italy, in the *Return of San Gioacchino* (school of Cimabue) in the Museo Civico of Pisa	85
Plate 18. The "thirty villas" and the "twelve castles" around the "small city" of the early communal period, in *San Cosmo and San Damiano* by Fra Angelico	90

LIST OF PLATES AND FIGURES

Plate 19. Plantations of the communal period in the *Garden of Olives* by Duccio di Buoninsegna — 93

Plate 20. Planted vines festooned on trees and the pressing of grapes in a miniature from the *Theatrum sanitatis* — 96

Plate 21. Individual clearings and an extensive systematization on a hill in the *Seaside* by Lorenzetti — 100

Plate 22. The suburban agricultural landscape in communal Italy in a detail from the *Buon Governo* by Lorenzetti — 102

Plate 23. Forms of landscape in the countryside of communal Italy from the *Buon Governo* by Lorenzetti — 105

Plate 24. The pastoral landscape of communal Italy in *San Gioacchino and the Shepherds* in the Pinacoteca Vaticana (attributed to Bartolo di Fredi or Andrea di Bartolo) — 108

Plate 25. The woods and hunting in communal Italy from a miniature in the *Theatrum sanitatis* — 111

Plate 26. Fields enclosed with live hedges in the Renaissance landscape, from a detail of the *Flight into Egypt* by Gentile da Fabriano — 115

Plate 27. Hillside systematization and *campi a pigola* in Renaissance Italy, from a detail of the *Adoration of the Magi* by Gentile da Fabriano — 120

Plate 28. The system of closed fields and systematization *a magolato* in a detail from the *Disposition* by Fra Angelico — 123

Plate 29. Live enclosure of meadows and pastures in the Renaissance landscape, from a detail of the *Meditation on the Passion* by Carpaccio — 128

Plate 30. The geometric landscape of improvement and irrigation in a drawing by Leondardo in the Uffizi Gallery — 132

Plate 31. An irrigated meadow of the Lombard Renaissance, in a miniature from *De sphaera* from the Biblioteca Estense in Modena — 134

Plate 32. The landscape of the *piantata* in the Po valley from a sixteenth-century map of the countryside around Mantova — 138

Plate 33. The "bel paesaggio" of the Tuscan Renaissance in a detail from the *Journey of the Magi* by Benozzo Gozzoli — 144

Plate 34. The "bel paesaggio" of the Venetian Renaissance in the *Adoration of the Magi* by Domenico Veneziano — 146

Plate 35. The Medici villa of Cafaggiolo in an anonymous decorative panel in the Palazzo Riccardi in Florence — 148

Plate 36. The integration of the Renaissance pastoral and agricultural landscapes in the *Virgin and Child* by Giovanni Bellini — 151

Plate 37. Clearings and plantations on the hillsides and mountains in a sixteenth-century panorama of Serravalle Veneta — 155

Plate 38. The deforestation of the hillside and mountainous landscape in *Gravinana Attacked by the Militia of Orange* by Vasari — 159

Plate 39. Arrangements in plantations on the hills in *Prayer in the Garden of Olives* by Barna Senese — 161

LIST OF PLATES AND FIGURES xv

Plate 40. Plantations on gradings in a detail from the *Penetent San Girolamo* by the pseudo Pier Francesco Fiorntino 166
Plate 41. Painterly style and graded arrangements (*ciglioni*) in the *Incidents in the Life of Pope San Silvestro* by Battista da Vicenza 167
Plate 42. Works of construction in the form of terraces in a Renaissance topographic landscape of the Gonzaga villa at Castiglione dello Stiviere 170
Plate 43. The road system and hillside arrangements in a detail from the *Martyrdom of San Giacomo* by Mantegna 172
Plate 44. The sixteenth-century landscape of irregular fields at Montepulciano in an anonymous painting in the Palazzo Ricci 175
Plate 45. Preserves and the Mediterranean garden in a Renaissance map of Nardò 178
Plate 46. The spread of new crops: one of the first iconographic representations of Indian corn in the second edition of the cosmographic work of Giovanni Battista Ramusio 182
Plate 47. The landscape of marshes and fens in a painting by Giulio Romano 186
Plate 48. The *Roman campagna* by Poussin 191
Plate 49. The Tuscan villa, in a detail of the *Paradise* by Benozzo Gozzoli 197
Plate 50. The degraded landscape of seventeenth-century Italy in the *Seaside* by Salvator Rosa 201
Plate 51. Industrial crops and agricultural systems of continuous rotation: the harvesting and retting of hemp in a painting by Guercino 208
Plate 52. The belt of plantations of trees around Palermo in a map of the city at the beginning of the eighteenth century 211
Plate 53. The *alberata* of Tuscany, Umbria, and the Marche from an seventeenth-century map of Fermo 215
Plate 54. The extension of the eighteenth-century *piantata* of the Po valley around Mantova from a map in the *Voyage en Italie* of de Beaurain 218
Plate 55. In the dawn of the age of Enlightenment: *Stormy Landscape with Fleeing Friars* by Magnasco 222
Plate 56. A villa and seigneurial enterprise in the mid-eighteenth century, in *The Harvest* by Gaspare van Wittel 229
Plate 57. The farm and the landscape of the *piantata* in Tiepolo's *Rest of the Peasant* 233
Plate 58. The expansion of the landscape of closed fields in an eighteenth-century view of Prato 238
Plate 59. Deforestations, clearings, and erosion of the hillside and mountainous landscape in a popular print of the late eighteenth century: the Calabrian coast of the Straits of Messina deforested, reduced to cultivation, and eroded in an enormous landslide 244

LIST OF PLATES AND FIGURES

Plate 60. Eighteenth-century improvements with *colmate* in the Valdichiana in a map from the *Memorie idraulico-storiche* by Fossombroni — 249
Plate 61. Systematization A: *a rittochino*; B: *a cavalcapoggio*; C: *a tagliapoggio* on a hill with a united two-directional slope; or D: divided into *ciglioni* (banks) from a plate in the *Dizionario di agricoltura* by Gera — 257
Plate 62. The landscape of a rice field in the region of Vercelli in a painting by Enzo Gazzone — 270
Plate 63. The operation of a hillside landfill (*colmata di monte*) in an illustration from the *Nuovo dizionario di agricoltura* by Gera — 278
Plate 64. The *Roman campagna* by Enrico Coleman — 285
Plate 65. The landscape of the Maremma in an engraving by Fattori — 286
Plate 66. The Imperial Regia Strada Ferrata Prinvilegiata Ferdinandea Lombardo-Veneta (Imperial Royal Lombard-Venetian Railroad), from a contemporary print — 292
Plate 67. The landscape of the atypical *piantata* in *Countryside at Monte Orobio* by Cesare Breveglieri — 299
Plate 68. An Emilian rice field in *Mondine 1940* by Aldo Borgonzoni — 307
Plate 69. The extension of the *alberata* of Tuscany, Umbria, and the Marche at Magione, in the zone of plain and hills near Lake Trasimeno, from a cadastral map of the early nineteenth century — 319
Plate 70. *Flocks in the Roman Countryside* by G. A. Sartorio — 326
Plate 71. A landscape of the Mediterranean garden with works of construction in *Caprile* by Rosina Viva — 326
Plate 72. Telemaco Signorini, *The Olive Grove* — 327
Plate 73. Filippo Palizzi, *Country Road* — 328
Plate 74. The landscape of the *larga* in the early twentieth century, in a map of the Gallare holding of the Istituto di fondi rustici — 342
Plate 75. Ernesto Treccani: *Melissa* — 350
Plate 76. *Occupation of Uncultivated Lands in Sicily* by Renato Guttuso — 351
Plate 77. The *piantata* of the Po in a painting by Aldo Borgonzoni — 360
Plate 78. *Branch Cutter* by Aldo Borgonzoni — 363
Plate 79. *Umbrian Landscape* by A. Salietti — 369
Plate 80. *Anacapri* by Renato Guttuso — 378

Figures

Figure 1. The lands of the temple of Athena Polias in Eraclea di Lucania, in the fourth century B.C — 20
Figure 2. The landscape of the Mediterranean garden in the *Tavola di Alesa*, in Sicily, in the first century B.C — 23
Figure 3. Old and new assignments in the *centuriatio* of Minturno, from a codex of ancient *gromatici* — 28
Figure 4. Assignments of lots and undivided lands in the *centuriatio* of Suessa — 28

LIST OF PLATES AND FIGURES   xvii

Figure 5. The Roman road system of the imperial age in the Italian
            agricultural landscape   30
Figure 6. The imprint of the Roman *limitatio* in the territory of Cesena   33
Figure 7. Lands assigned to *pascua publica* in the colony of Iulia
            Constantia   36
Figure 8. The *silva* and the *pascua publica* in a map made by geodesists   36
Figure 9. The lands of the *compascuo*, open to the use of nearby
            proprietors, in a map made by geodesists   36
Figure 10. A vine raised low-down in the system of closed fields of
            the medieval city and suburb, from an illustration of the *Martirologio
            di Adone* of the twelfth century   69
Figure 11. The traces of a plan of colonization of the twelfth century
            in the contemporary agricultural landscape of Villafranca Veronese   82
Figure 12. A farm and its clearings in the eighteenth-century landscape
            of Tuscany, from a sketch preserved in the Archive of Castel del
            Piano (Province of Grosseto)   205
Figure 13. Systematization *a ciglioni* (in banks) in the *Tableau de
            l'agriculture toscane* by Sismondi   255
Figure 14. The geometric schemata of irrigation channels and fields
            in the landscape of the irrigated *cascina* (from the illustrations in
            Note relative all'agricoltura milanese of 1784)   265
Figure 15. Rows of trees and arrangement in *porche* in the eighteenth-
            century *alberata* from an illustration in the *Tableau de
            l'agriculture toscane* by Sismondi   275
Figure 16. The work of systematization of the terrain for the Bolognese
            piantata in the early nineteenth century from the *Istituzioni di
            agricoltura* by Berti-Pichat   301
Figure 17. The relationship between zones of sowable land with trees
            and sowable land without trees in Italy [circa 1960]   358

FOREWORD TO THE SERIES
BY CHARLES S. MAIER

•••••

What is the value of reading modern Italian history? What lessons might Americans, and an English-language public more generally, derive from Italy's national experience and its historians' interpretation of that experience? What approaches to historical study can their outstanding works propose that we might not have already learned from the *mentalités* analyzed by the French, the documentation of class and elite politics contributed by the British, or the earnest archival scrutiny and national reassessment on the part of the Germans? To be sure, these questions presuppose an *American* reader or at least a reader oriented in his or her own national narrative. But the point is that Americans, including the historians among us, have had comparatively little opportunity to read the landmark postwar works of modern Italian history. A few books have made their mark in recent decades, for example the superbly erudite studies of the late Franco Venturi on the Enlightenment era, which this press has already begun to publish in translation. Studies of popular culture during earlier epochs, preeminently those of Carlo Ginzburg, have become celebrated as part of an international historiography that has reevoked the world of rural Europe. But the Italians' interpretation of their own experience as a unified nation, or their vigorous study of long-term and recent economic development have remained largely unknown outside specialist circles. For most Americans, Italian history has been mediated through gorgeous films: Visconti's *Leopard*, Olmi's *Tree of Wooden Clogs*, Fellini's *Amarcord*, Bertolucci's *Conformist* or *1900*. Historical prose, though, deserves its own chance.

The conviction that the Italian tradition of history writing deserves a wider international and English-language public has prompted the Giovanni Agnelli Foundation of Turin, together with Princeton University Press, to translate and publish some key works of modern Italian historiography. The Agnelli Foundation has long nurtured programs to diffuse knowledge of Italian culture outside Italy, including "La Biblioteca Italiana," the distribution abroad of important works in Italian. Princeton University Press takes pride in its continuing publication of monographs in Italian history by American authors and a limited number of translations by important authors. They have combined efforts to select, translate, and publish other works that promise to be especially rewarding. A committee of historians, Italian- and English-speaking, convened to decide on priority of titles and, further, to plan some thematic anthologies of outstanding articles and key primary documents. The series starts with three diverse works, revealing the broad range of historical discussion, and will, it is hoped, continue with others that introduce the history of Catholic and Socialist political movements.

The initial translations include the intellectual sketch of twentieth-century Italian political currents by Italy's preeminent post-war political writer—perhaps better described as public moralist—Norberto Bobbio. Bobbio (b. 1909) has brought his encylopaedic reading of Western political and legal theory to bear on the conditions of post-war politics, especially from the turbulence of the late 1960s to the recent sea change in Italian politics. The Press will also issue Emilio Sereni's historical survey of patterns of settlement in Italy since antiquity. Sereni's work is one of the most distinguished examples of the Marxian historiography that constituted a fundamental strand of Italian historical and political culture for over four decades after the Second World War. It also continues the notable tradition of Italian intellectuals' critical scrutiny of their rural society, which began in the Enlightenment and was carried on by conservative and democratic reformers after national unification. Finally this first "triptych" includes what may be the most luminous work in the Italian neo-historicist tradition: Federico Chabod's profound study of the aspirations and concepts of the founders of Italian statehood, preoccupied with establishing their fragile but ambitious new nation in the comity of rival great powers.

These works exemplify different disciplinary and ideological approaches to history. Bobbio began as a student of philosophy and law. In continental Europe that meant for his generation, who completed their studies between the wars, a far less technical education than American legal training suggests. Law faculties have sheltered the historians of public institutions and political theorists. History and political science departments have traditionally maintained specialists in the history of political doctrines and parties. Bobbio addresses those familiar with these diverse approaches as well as the general reader. He has focused increasingly on the long experiment in postwar Italian democracy of which he has been, simultaneously, a public critic and exponent.[1]

Emilio Sereni (1907–77), as R. Burr Litchfield's introduction to the *History of the Italian Landscape* points out, was one of the remarkable small, Italian Jewish intellectual elite, who have contributed so powerfully to culture, economy, and politics in the twentieth century.[2] Even

---

[1] In addition to Massimo Salvadori's introduction to the volume, see Perry Anderson's 1988 appreciation, "The Affinities of Norberto Bobbio," in Anderson, *A Zone of Engagement* (London: Verso, 1992), pp. 87–129.

[2] For an introduction to this milieu, see H. Stuart Hughes, *Prisoners of Hope: The Silver Age of the Italian Jews, 1924–1974* (Cambridge, Mass.: Harvard University Press, 1983).

more notably Sereni was trained in agriculture and concerned with the long history of the shaped environment in Italy over the millennia. His works exemplify the premises of Italian Marxism, which ever since the massive impact exerted by *Community Party theorist Antonio Gramsci*, has concentrated on the peculiar obstacles to Italian historical progress allegedly posed by the power of Southern landlords allied with Northern industrialists. Sereni, in fact, represents a longer tradition of critics, conservative as well as liberal or Marxist, who have confronted the agricultural problems of Italy as the major constraint on political justice and economic development. Despite Sereni's intense ideological commitment, his historical sensibility is not neatly contained in a political framework. Like Marc Bloch, the great French medievalist two decades his senior, Sereni communicates an emotional attachment to the shaped landscape— the interaction of dwellers on the land with the variable resources of soil and water, forest, plain, and uplands that they were given to exploit. Perhaps, too, both historians' intense patriotism as assimilated Jews strengthened the cathexis to their national terrains—an emotional pull that could, displaced in objective, lead Sereni's Zionist brother to Palestine. And if Sereni and Bloch might share conservatives' love of local countryside, as democrats and historians they remained far more clearsighted about the continuing patterns of social domination and hierarchy that organized settlement.

Finally the Agnelli series offers the principal modern work of Federico Chabod (1901–60), who must rank, along with Franco Venturi (1914–94), as one of Italy's preeminent academic historians of the twentieth century. Chabod produced important studies of Machiavelli and the Renaissance as well as the magisterial and original work of international history that we publish in this series. *Italian Foreign Policy* was crafted not as a diplomatic narrative, but as an exploration of the agendas for state-building among the national founders. Chabod focuses on their differing orientations toward potential allies, whether the conservative powers of Central Europe or liberal France and Britain, on their awareness of national vulnerability, and their concern, as democrats, conservatives, or would-be nationalist authoritarians, with the resources of civic culture and public opinion. If Marc Bloch serves as a non-Italian pendant for Sereni, Chabod might usefully be contrasted with the German historians of state-building, either Friedrich Meinecke of an older generation, or the critical researchers of the 1960s such as Fritz Fischer or Hans-Ulrich Wehler. Like Meinecke—indeed as an intellectual heir of German idealist historical culture—Chabod focuses on the national project as an expression of powerful developing ideas. But he never assigns the Italian nation-state a teleological role; neither does he grade his protagonists according to their national commitment; nor unlike the later Meinecke does he propose a chronicle of *raison d'état* that fatalistically dictates national *Realpolitik*. On the other hand, Chabod has little interest in exploring the economic and class variables that help to shape differing prescriptions for foreign policy and which historians influenced by social democracy or orthodox Marxism would emphasize in the 1960s and after. As with all neo-historicist accounts that seek to place the reader empathetically within the mind of the protagonists, there is danger of mystification. The apparently serene and lofty mastery of written texts can obscure a welter of personal rivalries and material interests. Nonetheless, Chabod is hardly uncritical; he must ask why state-building led, at least temporarily, to a fascist outcome; and his perspective remains so cosmopolitan and encompassing that it reveals far more than it might obscure. Without self-conscious rhetoric it renews international history with intellectual and cultural analysis.

It is fitting that two of these writers should have Piedmontese roots, given the Agnelli Foundation and the Fiat firm's own roots in Turin. Turin has long produced a rather proud community of intellectuals nurtured by the city's syncretic traditions. A dynamic industrial culture and a politically mobilized industrial labor force long allowed an interaction among market-oriented liberals and young democratic and Marxist intellectuals. But to what degree can these works be viewed as exemplifying a characteristically Italian national historiographical approach? Can one find categories to compare the historical output of the past half century in Italy with that, say, of France or Germany, Britain or the United States? It would be too facile to identify a single national tradition. The historians chosen here—who represent a generation that achieved its influence after the War but not in the most recent transformations—are less of a university-based cohort than some of the younger contributors. They present work less as a part of a group effort or enterprise than as monuments of a broader effort to define a national culture. How might we characterize their output, and that of related great scholars whose work is already translated, in comparison with scholars elsewhere?

For one thing, the works we have chosen are in effect in deep dialogue with key debates about the nature of Italian political society. Each of the writers in this first tranche of authors—Bobbio, Chabod, and Sereni—had to conceive his work in the aftermath of Italy's twenty-year acceptance, indeed invention, of fascism. No developmental trend might be discussed that did not pose the issue of what long-term vulnerability might afflict national politics, and conversely, what long-term strengths could be drawn upon. To be sure, German historians such as Hajo Holborn or Hans Rosenberg, who had migrated to the United States, or Theodore Schieder, posed similar questions with respect to Nazism. Historians and political commentators in both cultures had to ask to what degree the previous political cultures had been incompletely democratized and vulnerable to authoritarianism. But there were differences: unified Italy had been the creation of self-conscious liberals, some of whom admired Germany and others of whom feared its authoritarian tendencies. Most did not depart from a liberal or a conservative-liberal testimony of faith, whereas the exponents of German statehood had always been more explicitly divided between authoritarian and liberal currents. That is to say, the Italians could believe that Italian political society as created at the national level in the nineteenth century still had a positive, indeed redemptive agenda; this was harder for the Germans.

Of course, Italian historians and commentators—Massimo Salvadori cites Gaetano Salvemini in his foreword to the Bobbio volume—were sharply critical of Italy's insufficient democratization. And many historians, including the Englishman Denis Mack Smith, would develop a historiography of lost democratic opportunities. Conversely, conservative German historians would over the course of the last forty years endeavor to liberate the German national project from the retrospective shadow of having led up to Hitler. By and large, however, the Italians could suggest that national politics might play a positive role in overcoming backwardness, creating a civic community, and educating for citizenship. The framework of civic life was not so problematic from the outset as the Prussian-German traditional appeared to postwar historians of Germany. In a period when we have discovered again the pathologies of ethnic nationalism, the works offered here provide a revealing glimpse into how thoughtful, acutely responsible political historians envisaged a potential national society that might satisfy longing for community association without brutal chauvinism.

The historians represented in this initial series—like Venturi—were keenly aware that after fascism, the project of constructing an Italian national community must be built on a democratic basis. They differed to what degree Italy's Marxist tradition might play a constructive role in that endeavor. Bobbio has been preeminently concerned with drawing on liberalism and socialism and explaining the points of contact but confirming the differences. Sereni, of course, was a committed Marxist and never felt that his commitment must violate democratic norms. Chabod was less preoccupied by this issue although he too wrote a small volume about Italy's descent into fascism and was clearly pledged to a postwar liberal democratic order.

Each of these historians was old enough to be caught between the great division in Italian ideologies represented by Benedetto Croce on the one side, and Antonio Gramsci on the other. Although he lived long enough to participate in politics after fascism, Croce exemplified Italy's pre-fascist liberalism. In the steps of an even earlier generation of

Neapolitan Hegelians, he helped implant into Italian culture the currents of German idealism and historicism. In this framework, history constituted a clash of ideas between the liberals who gradually opened the frontiers of a civic community to a mass citizenry, and two sorts of demagogues. On the Right were the elitists and later fascists, on the Left the collectivists who would endanger this democratic progress—a threat of interacting extremes that preoccupied Bobbio, too, in the 1970s. Chabod's scenario of state-building also exemplified this grand narrative line according to which national politics must ultimately represent a contest between emancipatory and reactionary or demagogic concepts.

On the other side, Gramsci, whose prison writings oriented a generation of intellectuals after fascism, criticized Crocean idealism as a conservative mystification that served to uphold the regime of landed and industrial property. What distinguished his analysis from Marxist critiques elsewhere was its grounding in the fundamental constraints of Italy's peculiar history. His dualist conclusions, confirmed by the fascist outcome of the 1920s, envisaged a backward Mezzogiorno, enmeshed in clientelism and feudal legacies, undermining critical intellectuals and frustrating the forces of modern technology and proletarian democracy based in the industrial North. Sereni suggests some of the same duality. A Marxist like Gramsci, he understood that capitalism penetrated the countryside—but only imperfectly and not without drawing on the political resources of pre-market hierarchies. It is his affection for this long agrarian history, however, that keeps his work from becoming just a schematic narrative and delightful exploration of how men lived with the country. History is presented as a story of constraint, but the constraint imposed by the soil and elements as much as by landlords.

It is fitting that the Princeton-Agnelli series begins with these three authors, for together they exemplify the grand narrative lines in which Italians conceived their national history from the Risorgimento through the long era of political and economic modernization that followed the fall of fascism and the institution of the postwar republic. These authors emphasize the conflict between liberating ideas and constraining hierarchies; they envisage Italian history as a great contest between projects for civic emancipation and habits of domination, between politics as ideas and politics as ancient constraints. In part because Italian intellectuals imported these grand narrative lines from abroad, they took on an exaggerated clarity, less polarizing than in Spain, but dramaturgic nonetheless. This simultaneous sensibility for the force of contending ideas and the tenacity of material structures endows Italian scholarly historiography with a sense of drama that British and American authors summon more readily from biography. We tend to personify the contests that the Italian scholars translated in this series could incorporate in collective narratives.

Will the reader find such monuments of synthesis today? Contemporary Italian historians, like their counterparts in this country and elsewhere in Europe, have become more preoccupied by multiple voices, fragmentation, and the dissolution of overarching "metanarratives." Even the great national drama of the Resistance to fascism and nazism, which for almost half a century helped to orient the alliances of national politics, has become less clear and more confused. Still, historians writing today have understood how to transform even the new ideological and methodological uncertainties into various reassembled histories. What Italian historians continue to furnish even as their narratives decompose is an acute self-awareness of political ideas. If they can no longer offer us a confident footing either as liberals or Marxists, they continue to provide sophisticated self-awareness. Like the residents of Italo Calvino's imaginary city of Ottavia who live suspended over an abyss into which they know they must ultimately fall, they provide structures all the more remarkable for their vulnerability. The writers of the successive postwar generations—first Sereni and Chabod, then later Bobbio, the interpreter of the ideological conflicts of the years of upheaval and terrorism in the late 1960s and 1970s—wrote with more certitude than we are likely to find today. Perhaps, too, with more power and penetration. The English-speaking reader now has a valuable chance to judge.

INTRODUCTION TO THE ENGLISH TRANSLATION

• • • • •

Emilio Sereni was born in Rome on 13 August 1907 into an old established Jewish family of professional and intellectual tendencies. His father had been among the physicians who attended the Italian royal house, and members of his mother's family, from Pisa, were prominent in physics, film, and law.[1] His personal biography, his political career in the Italian Communist Party, and his outstanding intellectual accomplishments were enmeshed in the contradictions of Italian society, politics, and culture of his age: in the painful transition from the liberal monarchy and Fascism through the anti-Fascist resistance to the post–World War II republic, in the social contrast between urban and rural life and between the Italian North and South, and in the contrasting intellectual tendencies of philosophical and historical idealism and the more strictly economic characteristics of Marxism-Leninism.

Since not even the bare facts of Sereni's biography are widely known outside of Europe, it is worth pausing over them for a moment here. One of his elder brothers, Enzo Sereni, who had studied philosophy with Giovanni Gentile at the University of Rome, became a leading member of the Italian Zionist movement and emigrated to Palestine in the late 1920s. He was later executed in the Dachau concentration camp in 1944 after he had returned to participate in the Jewish brigade of the resistance and was captured by the Nazi SS in northern Italy. Emilio Sereni's career took a different direction.[2] After his early studies in Rome he attended a well-known agricultural institute, the Istituto Superiore Agrario in Portici, near Naples, where he studied with the noted

---

[1] In writing this introduction, I am grateful for the suggestions kindly offered by Emilio Sereni's daughters Lea and Clara in Rome in May 1992, and also for the assistance of Fulvia Di Giulio at the Istituto Alcide Cervi, where Sereni's private library is preserved. A complete bibliography of Sereni's works was published in 1987 (Istituto Alcide Cervi, *Bibliografia degli scritti di Emilio Sereni* [Florence: Olschki, 1987]). The archives of Sereni's political career are preserved in Rome at the Istituto Gramsci. I am also grateful for conversations about Sereni with my Italian friends: Giuseppe Ricuperati and Mauro Ambrosoli in Turin, Giovanni Levi in Venice, and Sergio Bertelli and Carlo Pazzagli in Florence.

[2] His eldest brother, who died in 1929, studied biology in Rome, and the next eldest brother also emigrated to Palestine. Sereni's family background is noted in Edoardo Volterra, *Emilio Sereni: Studioso dell'agricoltura* (Rome: Istituto Alcide Cervi, 1978). The only other biographical works on Sereni are *Dirigenti Communisti: Sereni* (Rome, 1946) and the article of E. Mannari, "Sereni, Emilio," in Franco Andreucci and Tommaso Detti, eds., *Il movimento operaio Italiano: Dizionario biografico, 1853–1953* (Rome: Editori Riuniti, 1978), 4:608–12.

agronomist Oreste Bordiga and received his laureal degree in 1927. At Portici, Sereni developed his lifelong attachment both to the history of agriculture and to the Italian South. His fellow student and close friend in this period, Manlio Rossi-Doria, later devoted himself to the agricultural development of the South, first in the Italian Communist Party (PCI) and then in the Socialist Party (PSI). But Sereni's upbringing gave him much broader cultural and intellectual interests than were contained within the study of agricultural development alone. He read philosophy passionately in Portici, not so much in the direction of the Neapolitan liberal idealist philosopher, Benedetto Croce—who had recently published his *History of the Kingdom of Naples*—as in the works of Marx and Lenin, which became the other pole of his cultural formation. His friends later retained warm recollections of his intensity in these years.[3] Moreover, the acquaintances and friends of Sereni and Rossi-Doria in Portici included such significant figures in the political resistance to Mussolini's Fascist government as the young Carlo Rosselli, who was later assassinated by Fascist agents in France in 1937, and Giorgio Amendola, son of the liberal deputy Giovanni Amendola who was murdered by a Fascist squad in Rome in the summer of 1925, who later became vice-secretary of the PCI and the righthand man to Palmiero Togliatti, replacing Pietro Secchia in 1955. As the events of his life unfolded, Sereni developed as an intellectual with deep political commitment.

Sereni's definitive conversion to Marxism-Leninism and his adherence to the Communist Party dated from 1926–28, and he returned from Rome to Naples in 1928 after his marriage to Xenia Silberberg, from a family of Russian socialist exiles. Her father had been hanged by the czarist government after the October revolution of 1905. In Naples during the next two years he worked along with Rossi-Doria to establish the Federazione Comunista Napolitana among anti-Fascist worker and peasant groups, and to make contact with the central direction of the PCI in Paris. Rossi-Doria recalls his awareness of Antonio Gramsci's "Some Aspects of the Southern Question" in this period—which posed the question of a collaboration between workers in the industrial cities of the North and peasants in the South—as well as Sereni's first economic writings, "Sulle cause della crisi agraria" (On the causes of agricultural crises) and "Il capitale finanzario in Italia" (Financial capitalism in Italy), which were published in 1930, under pseudonyms, in *Stato operaio*, the Parisian journal of the PCI in exile. These proved to be

[3] There are recollections of Sereni, who was nicknamed "Mimmo" as a young man, in Manlio Rossi-Doria, *La gioia tranquilla del ricordo: Memorie, 1905–1934* (Bologna: Il Mulino, 1991); Giorgio Amendola, *Una scelta di vita* (Florence: La Nuova Italia, 1978); and Clara Sereni, *Il Gioco dei Regni* (Florence: Giunti, 1993).

early studies for Sereni's later book on the economics of Italian agricultural development in the late nineteenth century, *Capitalism in the Countryside, 1860–1900*.[4]

In September 1930 Sereni and Rossi-Doria were arrested by the Fascist police, tried by the so-called Tribunale Speciale dello Stato that had been established in 1926 for political crimes, and condemned to fifteen-year prison terms. Sereni served successively in the prisons of Poggioreale, Regina Coeli, Lucca, Viterbo, and Civitavecchia, where he encountered others in the Italian Communist movement, among whom such later important members of the party as Pietro Secchia, Mauro Scocciamarro, and Umberto Terracini. In prison he further expanded his already ample linguistic abilities by studying Arabic, Chinese, and Japanese (in Naples he had read Lenin in Russian and Marx and Engels in German, and he also knew French, English, Spanish, Greek, Latin, Hebrew, Yiddish, and Aramaic). He studied higher mathematics in prison, and he also taught history, economics, and economic history. According to his wife Xenia he requested so many books "Because, he says, if he is not distracted a bit it becomes tiresome to always read the same things." A fellow prisoner at Viterbo recollected that "Sereni read three books a day," and another that "when he had read a book he knew it by heart." Gian Carlo Pajetta, later a member of the party's Central Committee, who had known Sereni in prison wrote: "I met him, not only because people studied the books he had read ('Sereni recommended this'), but also because in some way he had already become part of the party legend. . . . Sereni 'knew everything.'"[5]

Released from prison in September 1935 through a general amnesty, Sereni emigrated clandestinely to France and established himself in Paris, where he became the chief editor of *Stato Operaio* in 1936 and was co-opted into the Central Committee of the PCI. Although briefly suspected of a dangerous independence from Moscow, he was charged with the education of Communist youth.[6] He also edited the Parisian anti-Fascist daily newspaper *La voce degli Italiani*. To support himself, he worked as a machinist in a Paris suburb. With the German occupation of Paris in 1940 he moved successively to Tolouse, Toulon, and Nice, where he worked with partisan groups and edited the journal *La parola del soldato*. He was to recall his partisan experience in the countryside of

---

[4] *Il capitalismo nelle campagne, 1860–1900* (Turin: Einaudi, 1947).

[5] Gian Carlo Pagetta, "Vita di un communista," *Rinascita* 34, no. 12 (25 March 1977): 7. Sereni's activities in the 1930s and 1940s are recounted in Marina Sereni (his wife Xenia Silberberg), *I giorni della nostra vita* (Florence: La Nuova Italia, 1956).

[6] On some figures in the PCI in this period, see Sergio Bertelli, *Il gruppo: La formazione del gruppo dirigente del PCI, 1936–1948* (Milan: Rizzoli, 1980); and, more generally, Paolo Sprano, *Storia del partito communista Italiana*, 8 vols. (Turin: Einaudi, 1967–75).

the Côte d'Or and the Ligurian coast later in his study of rural communities in pre-Roman Liguria.[7] He also wrote a book at this time on more recent developments in the Italian countryside: *The Agricultural Question in the Italian National Rebirth*.[8] Again arrested and imprisoned, first in San Remo in the summer of 1943 at the moment of Mussolini's fall from power in Rome, he barely escaped execution by the Nazi SS, was sentenced to prison, again escaped, and reemerged in liberated Milan in 1944 in the Comitato di Liberazione per la Lombardia and as president of the Giunta di Governo.[9] These were the years when the PCI formulated what Palmiero Togliatti would call the "Italian way to socialism," that is, a progressive program—of which the tactics were debated in the years that followed—appropriate to Italian particularities and traditions and guided by a popular front of socialist groups, to eradicate vestiges of Fascism and promote social development in the interests of the working class.

Sereni's stature and his efforts in the North led to his selection for the Central Committee of the PCI at its fifth congress in December–January 1945–46, a position he continued to hold until the fourteenth congress in 1975. As a member of the Central Committee he was charged successively with party organization in Naples and the South, and with the cultural activities of the party. He also served in the constituent assembly for the Italian republic, after rejection of the monarchy, and as minister for postwar assistance and public works in the second and third De Gasperi governments in 1946 and 1947—an experience recounted in his *The South in Opposition: From the Notebook of an Ex-Minister*.[10] He was made a member of the Italian Senate 1948, where he remained until 1963 when he was elected to the legislature for two terms. He continued to write and speak on politics and agricultural policy, was active in the Communist international peace movement, had four of his earlier books translated into Russian, and in 1955 succeeded Ruggero Grieco, who had headed the agricultural section of the PCI since 1945, as president of the newly founded Alleanza Nazionale dei Contadini. Generally Sereni remained in the more traditional circles of the PCI during the upheavals of the 1960s and 1970s: the period of Amintore Fanfani and the Christian Democratic Party's "opening to the left," the *Manifesto* of Pietro Ingrao, and Enrico Berlinguer's "historic compromise" of col-

---

[7] *Comunità rurali nell'Italia antica* (Rome: Edizioni Rinascita, 1955).

[8] *La questione agraria nella rinascita nazionale Italiana* (Rome: Einaudi, 1946)

[9] These activities are reflected in Emilio Sereni, C.L.N.: *Il comitato di liberazione nazionale della Lombardia al lavoro Nella cospirazione, nella insurrezione, nella ricostruzione* (Milan: Percas, 1945).

[10] *Mezzogiorno all'opposizione: Dal taccuino di un ministro in congedo* (Turin: Einaudi, 1948).

laboration between the PCI and the Christian Democrats. Despite the relative decline over time of some of his older party associates he was made editor of *Critica marxista*, the theoretical journal of the PCI, in 1972.[11] Finally, in 1975 he became director of the scientific committee of the Istituto Alcide Cervi, an institute in Rome devoted to the history of peasant movements and agriculture, where he remained until his death on 20 March 1977.

Sereni's works on the history of agriculture—and his most important synthesis, *History of the Italian Agricultural Landscape*, written in the mid-1950s but not published until 1961[12]—can best be understood in the context of these personal and political developments. For all of their academic originality and validity, and their impressive sweep through time, his historical works belong to the genre of politically committed history, and one might add, to the particularly Italian formulation of philosophical, ideological, and political discourse in the pre– and post–World War II period. This is what most truly distinguishes Sereni's work from Marc Bloc's *Caractères originaux de l'histoire rurale française*, with which he polemicizes in the preface to the *Agricultural Landscape*, or from Philip Jones's chapter on medieval agricultural history in the *Cambridge Economic History*, to which he also alludes. The *History of the Italian Agricultural Landscape* also reflects some of Sereni's other historical, theoretical, and political writings, which we will examine briefly.

Sereni's first work on the economic history of agriculture—*Capitalism in the Countryside, 1860–1900*, published in 1947—was devoted to the effects in the Italian countryside of the establishment of a national market after unification, through unified tariffs, the building of the railroads, the postal system, the changing relationship between rural and urban population, the increasing dissociation of agriculture from industrial activity, and the complications that residues of the feudal system posed for this process, particularly in the South. He wrote further works during the 1950s and 1960s on subsequent developments in the countryside.[13] But Sereni also had strong and more strictly academic interests in the ancient, medieval, and early modern history of Italy that emerged

---

[11] Some of Sereni's earlier political essays from the 1940s were republished in Emilio Sereni, *Scienza, marxismo, cultura* (Milan: Le Edizioni Sociali, 1949). Examples of political speeches in the late 1950s and early 1960s are *Le trasformazioni colturali e la controriforma agraria dell' on. Fanfani* (Rome: Alleanza Nazionale dei Contadini, 1958) and *Da Bonomi a Fanfani: La crisi nelle campagne ha nome D.C* (Rome: Sicca, 1960). Some later essays from *Critica marxista* are collected in Giuseppe Prestipino, ed., *La rivoluzione Italiana* (Rome: Editori riuniti, 1978).

[12] *Storia del paesaggio agrario Italiano* (Bari: Laterza, 1961).

[13] *Vecchio e nuovo nelle campagne* (Rome: Editori Riuniti, 1956), *Due linee di politica agraria* (Rome: Editori Riuniti, 1961), and the essays collected in *Capitalismo e mercato nazionale in Italia* (Rome: Editori Riuniti, 1966).

through a number of works written before or during the composition of his *Agricultural Landscape*. These showed his underlying linguistic, literary, and cultural interests and associated him in significant ways with the then reigning Italian school of historical idealism.

There is a certain fluidity in Italy between politics and academic life, and although Sereni never held a university position, he had a vast historical erudition, and he received a state *libera docenza* in the history of agriculture in 1960, which might have opened to him an academic career. Some of his historical articles are reminiscent of Croce's essays on the popular culture and folkways of early modern Naples. Traces of his Crocian orientation remained in the significance he assigned to "rethinking" human activity in the past, which was typical of Crocean methodology. There was, for instance, his essay "The People and Popular Poetry in Italy in 1848" (1948), on the folksongs of the 1848 revolution, and also his "Notes on the Traditional Folk Songs of the People of Umbria" (1959) on Umbrian folk songs, that extend in subject matter backward to the time of Dante and to Fransciscan and goliardic poetry.[14] Also in a Crocean mood was his splendid essay "Notes on the History of Diet in the South: Neapolitans from 'Leaf-Eaters' to 'Maccheroni-Eaters'" (1958), which traced through literary references the sixteenth-to-eighteenth-century shift in diet of the popular classes of Naples from consuming mainly leafy vegetables to consuming pasta.[15] Closer to the subject of the *Agricultural Landscape* were his "Toward a History of the Most Ancient Techniques and Nomenclature of Vines and Wine in Italy" (1964) on the ancient, and pre-Roman, nomenclature of viticulture,[16] and his posthumously published essays on the old peasant mythology, and vocabulary, of clearing new forest land when already cultivated clearings had become infertile, and on the still more ancient mythology and nomenclature of the horse in Eurasian societies.[17]

---

[14] "Popolo e poesia di popolo in Italia attorno al '48," in *Il 1848: Raccolta di saggi e testimonianze* (Rome: Quaderni di Rinascita—1, 1948); and "Note sui canti tradizionali del popolo Umbro," *Cronache Umbre* 2, 4–6 (1959), 19–51, 15–40.

[15] "Note di storia dell'alimentazione nel Mezzogiorno. I napoletani da 'mangiafoglia' a 'mangiamaccheroni,'" *Cronache meridionali* 4–5 (1958), 272–95, 351–77 republished in Emilio Sereni, *Terra nuova e buoi rossi e altri saggi per una storia dell'agricoltura Europea* (Turin: Einaudi, 1981), 292–371.

[16] "Per la storia delle più antiche tecniche e della nomenclatura della vite e del vino in Italia," in *Atti e Memorie dell'Accademia Toscana di scienze e lettere La Colombaria* (Florence, 1964), republished in *Terra nuova e buoi rossi*, 101–214.

[17] "Terra nuova e buoi rossi: Le tecniche del debbio e la storia dei diboscamenti e dissodamenti in Italia," and "La circolazione etnica e culturale nella steppa eurasiatica: Le tecniche e la nomenclatura del cavallo" in *Terra nuova e buoi rossi*, 2–100, 215–291.

These essays all display the extraordinary philological and linguistic abilities that appear prominently in Sereni's *Agricultural Landsape*, as well as his early training in agricultural techniques at the Istituto Agraria in Portici, where he undoubtedly studied nineteenth-century agronomists who used many of the old technical terms he employs of Italian rural nomenclature. Another important element in Sereni's historical thought, which appears in the articles he published while writing the *Agricultural Landscape*, was his deep interest in agricultural technology, and in the individual technicians and early theorists of agronomy who chronicled the shaping of agrarian activity in new ways: medieval and Renaissance figures such as the Bolognese Pietro de' Crescenzi and Paganino Bonafede, the Tuscans Michelangelo Tanaglia and Luigi Alamanni, the Neapolitan Luigi Tansillo, the Brescians Agostino Gallo and Camillo Tarello, and the many other later technicians and agronomists who appear in his *Agricultural Landscape*.[18] Sereni conceived of agriculture as a practical, active, reasoned response—which often emerged among the cultivators themselves, but was reflected in the basic processes and implications of technology—to the problems posed by the material and economic environment of the rural landscape. For this reason, throughout this translation, I have utilized the English word "agricultural," rather than the weaker terms "agrarian" or "rural," to emphasize the practical active element in technology that Sereni thought was such an important factor in historical adaptation and change.

Sereni's most substantial academic book of the 1950s was his *Rural Communities in Ancient Italy* (1955).[19] This introduced and developed another significant aspect of his distinctive approach, which complemented his cultural, literary, linguistic, and technical interests: the very concept of historical "agricultural landscapes." The intersection be-

---

[18] "Pietro de' Crescenzi e la tecnica agraria d'avanguardia" in *Riforma agraria* 3, nos. 11-12 (1955); "Paganino Buonafede e l'agronomia popolare in Italia nell'età dei Comuni," *Riforma agraria* 4, no. 1 (1956); "Michelangelo Tanaglia (1437-1512) e l'agronomia dell'Umanesimo in Toscana," *Riforma agraria* 4, no. 2 (1956); "Luigi Alamanni (1495-1556) e il bel paesaggio agrario in Toscana," *Riforma agraria* 4, no. 3 (1956); "'Il podere' del Tansillo (1510-1568) e l'agronomia del Rinascimento nel Regno di Napoli," *Riforma agraria* 4, no. 4 (1956); "Agostino Gallo (1499-1570) e la scuola agronomica bresciana," *Riforma agraria* 4, no. 5 (1956); "Il 'Ricordo d'agricoltura' del Tarello (1567) e gli inizii della rivoluzione agronomica nella Padana," *Riforma agraria* 4, no. 6 (1956).

[19] *Comunità rurali nell'Italia antica*. See also his "Il sistema agricolo del debbio nella Liguria antica," *Memorie dell'Accademia lunigianese di scienze, lettere ed arti Giovanni Capellini* 25 (1953): 11-29, and "La comunità rurale e i suoi confini nella Liguria antica," *Rivista di studi liguri* 20 (1954). Sereni later returned to this subject, but more from the point of view of "structures" than landscapes: "Villes et campagnes dans l'Italie préromaine," *Annales· Economies, sociétés, civilisations* 22, no. 1 (1967): 23-49.

tween geography and history had been a chiefly French concern in the 1950s, one that referred back to geographers such as Paul Vidal de la Blache, and to historians such as Lucien Febvre or more recently Fernand Braudel, although one might note that a somewhat similar discussion in Italy extends backward into the nineteenth century, at least to Carlo Cattaneo. Sereni's *Rural Communities* is a linguistic, ethnographic, economic, and geographic study of the Ligurian coast in the second century B.C., on the eve of the Roman colonization of this region, considering its peoples, tribes, and territories with their diverse nomenclatures, customs, and rural economies—a diversity of what might best be called "cultural landscapes": human aggregations within the physical environment of a regional geography. His discovery in ancient Liguria of historically concrete but culturally distinct rural "landscapes" in this sense—and not just the imperfect reflections of landscapes in the pictorial works he later chose as illustrations of agricultural customs and techniques for subsequent periods—was to give Sereni's work some of its greatest originality. In his *Agricultural Landscape* Sereni revealed the contrasts in environment, society, and agricultural technique among the regional landscapes that are essential to comprehending diversity within Italian history. Sereni was also a pioneer in studying environmental history, and the forest burned spots and clearings of ancient Liguria that he studied in his *Rural Communities* called attention to the first signs of ecological deterioration, which he later developed as a much more significant theme, one that is very timely for today. The reader may be surprised to discover from Sereni's *Agricultural Landscape* what a serious problem deforestation and erosion were in early periods of Italian history, and how important technology was in the late Middle Ages, Renaissance, and early modern period for conserving water and soil, and for giving terraces to Italian hills, and networks of canals to its plains.

The cement that holds together Sereni's synthesis in the *Agricultural Landscape*, and permitted him to construct his insightful analysis, was provided by his Marxism-Leninism, but even this was sifted through and colored by the Italian cultural and political environment in which he worked. The relative independence of the PCI in the post–World War II period made Italy a hotbed of theoretical debate within the Communist movement, a debate that extended in the Italian party back at least to the confrontation between Amadeo Bordiga and Antonio Gramsci at the Lyon conference in 1926. The debate involved traditional Marxists-Leninists, and others who took a new direction in postwar Italy. Through the events in Hungary in 1956, the death of Togliatti in 1964, and the Prague Spring in 1968, the PCI gradually moved toward Eurocommunism.

The new left often drew strength from theoretical works of Gramsci. In several of his works, Sereni cites Gramsci's 1926 essay "Some Aspects of the Southern Question."[20] In this essay Gramsci had posed the question of a collaboration between workers in the northern Italian industrial cities and the disorganized peasants of the South to oust the bourgeoisie from state power and swing the weight of the bureaucracy behind the peasants in their struggle with the southern landowners. Gramsci also emphasized the great difficulties that beset this possible alliance through the ideological backwardness of the South and the failure of southern intellectuals to assume a hegemonic function in giving the inchoate southern masses an awareness of their political role. This was the essence of the theoretical insight that Gramsci developed further in his *Prison Notebooks*, and that was so significant in subsequent debates within the PCI. It involved the importance of ideological hegemony created by "organic" intellectuals through a mass party to effect cohesion for political action, on one hand, and historical, class-derived, ideological, and political alliances, on the other: in this case between northern workers and southern peasants. Gramsci's theory of social and historical change in this and his later works was an idealist offshoot from traditional Marxism-Leninism, which stemmed from his own early Crocean orientation and led to his emphasis on the primacy of ideology and politics—that is, the superstructure of thought and action over the infrastructure of capitalist development and the dialectic of social classes.

Sereni later wrote that in his early works he was unaware of much of Gramsci's thought,[21] and in many of his writings he retained a certain distance from the Gramscian tendencies of Italian communism. He was far more an "economist" than Gramsci, and he had a somewhat different view of how historical change articulated itself.[22] He doubted, for instance, that the hegemonic influence of intellectuals would suffice to resolve the problems of the Italian South, and he assigned less importance than Gramsci had done to the shifting ideological alliances of party groupings in the state. More impressive in his estimation was the

[20] "Alcuni temi sulla questione meridionale," published in *Stato Operaio* in 1930. The essay is available in English in Antonio Gramsci, *Selections from Political Writings, 1921–26*, ed. and trans. Quintin Hoare (Minneapolis: University of Minnesota Press, 1978), 441–62.

[21] In his introduction to the second edition of his *Capitalismo nelle Campagne* (Turin: Einaudi, 1968). Sereni's 1948 essay on Gramsci ("Gramsci e la scienza d'avanguardia," *Società* 4, no. 1 [1948]: 3–30) is little more than a review of Gramsci's *Il materialismo storico e la filosofia di Benedetto Croce*.

[22] See Giuseppe Prestipino, "Il marxismo militante di Emilio Sereni," *Critica marxista* 14, nos. 5–6 (1977): iii–xv, and Renato Zangheri "Emilio Sereni e la questione agraria in Italia" in *Emilio Sereni e la questione agraria in Italia*, by Renato Zangheri, Pasquali Villani, and Attilio Esposto (Rome: Editori Riuniti, 1981), 15–44.

direct practical action that southern peasants took in the immediate post–World War II years to occupy the uncultivated lands of latifundia, the great southern estates, and to force the government to accede to their immediate interests. His *Capitalism in the Countryside*, which was published the same year (1947) as Gramsci's posthumously published *Prison Letters*—well before publication of the *Prison Notebooks*—took its theoretical inspiration instead from Lenin's *Development of Capitalism in Russia*, and from Lenin's writings on agriculture in the period 1901–15, which Sereni had read in Russian. Lenin had observed that the agricultural question involved, above all else, the development of capitalism in the countryside. He also distinguished typologies in the rural development of capitalism between what he called the "Prussian type," where the large feudal estates of the junker were gradually transformed in a capitalist direction while reducing the peasant population to the situation of landless day laborers—which Sereni appropriated also as the "Italian type" of capitalist development in the countryside—and what Lenin called the "American type," based on the operations of small free farmers, where capital accumulation and market conditions in different sectors of the agricultural economy would eventually lead, in a different way, to a similar outcome.[23]

Sereni developed these insights from Lenin in his *Capitalism in the Countryside* for the period 1860–1900, and then further in his subsequent works that brought his analysis of conditions in Italy up through the 1950s.[24] He continued to emphasize the degree to which the residue of feudalism in central Italy, the South, and the Islands and the newer monopolistic tendencies of the northern Italian industrial elite complicated and frustrated agricultural development. Chapters 8 and 9 of his *Agricultural Landscape* summarize the discussion in his *Capitalism in the Countryside* and his other works on these questions. Sereni's later position on agricultural reform was consistent with the central program of the PCI in the 1950s and 1960s, which he helped to elaborate. The Gramscian position of an alliance including southern peasants continued. There was no question in Italy of collectivization of landownership: the aim was to assure "the land to those who work it," in line with the decrees of Communist Minister of Agriculture Fausto Gullo in 1944 and the peasant movement to occupy the southern latifundia that came to a head in the autumn of 1949. Subsequent policy was to complete the elimination of latifundia, to improve the conditions of labor contracts

---

[23] For a further discussion of Lenin, which also shows the continuing interest in him among Italian agricultural historians, see Giorgio Giorgetti, "I quaderni di Lenin sulla questione agraria" in his *Capitalismo e agricoltura in Italia* (Rome: Editori Riuniti, 1977), 352–79.

[24] The works cited in nn. 8 and 13.

for sharecroppers and agricultural renters throughout Italy, to sponsor public investment and technical assistance, and to encourage the formation of peasant voluntary associations. Sereni took the position that the real problems of agriculture arose from the monopolistic control of markets, and from the failure of the Italian industrial elite to provide for adequate infrastructural support for agriculture, particularly in the South.

Thus Sereni and Gramsci both began their intellectual quests in the environment of philosophical idealism and both then emerged into the political environment of the anti-Fascist resistance, but they followed different courses and elaborated somewhat different individual positions. Sereni differed from Gramsci in the greater emphasis he placed on the underlying economic process of the spread of capitalism in the countryside, which was complicated by vestiges of feudalism and by monopolistic industrial interests more than by the contradictions arising from competing ideologies, and he also more fully explored Gramsci's concept of political alliance, what was known as a *blocco storico*. In some of his articles in *Critica marxista*, and his historical works, he interpreted this concept in a fuller "economic" sense, less as a system of alliances among hegemonic party groupings than as a historical unity arising in both economic interests and consciousness.[25] This was a sense more complex than the type of alliance contained, for instance, in the term "military-industrial complex" from the United States of the 1960s, which might seem at a first reading similar to what he suggests in the last chapter of the *Agricultural Landscape*. However, Sereni differed from Lenin too, not only in the phrase "the land to those who work it," but also in the lesser emphasis he placed on industrial workers as the exclusive vanguard social class. There was also his greater emphasis on the innovative possibilities of practical applications of technology and planning to overcome obstacles in the natural environment, through improved agricultural techniques, and in the capitalist environment, through innovative cooperative planning. Sereni was similar to Gramsci in emphasizing the democratic way to socialism. As he capsulated his own ideological position in *Critica marxista*, echoing his preface to the

---

[25] See his "Blocco storico e iniziativa politica nell'elaborazione gramsciana e nella politica del PCI," *Critica marxista* 5 (1972): 3–20 In a 1967 article he had called the "bloc historique" a "new type of intimate and strong relationship between structure and superstructure." "Villes et campagnes dans l'Italie préromaine," 29. There is a further discussion of Gramsci, in the context of a more general discussion of the theory of agricultural development, in his "Agricoltura e sviluppo del capitalismo: I problemi teoretici e metodologici," *Studi Storici* 9, nos. 3–4 (1968): 477–530. Here Sereni also specified his differences with the more capitalist historians, A. Gerschenkron, P. Bairoch, R. Romeo, and others.

Agricultural Landscape: "Thus, structural or superstructural approach? or rather, perhaps, the Faustian, Marxist, Leninist, Gramscian approach of *Im Anfang war die Tat*, 'In the beginning was action'? an action that invests not only structures (a great syndicalist class movement), or superstructures (the development of class consciousness), but also, and above all, their dialectical, deep, active, and most intimate connection: liberty, democracy, politics."[26]

Sereni's modified Marxism-Leninism appears most clearly in the later chapters of the *History of the Italian Agricultural Landscape*, on the period since the mid-eighteenth century, but it also deeply colors his treatment of ancient, medieval, and Renaissance history. Indeed, the synthetic sweep of his work through centuries, for one who, as he says in his preface, wanted to see fragmentation recompose itself, "to become history again," would have been difficult to achieve without it. His erudition, and his awareness of the monographic literature for all periods—although the form he chose for his work omits its specific citation—always nuances the more schematic aspects of his exposition. In fact, the theoretical approach goes well beyond the monographic literature and continually suggests to the informed reader unexpected connections, problems, and possibilities of research.

Some idiosyncrasies, however, emerge from Sereni's theoretical imposition. One notes, throughout, from his discussion of Hippodamus of Miletus and the layout of ancient cities, through Roman centuriation, the planned colonies of Italian towns in the age of the communes, his ranking of Lombard over Tuscan agriculture in the Renaissance, up even to the planned fields of the current great agrobusinesses of the Po valley, Emilia, and Puglia, Sereni's preference for a planned and organized rural landscape. The antithesis to this, throughout his work, was the spontaneous, unplanned, irregularly shaped field of the individual squatter or entrepreneur, the Provinçal *Marrello*, whose function he disputes with Marc Bloch, or rather the *Campo a pìgola*—the term he prefers from the agricultural nomenclature of Volterra and that, despite his deep understanding of regional differences, he tends to generalize through the regions and centuries of Italian history. One notices also that Sereni imposes a certain unity of mentality through time and space—for instance, the taste for a "bel paesaggio" that in Renaissance Tuscany was the same for peasants, the painter Benozzo Gozzoli, and Giovanni Boccaccio in his poetic passage from the *Ninfale fiesolano*. There is not among Sereni's peasants, as Pasquale Villani has pointed out, any "moral economy," to use the English Marxist historian E. P. Thompson's term, that

---

[26] "Democrazia e socialismo nel pensiero e nell'azione di Togliatti," *Critica marxista* 4, no. 2 (1966) 48.

made the mentality of peasant communities different from those of artists, city dwellers, or technicians.

But it is not hard for the reader to see through such psychological limitations to the greater merits and complexities of Sereni's work. As Villani—an important historian of the Italian South—put it, referring to Sereni's striking characterization of the economic effects of the late nineteenth-century juridical dismemberment of the feudal system in the Kingdom of Naples:

> He discusses the sale of ecclesiastical lands, their dispersion among different social classes, and the financial aspects of the sales; he discusses the alienation and allotment of demesne lands, and assesses the weight of ex-feudal property and latifundia in the agrarian structures of the South; he studies the development of the peasant bourgeoisie, estate agents, the strata of the small and middling bourgeoisie above the peasants, and their intermediary political and social function; he examines the problem of the specialization of crops, agricultural change, the progress of the expropriation of peasant property, and the spread of subdivisions; he confronts the large question of mortgage obligations; . . . he discusses the process of proletarianization and the problems of emigration: he discusses each and all of these problems. The indications and suggestions that come from the work of Sereni are almost obligatory points of departure for any further research.[27]

Sereni's *History of the Italian Agricultural Landscape* has had a history since the publication of the first edition by Laterza in 1961 (the fifth Laterza edition was published in 1989). There was a French translation in 1965.[28] Sereni restated the main lines of his argument, with some further clarifications, in his essay "Agriculture and the Rural World" in the Einaudi *Storia d'Italia* in 1972.[29] English-language scholarly journals did not give much attention to Sereni's book, but it was reviewed most favorably and at length by Georges Duby, who did not find the chapters on medieval history the best part of the book, in the *Annales: Economies, sociétés, civilizations* in 1963.[30] As I have noted, Sereni did not hold an academic position or have students; he was thus

---

[27] Pasquale Villani, "Emilio Sereni, storico dell'agricoltura," in Zangheri, Villani, and Esposto, *Emilio Sereni e la questione agraria in Italia*, 66.

[28] Emilio Sereni, *Histoire du paysage rurale italien*, trans. Louise Gross (Paris: René Juliard, 1965)

[29] "Agricoltura e mondo rurale," in *Storia d'Italia*, Vol. 1: *I caratteri originali* (Turin: Einaudi, 1972), 135–252.

[30] Georges Duby, "Sur l'histoire agraire de l'Italie," *Annales: Economies, Sociétés, Civilisations* 18, no. 2 (1963): 352–62. In a review of Sereni's *Capitalismo e mercato nazionale in Italia* in the *Economic History Review* (22 [1969]: 385–86), Luciano Cafagna called attention to the *History of the Italian Agricultural Landscape*.

not able to create a school of historical writing in this way. But his presence has continued to be felt in Italian historiography. One might note as indicative his exchange with the economic and agricultural historian Carlo Poni in 1963–64, which stemmed from Sereni's review of Poni's book *Plows in the Agricultural Economy of Bologna in the 17th–19th Centuries* in *Studi storici*,[31] where Sereni chided Poni for having provided incomplete etymologies for agricultural terms, for having underestimated the rotation between grain and hemp, and for slighting iconographic evidence. Poni replied[32] that Sereni had used an agricultural vocabulary inappropriate to the Bologna region, had miscalculated the extent of hemp production from the secondary sources he had used, and had misunderstood the impact of plowing on the systematization of fields. As for the iconographic evidence, a painter in question had depicted an entirely imaginary plow, with wheels mounted behind the plowshare in such a way that it could not have been used by anyone at any time to plow a field. When these facts were brought to light Poni came out rather better than Sereni in the exchange, but Poni still held that Sereni's work was the "valid and obligatory point of reference" for further discussion.

The iconographic evidence, the works of art that Sereni chose to illustrate his *Agricultural Landscape*, add an important dimension to the work, and he would have included further illustrations if publication costs had permitted. It must be admitted that art historians have not responded enthusiastically to the rather direct relationship he hypothesized between the expression of artists and the appearance of their rural environment. "Artifex ex artificio aliquo viso concipit formam, secundum quam operare intendit" (the artist conceives the image on which he plans to work on the basis of another work of art he has seen before) reproved one critic in 1976—Giovanni Romano, citing Saint Thomas Aquinas's *De veritate*.[33] Romano, who otherwise called Sereni's book a "classic of interdisciplinary research," provides a full discussion of its illustrations: the medieval calendars of the months, the *Theatrum sanitatis*, the figuration of saints, Andrea Pisano, Giotto, Ambrogio Lorenzetti's *Buongoverno* in Siena, Leonardo da Vinci, the great Venetian

---

[31] *Gli aratri e l'economia agraria nel Bolognese dal XVII al XIX secolo* (Bologna Zanichelli, 1963), in *Studi Storici* 4, no. 4 (1963): 783–92.

[32] Carlo Poni, "Aratri e sistemazioni idrauliche nella storia dell'agricoltura bolognese," *Studi Storici* 5, no. 4 (1964): 633–74, republished in his *Fossi e cavedagne benedicon le campagne: Studi di storia rurale* (Bologna: Il Mulino, 1982), where Poni cites Sereni frequently.

[33] Giovanni Romano, "Documenti figurativi per la storia delle campagne nei secolo XI–XVI," *Quaderni Storici* 31, no. 1 (1976): 130–201, republished in his *Studi sul paesaggio* (Turin: Einaudi, 1978).

painters, botanical illustration, and the gradual emergence of technical representation. He concludes that Sereni should rather perhaps have paid more attention to archival evidence from the more prosaic agrarian cadastres that had begun to appear in the fifteenth century than to illustration. But one might respond that Sereni did not really confuse the painterly with the "real" agricultural landscape, whose internal relationships remained the discovery of agricultural, economic, and social historians. And in this sense there has been a continuing interest in Sereni's innovative concept of an "agricultural landscape," as evidenced certainly by the younger historians who participated in the international conference on the "European Agricultural Landscape" at Cesena in October 1987, and whose papers were published in the *Annali* of the *Istituto 'Alcide Cervi'* in 1988.[34]

The most recent book on all aspects of the modern history of Italian agriculture is the impressive three-volume work edited by Piero Bevilacqua.[35] The Italian and foreign specialists on different subjects who contributed the sixty-two detailed articles that make up this work cite Sereni frequently and respectfully. The details of knowledge of the Italian agricultural landscape have increased enormously in recent years, but Sereni's work remains, as Carlo Poni said in 1964, a "valid and obligatory point of reference" for the specialist; and it is still the best introduction to Italian agricultural history for the general reader.

The final part of Sereni's *Agricultural Landscape*, Chapter 10, provides an "Agricultural panorama of contemporary Italy" as it appeared from the standpoint of his last revisions of the book in 1961. At that time, the outcome of the Italian post–World War II "economic miracle" was as yet unclear, and Italy had only recently signed the Treaty of Rome that established the European Economic Community (EEC). Readers may wonder to what extent Sereni's picture is still valid, given the rapidly changing situation of the past three decades. In the EEC, Italy, like Great Britain and Germany but unlike France, Holland, and Denmark, has been a net importer of agricultural products. The effects of the policies of the Fascist period that aimed at agricultural autarky disappeared very rapidly in the 1950s, when the Italian government also made the strategic choice of favoring Italy's industrial potential for producing modern consumer goods over the more traditional sector of agriculture and, given Italy's scarce resources of industrial raw materials, also

[34] *Annali dell'Istituto Alcide Cervi* 10 (1988), with an introduction by Rosario Villari. The participants were J. Marino, R. A Butlin, H.-J. Nitz, J. Marterné, M. Goossens, C. Pfister, V. Zimányi, P. Sereno, R. Comba, G. Chittolini, F. Cazzola, C. Pazzagli, and P. Bevilacqua.

[35] Piero Bevilacqua, ed., *Storia dell'agricoltura italiana in età contemporanea*, 3 vols. (Venice: Marsilio, 1989–91).

favored free trade. Italy, within the system of the EEC, has not had such protective policies favoring agriculture as France in recent decades, and the agricultural component in Italian GNP has fallen steadily (from 22.8 percent in 1951 to 6.5 percent in 1981). The "economic miracle" of the late 1950s and the 1960s produced a vast internal migration: between 1955 and 1971 more than a fifth of the population relocated itself, moving chiefly from agricultural zones of the South, Islands, and northeast to industrial zones of the northern center and northwest. Whereas in 1951 43.9 percent of the active population (56.7 percent in the South) was engaged in agricultural pursuits, this proportion had fallen to 12.1 percent in 1981 (21.6 percent in the South). The migration has subsided from its frenetic pace in the 1960s, but this has not stopped the abandonment of use of land for agriculture. Between 1961 and 1971 alone, the extent of cultivated land decreased by 1.5 million hectares.[36]

These changes, and the political economy of the EEC, have produced a new and different market situation—about which Sereni would undoubtedly have had much more to say. Italian agricultural production grew most since the 1950s in areas of export of "Mediterranean products" to the EEC—such as olive oil, wine, fruit, and citrus fruits—at least until entry into the EEC of the other Mediterranean countries: Greece, Spain, and Portugal. Italy is more self-sufficient in agricultural products than might at first appear, but in general the system of internal compensations and subsidies by the EEC has benefited the "beef and butter" interests of northern Europe more than Italian producers. For urban consumers, the tempting tomatoes, lettuces, and zucchini-flowers in Italian markets are still local products, but it often seems that frozen orange juice comes from Spain, and beef from Hungary—by way of Bavaria. The moderate prices favor city dwellers rather than Italian farmers. As well, with the disruption that has resulted from the rural exodus, and the decreasing component of agricultural production in GNP, the political articulation of agricultural interests has weakened, so that there are not very strong cries for agricultural reform from the Italian countryside at the present moment. Still, Italian production of most agricultural items has continued to grow, and partly through encouragement by the government, although alongside some successes of the (now defunct) Cassa per il Mezzogiorno in the South—one might mention the construction of the aqueduct of Puglia—there have also been some conspicuous failures. As well, a policy of industrializing the South began in

[36] On developments since the 1950s, see Paul Ginsborg, *A History of Contemporary Italy: Society and Politics 1943–1988* (Harmondsworth: Penguin, 1990), Guido Fabiani, *L'agricoltura Italiana tra sviluppo e crisi, 1945–1985*, 2nd ed. (Bologna: Il Mulino, 1986), and the essays in Bevilacqua, *Storia dell'agricoltura italiana in età contemporanea*.

1957, so that, although the problem of the South continues, regions once exclusively agricultural now have a more mixed economy.

Another serious problem, ecological deterioration, has worsened in recent decades. Sereni would have had more to say about this as well. The tractors and motor vans that Italian industry has sold to farmers in such numbers, which have helped to supplement the much reduced agricultural work force, and other new modern techniques, have contributed to the ecological deterioration. The delicate terracing of Italian hills was designed to withstand the lighter and surer tred of horses, oxen, and mules; and an unwanted by-product of the chemical fertilizers used in such quantities throughout the Po valley has been the thick smelly growth of algae that now clogs the lagoon of Venice.

Sereni had already perceived effects of most of these trends when he wrote his "Agricultural Panorama," and thus his picture still has much validity. In the South, it is true, land is now cultivated that once was waste, and, overall, Italy with its typical "Mediterranean gardens" still has by far the largest proportion among the EEC countries of very small farmers with five hectares or less of land (76 percent of the farms and 20.5 percent of the farmed surface in 1961, and 75.4 percent of the farms and 16.3 percent of the surface in 1981). Large concerns though, with a hundred hectares or more—like the agrobusinesses of the Po valley and Puglia—have increased in number and size (from .5 to .7 percent of the farms and from 29.1 to 35.6 percent of the surface during the same period). From the train window traveling from Bologna to Milan and on to Turin in the North, one still sees ruins of the famous "Piantata" of the Po valley—although more in the stretch from Bologna to Milan than between Milan and Turin, where rice fields predominate. One also sees another kind of field planted with trees that Sereni had noticed less, scattered tree farms with thick, regularly planted stands of poplars. Sereni did not perceive so clearly the rapidity with which the millennial, hereditary, and culturally stable social class of peasants would fade in the postwar decades as an occupational category, or that there could ever in Italy be agriculture without peasants. But he did perceive clearly the also millennial problem of Italy's ecological deterioration. It remains now for the splintered and re-formed remnants of the PCI, for Italy's small *Partito Verde*, and for others to spur on better planning, or rather orchestration of private interests and public institutions, so as to make more effective use of the Italian countryside for agriculture, and to innovatively restore its truly splendid "bel paesaggio."

IN MAKING this translation—from the 1989 Laterza edition—I have tried to remain as close as possible to Sereni's prized agricultural vocabulary, and thus I have included a brief glossary to help make the translation of agricultural terms clearer. I have also tried to disentangle Sereni's somewhat complex style so as to make the exposition more transparent for readers of English. Sereni published his book without footnote references, to make it more palatable for general readers. But he did make occasional bibliographical references in the text, which Italian experts might recognize but readers of English might not. Thus, when possible, I have added a few bibliographical footnotes to the English text where they seemed most needed to make these references more evident.

R.B.L.
Providence, R.I.
August 1996

History of the Italian Agricultural Landscape

•••••

PREFACE

. . . . .

In the sketch of a history of the agricultural landscape of Italy that we present here to readers, we have tried to collect and express in a summarized and unspecialized way, without erudite apparatus, the results of an investigation that we pursued for long years, up to 1955. In fact, this essay was drafted in 1955, although various events retarded its publication, so that the last chapter on recent developments in the forms of the Italian agricultural landscape has been brought up to date by the author.[1]

This distance of time has produced a fuller awareness of the limits and gaps in our essay; and for this reason particularly, when we finally decided to publish it, we felt the need more than ever to introduce and defend it. "In the development of a discipline," Marc Bloch wrote in the introduction of his *Les caractères originaux de l'histoire rurale française*, "there are times when a synthesis, even if seemingly premature, can contribute more than many analytical studies; when, in other words, it is more important for the moment to formulate questions well than to seek to answer them. French rural history seems now to have reached one of these moments. All I have tried to accomplish here has been the general survey of the horizon made by an explorer before plunging into the undergrowth, where a broader view is no longer possible."

After what we have said of our awareness of the limitations and gaps in our work, we hope that it does not seem presumptuous to refer to this justification of Marc Bloch, who was the first to state the essential themes of historians of the agricultural landscape in his now famous work of 1930. Indeed, he rightly should and must be considered not just the leader but also the founder and pioneer of this new discipline. The pains we took to collect the materials for this essay and define its methodology, and even our awareness of the limits and weaknesses of the results of our labor, have persuaded us more than ever of the need for such an attempt at synthesis, even if premature. Given the state of Italian agricultural historiography, this seems to us even more urgent than it must have seemed to Bloch in France during the thirties.

It is not that Italy lacks worthy students of agricultural history and customs, or human geography, who have delved into individual problems of the typology and development of the Italian agricultural landscape. Through more or less specifically focused works, these have made contributions that we do not wish to belittle but, instead, to utilize to

---

[1][To about 1961, when the first edition was published.—Trans.]

the fullest extent in our work. Precisely the existence and authority of these contributions make it difficult to mention here by name the numerous authors whose pioneering work was so valuable in providing our first orientation into this new and complex subject matter. A young Italian scholar, Gambi, in March 1958, when reporting in the *Rivista geografica italiana* on the results of the first international congress on history and rural geography held at Nancy in September of the previous year, stated rightly the congress's unanimous view "that Mediterranean rural structures, and especially those of Italy (excluding the recent work of Sereni, Zangheri, and Ortolani for Emilia, of Dal Pane, Masi, and Ricchioni for Puglia, of Villari for the Cilento region, and of Le Lannou for Sardinia), had been little studied, and that the most important detailed works (like those of Curis and Cassandro on *usi civici*) have been limited to juridical problems."[2]

In fact, if he does not turn directly to the sources, a scholar most often turns to a broad (and often valuable) reading of the juridical literature when he seeks an initial orientation and wants to profit from a preliminary collection and elaboration of information about the history of the Italian agricultural landscape, even if it is only for some particular region or sector. But in this literature, he inevitably finds the attention of authors focused on juridical institutions: and it is not always clear from the occasional or marginal notes how this juridical institution might be reflected in the reality of the agricultural landscape of the time, or what above all might be its relationship to the technological, productive, or social reality of that age or previous ones. On these themes, to be sure, precious light is shed for the researcher by the study of toponymical and historical linguistics, for which Italy boasts one of the largest and most valuable bibliographies. But to advance his first orientation into the history of the agricultural landscape, the researcher must himself make the difficult combination (so to speak) of the facts of juridical institutions with the results of toponymical or historical linguistic inquiries. Even greater difficulties present themselves if he wishes to examine more deeply the relationship between the reality of the landscape in a certain period and region and the level of agricultural technology of that period or preceding ones. Although the literature on agricultural technology provides abundant information about the modern world, the history of agricultural technology in Italy has still today attracted very few studies. Thus, in many still valuable studies of human geography, the researcher often finds pertinent descriptions and characterizations of contemporary agricultural landscapes, as well as useful considerations on

---

[2] [Lucio Gambi, "In margine al primo Convegno internazionale di storia e geografia rurali," *Rivista geografica italiana* 65, no. 1 (March 1958): 57.—Trans.]

the relationship between these and the technological, productive, and social realities of our own time. But rarely in such studies—which have more a geographical than a historical perspective—will he find direct answers to questions he may have posed about the *history* of the agricultural landscape.

Otherwise, in confronting this specific but complex historical subject, the researcher must not only confront the difficulties just mentioned; no less are those of methodology and even terminology. His first spontaneous impulse will lead him, almost inevitably, to fall back on the formulations of problems and the nomenclature elaborated and adopted by Bloch and the French school, which has the undoubted merit of having opened the path to research on the agricultural landscape. The fact that these formulations of problems and this nomenclature were elaborated, naturally, with an eye turned particularly to French historical reality has not been an obstacle to their progressive imposition on an international scale, particularly in lands like those of central Europe, which have had, and have, a general similarity to France in landscape, agricultural regimes, and forms of property. However, already at the first international congress on rural history and geography it was rightly observed that these formulations of problems and nomenclature, which have now become traditional and almost obligatory in this type of historical inquiry, are inadequate to grasp the reality and history of a landscape of the Mediterranean generally, and of Italy in particular. It is enough to consider, for instance, the almost exclusively "horizontal" orientation, one might say, of the terrain in which agricultural landscapes developed in places like France or Germany; and the decisive importance assumed instead in a place like Italy—with its cultivated land that climbs up more than a thousand meters in altitude, its terraces, and all the variety of its hills and mountains—by what one might call the "vertical structure" of the agricultural landscape. But even if we limit ourselves to the "horizontal structures" of our landscape, we must consider the importance assumed in regions like the Po valley, for instance, by improvements through systematization of waterworks and irrigation, whose quality and extent cannot be explained only by a difference of climate and soil with respect to France or Germany, but instead reflect and express a much more complex nexus of conditions and natural, technical, demographic—in short, historical—agents.

Or consider as well, to cite one further example, the terminological as well as methodological difficulties confronting a researcher trying to describe a type of landscape with *closed* fields, which was quite common in the Italian past, which was characterized by a frequency of contiguous parcels of small size and irregular outline, or might be defined as a varied number of rectangular segments of different lengths and proportions.

The closest approximation of this kind of landscape that we have been able to find in nomenclature used north of the Alps is the term *marrello*, used (in a sense that roughly expresses what we have just described) in some dialects of Provence; while in others it meant a similar landscape, but one characterized by fields *open* to pasturage after the harvest. This last meaning of the word is certainly the most ancient, and in the environment of Provence it genetically and historically explains the first. But so far as we can see, the scientific vocabulary of the French school has not made use of the term *marrello*, which is not even found listed, for instance, in the recent volume of Georges Plaisance, *Les formations végétales et paysages ruraux (lexique et guide bibliographique)*, of 1959, which nonetheless, with its 322 packed pages of rural nomenclature, discharges well the lexographic function expressed in the first part of its title. In fact, French scholars have not paid any particular attention to the specific type of landscape of closed fields in question, which they seem to confuse wrongly with open fields through their general and comprehensive nomenclature. They have thus called it, and diffused the term in international parlance ("faute d'un meilleur nom," as Marc Bloch remarked), a *paysage à champs irréguliers* or *à champs en puzzle*. This last term might be considered, basically, a translation of the Provençal *marrello*. But certainly, at least in an Italian context, where this type of landscape is often not even genetically or historically derived from a landscape with open fields, a term like the French one risks confusing it with other quite different types of landscape with irregular fields, like that of the "Mediterranean garden," for example, or that of "closed pastures." To avoid such confusion, we have been led to revive terms from Tuscan usage, such as *paesaggio dei campi a pìgola* (landscape of irregular fields), which have not seemed to us lacking in expressive vigor, and we have followed the example of such eminent writers on agronomy as Ridolfi and Cuppari in the past century in utilizing them in this essay.

But it is clearly not only through questions raised by terminological difficulties, or necessarily by a new formulation of questions, that the objective environmental and historical conditions of Italy emerge in comparison with other lands where the historiography of agricultural landscapes has a longer tradition. Instead, precisely the inadequacy of certain formulations (and nomenclatures) in the French school's attempt to express an Italian reality arouses critical awareness in the researcher as to the limits and gaps in the formulations themselves. These limits and gaps were often overcome in Marc Bloc's work by the genius and vivid historical sensitivity of the author, but they appear more openly in the orientation of other scholars of the French school, whose very Cartesianism, that linear *esprit de clarté* that adds so much to the effectiveness of their writing, has not always, it seems to us, made it easier to deepen the internal dialectic of historical reality in the agricultural landscape.

Marc Bloch himself was not free from the problem of a kind of hypostatization, one might say, in the typology of the landscape; and even in his magisterial work, what should and must remain at most a classificatory schema, auxiliary to the historical content, sometimes risks becoming a substitute for the substance of the content itself. From this is born, perhaps, that uneasiness which an Italian scholar feels when he tries to contain the reality of a landscape like ours within the schema diffused by the French school. And this uneasiness and difficulty is born not only, or even, from an *ignoratio elenchi* of the facts of our reality, which is inevitable in a classificatory schema conceived for a different subject matter, as much as from something deeper that invests the very notion of agricultural landscape and its internal historical dialectic.

It could be said that the need to deepen this dialectic, which is a characteristic of the whole Italian historiographical and cultural tradition, makes the writer particularly sensitive to the nature of his own personal background; and some details of this will perhaps be useful to clarify the methodological criteria that we have attempted to apply to our work. The particular interest of the present author for the problem of landscape developed through research along two different and distant lines, in appearance quite distinct, and orientated, on one hand, toward Italian prehistorical and early historical agricultural techniques and institutions, and on the other, toward the agricultural history and politics of contemporary Italy. There are thus two levels of investigation, tied to ages that seem irredeemably distanced and separated by millennia for those who think the abstract flow of time devoid of history, but that instead, precisely history, as the continuity of customs shared by a common humanity, nears, ties together, and directly confronts in a kind of eternal "quarrel between ancients and moderns." And, truly, each new generation of humanity cannot take up its own living and current activity except from a reality that the work of past generations has painfully elaborated, imposing on it forms, contours, and well-defined limits. But by grounding itself only on a concrete and well-defined historical reality, human action loses its effectiveness and remains within old limits, in a tired replication of already established forms, always failing to go beyond or to overcome these historical limits so as to infuse present reality with new and original contents and forms.

For the tasks of a historian of contemporary agricultural reality, as for the activity of a reforming politician, the problems of the landscape present and impose themselves first of all as problems of historically determined *facts* to which he cannot help but respond. But inasmuch as these are also problems, and for that reason alone, they are a barrier or *limit*, before which he can by no means stop without the risk of abandoning at the onset the very reasons for his historical investigation and the possibility of innovative action. *Le mort saisit le vif* was always a principle of old French law, which has not lost any of its effectiveness in Italy. Any reformer comes up against it, for example when—in seeking to discern the route of a canal, or a road between farms, or even only a line of trees in the Po valley—he sees himself obliged to follow (or somehow cross with difficulty) some boundary fixed by the grid of a *centuriatio*: that is, by the form that Roman peasants centuries or millennia ago, following their *own* productive and social needs and their *own* traditions, imposed on the landscape of so many parts of Italy. In the Pontine Agro, or in the plain of Crotone, the politics of reform have encountered, and still encounter, the disaggregation of forms of the agricultural landscape and the dislocation of its inhabitants, who go back to generations even more distant from us in time. Somewhere else, it is a farm from the communal period, or an even more recent attempt at improvement from the Fascist period that still leaves its mark on the landscape. These forms present themselves as determined facts, and at the same time as a historical barrier, or limit, to the work of the political reformer, or even to that of some private economic operator who sets out, for instance, to adjust the organization and dimensions of his enterprise to new technological or economic needs.

To resolve this internal contradiction of the landscape, as a determined fact and as a barrier or limit set by the historical process, the political reformer, like the private economic operator, resorts—and can resort—to his own current living action, which overbearingly affirms his own rights against the acts and rights of past generations that are now crystallized and solidified in the form of the landscape. But should not the historian of present-day agricultural reality respond similarly to this Faustian *im Anfang war die Tat*, to this "in the beginning was action," if he does not want to halt before the pure and simple fact of the form of the landscape, but wants instead to find out its logic and historical dynamics. He can only do so by referring to the actions of distant or more recent generations, which he succeeds in making live again for us as a living and current activity, as something *to do* or *to be done*, in effect, rather than as a fact.

At first sight, the position taken toward a certain landscape by a scholar of the pre- or early history of Italy might seem different. For research directed to a distant age, with few epigraphic and archival sources and with few and uncertain literary sources, the landscape in question presents itself as accompanied by and closely connected to archaeological and linguistic sources, as a fundamental historical *document* more than as a fact. The study of toponymics, along with certain very ancient forms that still today persist in the landscape, can provide a precious documentation for the uprooting and migrations of a given population, or its type of settlement, which often is all that lets us give a name and a voice to archaeological evidence that might otherwise remain anonymous and mute. Equal, and not less valuable data can be provided about the agricultural and natural environment in which this population lived, and its productive activity, by archaeological finds and their concentration in given strata, or in a given soil and vegetative environment, and so on. Or, from another point of view, data are provided by historical linguistics: the systematic study, for instance, of the relics of pre-Roman words in regional Latin or in a certain group of Romance tongues. The fundamental importance of the physical landscape as a document of pre- and early historical reality should not be undervalued, and we ourselves have experimented with the use of this kind of material more widely than is generally done these days in the study of pre- and early historical Italy.

But so much for that. More than anything else, perhaps, experience confirms, if there were need, that no document can become a source for the historian without being read and interpreted in the light of philological criteria: not, that is, merely as a pure and simple fact outside of the historical process, but as an integral part of the historical process itself. And thus, when attempting to identify that complex of linguistic toponymical elements in which a certain document of the landscape of an early period of Italian history evolved, we cannot dispense with an investigation of "the who and what" of the people who utilized that form of words, which has no other documentation; and the way that form of words, for millennia silent, could express in a living way, and name, hills and marshes, wastelands and cultivated fields, and the settlements of men and their well-provisioned strongholds. A fact of the landscape can become a historical source for us thus only if we succeed in making it not merely a historical given, or *fact*, but again something those living people *did*, or had *to do*: in their productive activity, in the pattern of their life together, with their struggles, and with their language, which of those productive activities, that life together, and those struggles was the living, productive, and always innovative interpreter.

This indicates the danger—not only for the political reformer and historian of contemporary agricultural reality, but also for the scholar of a historical reality much more remote in time—of a tendency to hypostatize forms of the agricultural landscape that overemphasize their consistency and geographical persistence, shall we say, at the expense of the process of their living and continual historical elaboration. From this arises the particular difficulty of circumstances in which study of the history of the agricultural landscape has developed in Italy, where, if by any, more direct attention is given to the landscape by students of geography (and it is to their indisputable merit) than by students of agricultural history. As well, one cannot say that an early crop of works on the history of agricultural labor and technology has, as yet, offered Italian scholars the quantity of materials needed to effect this change in emphasis of research—that is, to study more closely the process of *elaboration* of the landscape through past human collective action, which was current and living, and which aimed at overcoming those limits imposed on itself which we have emphasized in the preceding pages.

In these conditions—which have only begun to change in recent years, as Gambi properly stated in the passage we cited, through the more direct and specific dedication of a few authors to the history of the Italian agricultural landscape—we have felt clearly the need for a first work of synthesis, like what Bloch, with more authority than ours, called for in France during the thirties; and it has not seemed to us that awareness of the modesty of our means (if not of our dedication to research) should exempt us from an effort to fill this need. On one point, at least, our research, which has long been devoted to the history of agricultural technology, suggested that we should undertake a more complex project on the history of this technology, whose realization would reveal the chief gaps in the field and permit discussion of its most difficult questions. But we were encouraged to publish the substance of that work in this more organic, although summary, form by the interest aroused by some of our early findings, both among the larger public of readers and listeners and among Italian and foreign researchers in quite varied disciplines, who where inspired by these findings to begin new and more specific research projects on Italian agricultural history. Already one of our early essays, "La comunità rurale e i suoi confini nella Liguria antica" (The rural community and its boundaries in ancient Liguria), and a more developed treatment of the natural and agricultural landscape of prehistoric, Roman, and medieval Italy in our volume *Comunità rurali nell'Italia antica* (Rural communities in ancient Italy) had the good fortune to attract the attention of eminent scholars in the history of law, such as De Francisci, Volterra, Chevallier, Pajakowski, of experts in historical linguistics and toponymy, such as Battisi, of archaeologists, such as Raymond Bloch, and were cited in the work of Italian and foreign researchers specializing particularly in the history of the Mediterranean landscape, such as Parain, Champier, Desplanques, and D'Elia.

The new interest in the problems of the history of the Italian agricultural landscape that we had sensed in our research was otherwise confirmed by the flattering reception to the publication of our "Note per una storia del paesaggio agrario emiliano" (Notes for a study of the agricultural landscape of Emilia)—which was the text of a paper read at the congress on "Le campagne emiliane dal Risorgimento ai giorni nostri" (The Emilian countryside from the Risorgimento to the present) held in Bologna in February of 1955—and by a lecture on the history of the Tuscan agricultural landscape given in Florence at the invitation of the Circolo Leonardo. Indeed, a year later, the reception to this lecture gave us the pleasure of putting the typewritten text of the present work at the disposition of Professor Philip Jones of the University of Leeds, who had been entrusted with writing the chapter on the agricultural history of medieval Italy for the *Cambridge Economic History*, and who in Florence meanwhile had expressed the desire to meet us to discuss themes of common interest that were touched on in that lecture. We must admit that the appreciation of our work by such a specialist as Professor Jones, and his courteous insistence, played a decisive role in the decision to publish this essay, even though we are aware, we repeat, of its defects and blemishes.

Among these defects and blemishes is also the fact that the reader—whether a specialist or not, as he may be—will have to seek out the meaning of the style and form chosen in our exposition, which is not the usual one for works of a specialized and erudite nature, but is more suitable instead for works that try to reach a larger public and circle of influence and to satisfy a more casual scientific curiosity. We think there is a large place for curiosity of this kind in a place like Italy, where at every step the variety of agricultural landscapes, and their combination and historical stratification, arouse fascination and a whole series of questions for cultivated but nonspecialized observers, whether they be Italians or foreigners. But it is even, and above all, out of modesty and scientific scrupulousness that we have chosen a form of exposition that is free from any pretense at erudite apparatus. Our research, in fact, was carried out—although with the valid support of the first historical, juridical, agronomic, geographic, toponymical, and linguistic works that we have mentioned—without a broad and direct reference to epigraphic, archival, archaeological, literary, iconographical, or other primary sources. It would not have been difficult for us, though (aside from editorial considerations), to translate the thousands of pages of notes in which we collected the results of our direct use, and our own elaboration, of these sources into an erudite apparatus. But precisely the need we felt to make wide and direct use of the sources, given the current state of literature on the history of the agricultural landscape, has kept us from slipping into what would have been, in reality, no more than an erudite temptation. However wide, in fact, given our own resources and abilities, that direct reference to the sources was, it inevitably had to be articulated through soundings made here and there (although with some criteria of selection) among the enormous masses of usable materials, rather than through a systematic exploration, which would naturally have to be entrusted to a larger group of scholars. Nor was it possible, without outstepping the bounds of space imposed on us, to give our exposition a more problematic character, punctuating it with questions, or with proposals for alternate solutions. And in this way, reference to an erudite apparatus would have risked attributing a definite (if not definitive) certainty to our conclusions, which they are far from being able to support.

Somewhat similar considerations have guided us in other aspects of the selection of the form of exposition, and in the choice of illustrations for this volume. It will perhaps seem strange, even to our most well disposed critics, that we have made only exceptional use of maps drawn from cadastres. These would, undoubtedly, not only be the most pertinent illustrative material, but also one of the fundamental documentary sources for a study of this type. It should be indicated, however, that Italy is in a backward state in precisely this area, not only in the publication and study, but even in the location of usable materials. Thus, in this direction more than ever, our personal research, except for the most recent periods, has had to take on the character of a limited sampling, and it would be misleading to ascribe an exaggerated significance to the results when they are not supported by firm evidence from other sources. On the other hand, it has seemed that a collection of iconographic sources of a quite different type—that is, artistic expressions, with that salient representativeness and intuition of the "typical" that a work of art offers—might not only provide more suggestive illustrations to the reader, but would also be more pertinent to the character and limits of our enterprise. Our review of these iconographic sources, more than two hundred thousand reproductions of artworks or their details from every period, has been, we believe, relatively exhaustive, and it was quite instructive. We were able to select, besides abundant materials relating to the history of Italian animal breeding, cultivation of crops, and agricultural technology and labor, at least several hundred iconographic details relevant to the specific object of our study, and we only regret that the editor's wish, and our own, to make this volume accessible to the largest possible number of readers by not encumbering it with excessive expense did not permit us to include here more than a small part of this illustrative material. It should be made clear, however, that we have used the illustrations not so much for documentation as—where their representativeness is guaranteed by other sources—for materials simply to illustrate our exposition.

For the purpose of this exposition, as well, the specialist will perhaps be surprised that we have often cited poems, georgic or other, or the more or less casual testimony of some Italian or foreign traveler, rather than referring to an archival document. This reflects not only our effort to make reading this volume less unpalatable for the nonspecialist, but also our just-stated concern to avoid erudite apparatus in the exposition of the results of research with the limits of ours. Here what we have just indicated about the choice of illustrations also applies. Where, for objective or subjective reasons, it is not possible to sift completely and systematically through the sources, an "involuntary" literary or artistic testimony, when supported and confirmed by other sources, can assume, for its ability to express the "typical," a representative value that would otherwise be sought only in the severe probability of a statistical analysis.

But there is perhaps something further that we should not conceal from readers. For scholars in individual disciplines who are aware of the unity of the historical process, there is always an uneasiness about the nonetheless necessary specialization of research, which risks fragmenting this oneness into many distinct strands: parallel, certainly, but for that matter also destined to come together again as a unitary process. Through the course of our research in agricultural history, what has particularly engaged our interest in the history of the landscape is precisely the fact that, in this discipline, fragmentation tends, at least partially, to recompose itself, to become history again. Thus, for example, it would be impossible to make any sense out of the landscape of Tuscany by referring only to the history of the technology or agricultural relationships of the region, without turning also to the social and economic development of the society of the communes, with their citizen life, their trade and products, their internal political struggles, and so on. But even by referring to this broader reality, we would not be able to explain fully the agricultural landscape of Tuscany, in what made it different from Lombardy, for instance, if we considered its formation apart from the historical reality of Tuscan *culture*, where the taste of peasants for a "bel paesaggio" in agriculture was born from the same root as the taste of Benozzo Gozzoli for a "bel paesaggio" in painting, or that of Boccaccio for a "bel paesaggio" in poetry in his *Ninfale fiesolano*. Thus, even in this way, we have assumed a task and adopted a form of exposition that reflects, to the extent of our abilities, the need for and awareness of unity in the historical process.

Only the reader will be able to judge to what extent these intentions are successfully realized in the pages of this essay, and thus to what extent it fills the need for unity that has imposed itself on our research. But however severe his judgment, we can assure the reader that more severe certainly is the judgment of one who, like ourselves, has been, and remains, directly involved in this research. What we are presenting here as first deductions and conclusions have already for some time seemed merely to be the source of new critical doubts and questions, and points of departure for new investigations. And if this could be the telling point of our essay for the larger public of readers, scholars, and critics; if this volume could succeed in arousing a more lively critical interest in the problems of the history of the agricultural landscape and a broader commitment to research in this area, we would be led to believe that the fruits of our commitment and labor were not useless.

. I .

NATURAL LANDSCAPE AND AGRICULTURAL LANDSCAPE

IF agricultural landscape means, as it does, *the form that man, in the course and for the ends of his productive agricultural activity, consciously and systematically imposes on the natural landscape*, it does not seem that one can speak of an Italian agricultural landscape, in the proper sense of the term, for periods before the Greek colonization and the Etruscan urban expansion.

To be sure, from the late Bronze Age onward, one can find traces on the soil of the peninsula and islands of the first agricultural activity of ancient populations; and these traces assumed an increasing significance in the Bronze and Iron Ages with the passage from primitive agricultural systems "of the hoe" to those of cultivation with the plow. From that time onward, in fact, the new productive activities of man must have imprinted themselves more deeply than had the more ancient activities of hunting and gathering on the natural landscape of Italy: its woods, thickets, and meadows began to be mottled by burned spots and clearings. Still, on the eve of the Greek and Etruscan colonization (and, in many parts of Italy, up to a much later period), the extent of areas where the new agricultural activity was exercised remained quite limited. And still more: even within these limits, the dominant agricultural system remained one of temporary clearings (*campi ed erba*) or burned spots (*debbio*—the equivalent on woodlands) where, as is well known, cultivation remained precarious and was practiced on virgin terrain, which, once its natural fertility was exhausted after one or a few productive cycles, was abandoned to be overrun again by spontaneous vegetation.

In these conditions, one clearly cannot yet speak of an agricultural landscape, of a form, that is, which the agricultural activity of man had consciously and systematically implanted onto the natural landscape. The very sporadic nature of cultivation prevented it from assuming a systematic stability and continuity, so that the new and precarious forms of landscape, which were nonetheless created by the first agricultural activity of man, can be considered at most to have been only elements of an agricultural landscape whose more solid lines would become precise and fixed only at a later period.

The possibility is not to be excluded that already before the Greek and Etruscan expansion the indigenous populations of peninsular Italy and the Islands had begun to develop, here and there, less precarious forms of fallow (*maggese*) from the traditional agricultural system of temporary clearings, where—it is important to emphasize—the cultivated land, once cleared, was permanently protected from the invasion of spontaneous vegetation, while its renewal of fertility was assured after each cycle of cultivation not only by a year of rest but also by regular tillage of the fallow, and eventually by more or less plentiful deposits of dung or manure.

There is no doubt, however, of the decisive role played by the Greek colonization and Etruscan expansion in the diffusion and final predominance of the system of fallow. Only with this development, which separated cultivated land from land abandoned to spontaneous vegetation, did an agricultural landscape begin to differentiate itself from the natural landscape of Italy. The new and lasting forms that agriculture implanted on this landscape became all the more precise and significant in this period because the spread of the system of fallow was also closely tied to the technical, economic, and political expansion of the Greek and Etruscan cities, which—in contrast to the indigenous populations who were left behind, for the most part, in the typical social forms of the age of tribal hunting and gathering—had already elaborated, in their lands of origin or in Italy itself, new forms of the private ownership of land, a territorial and state organization, and the productive, social, and political relationships inherent to these.

From that moment, the continuity of forms assumed by the agricultural landscape expressed not only the basic facts of geographic and climatic reality but also the new technological relationship between man and nature that was elaborated in the system of fallow. At a single blow, one might say, from this new relationship arose other aspects of the associational life of men: new social, political, and religious forms of property that also expressed themselves and were reflected in the form of the agricultural landscape.

. II .

ANCIENT ITALY

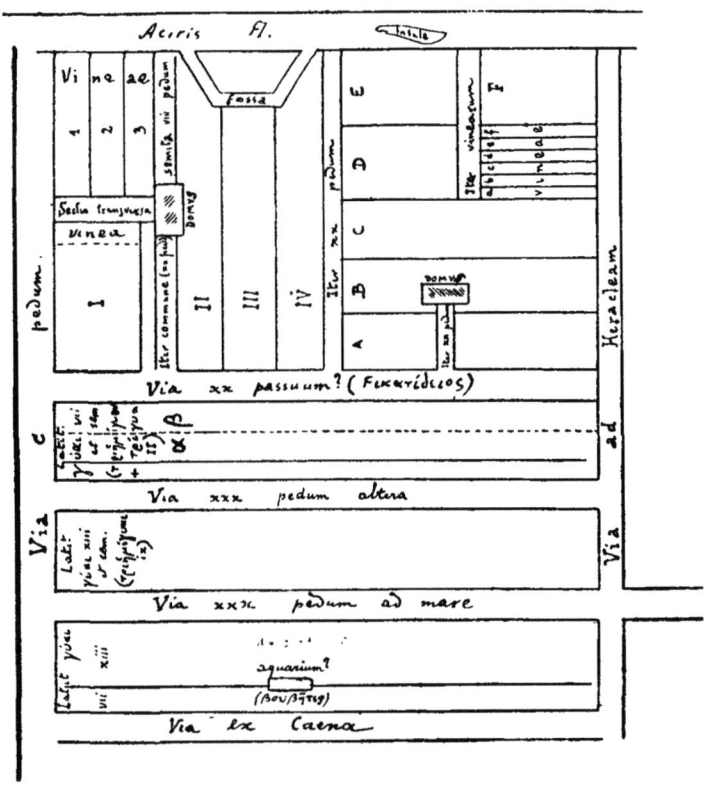

Figure 1. The lands of the temple of Athena Polias in Eraclea di Lucania, in the fourth century B.C.

· 1 ·

The Agricultural System of Fallow and the Landscape of Greek Colonization

Diodorus Siculus tells us in his *Histories* how, after the destruction of Sybaris (510 B.C.), the new Panhellenic colony of Thurii was founded (in 446 B.C.) according to a regular geometric layout, and that the land near the city was distributed to the first colonists according to the same orthogonal plan, utilizing parallel lines. It does not seem, in the foundation of Thurii, that this rigorous geometric plan was inspired by the rational urbanistic doctrines of Hippodamus of Miletus. Well before Hippodamus, at any event, although not in such a rational or rigorous way, orthogonal plans were adopted when colonies were founded, not only in the urban layouts, but also in the distribution of plots of land to the colonists for cultivation.

We have already indicated how the continuity and stability of geometric forms that the agricultural landscape now assumed for the first time were strictly inherent in the agricultural system of fallow, and the new property and productive relationships that this produced. Unlike what occurred in the old system of temporary clearings, cultivated lands were now firmly separated from uncultivated lands, or pasture, and were divided into *fields*, which had become the object of stable appropriation and had to be defended along their boundaries against abusive grazing or other usurpations. Hedges, walls, ditches, rivers, main or local roads thus designated the outlines of the agricultural landscape, where frequently, at least in the plain, the fields assumed regular geometric forms.

A document of inestimable value in the form of an inscription, the *Tavola di Eraclea*, permits reconstruction of the outlines of this agricultural landscape exactly as it presented itself toward the end of the fourth century B.C. on lands belonging to the temple of Athena Polias in Eraclea di Lucania. One notices particularly, in Figure 1 (which reproduces a plan drawn by Kaibel from the inscription), the regular geometric form of the fields, served by main and traverse roads, and the smaller divisions utilized for the cultivation of vines.

· 2 ·

*Greek Colonization and the Agricultural Landscape
of the Mediterranean Garden in Sicily*

In the foundation of a colony, like that of Thurii, or in the arrangement of the lands of a temple, like that shown us by the *Tavola di Eraclea*, the elaboration of the agricultural landscape in geometric form did not result from the spontaneity of individual initiative. The distribution of plots to colonists, or the arrangement of the fields of Athena Polias, was regulated by designated magistrates who operated on the basis of a *plan*. Whenever, even in later periods, we are confronted by an elaborated plan of colonization, as in the period of great clearings of the tenth through thirteenth century or with modern improvements, we find elements of a landscape elaborated according to regular geometric forms as in the *Tavola di Eraclea*. But from the time of Greek colonization, as in later periods, precisely the private appropriation of land, which was inherent in the agricultural system of fallow, brought even to planned colonies an individual, casual, and arbitrary element, which was further complicated by hereditary divisions, and by exchanges of property through sale or rental. The geometric regularity of the agricultural landscape thus became fragmented, minced, and so contorted in the arbitrary configurations of parcels that its form seemed to escape any norm, if not that of individual initiative.

Wherever, as in Sicily and in Magna Graecia, particularly close to cities and on the slopes of hills, the spread of cultivation of trees and shrubs assumed a greater significance relative to grass and fallow, this minced and contorted agricultural landscape acquired an appearance that remained typical, from the age of the Greeks up to the present, as the landscape of the so-called Mediterranean garden.

In Figure 2, the map that Sicca was able to redraw from another precious Greek epigraphic document, the *Tavola di Alesa*, shows us the elements of this landscape as they presented themselves in the first century B.C. in the suburbs of the city of Alesa (near present-day Tusa in the province of Messina). One sees, on a slope watered by streams, the irregular form of pieces of land, divided by small walls, ditches, and so on, and studded with various types of buildings. The landscape of the Mediterranean garden was a landscape of *closed* irregular pieces of land, dominated by the necessity of protecting the trees and shrubs from grazing animals, and their fruits from rural pilferage.

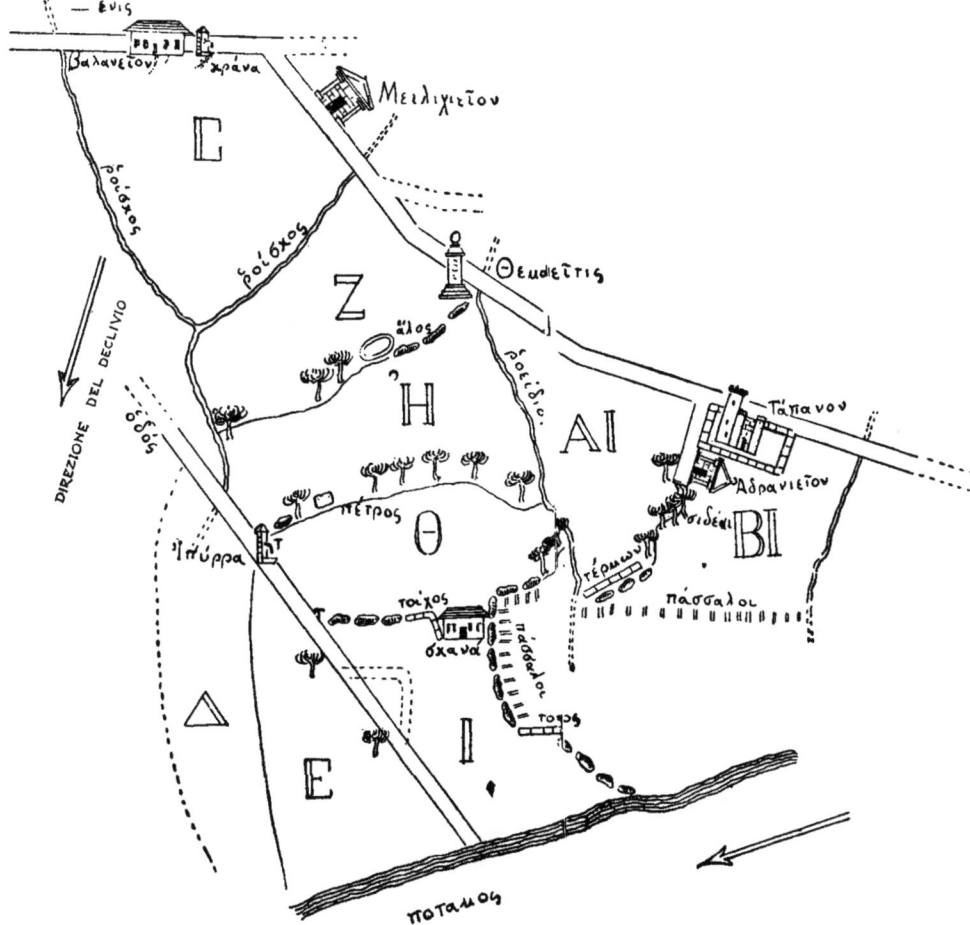

Figure 2. The landscape of the Mediterranean garden in the *Tavola di Alesa*, in Sicily, in the first century B.C.

· 3 ·

The Etruscan Urban Expansion, the Gallic Invasion, and the Landscape of the Piantata *in Central and Northern Italy*

The geometric scene of the landscape that we have found in the urban layout and distribution of plots of land for cultivation in the Panhellenic colony of Thurii, is found again, although with some variants inherent in a different tradition and a different elaboration of the same scene, in the Etruscan urban expansion: for instance, in the plan of an unnamed Etruscan city whose remains were discovered near the present-day Marzabotto. In the elaboration of the landscape of central and northern Italy, however, already in the time of the Etruscans, an element of the landscape began to take on particular significance that was lacking in the area of Greek colonization and that would be decisive in the area of Etruscan domination—that is, a system for cultivating vines in the cooler and richer soil of central and northern Italy, from the region of Capua to the Po valley, that differed from the Greek one, and left more scope to the robust vine branches that were allowed to run in long festoons high above the ground, and were eventually given a living support. Whereas in the area of Greek colonization, vines cultivated in a specialized way on low trees or dry stakes (*a palo secco*) made their typical mark on the landscape of the Mediterranean garden, in Etruscan territory, instead, a different system permitted mixed cultivation. Here, with vines raised high up, and eventually bound to poplars, maples, and elms, there are also examples of the cultivation of cereals in the same fields.

It is difficult, in the present state of research, to know whether this system of cultivation of vines was already practiced in the period before Etruscan colonization by the Paleolithic and other indigenous populations of the Po valley, who certainly, in any case, picked and used *labrusca* grapes: a type of wild vine, whose long vine branches in that climatic zone must have spontaneously laced themselves in the tops of elms, maples, and poplars. Up to our own time, the name *lambrusco* continues this ancient cultural and linguistic tradition. As for the techniques for cultivating vines, one cannot exclude the possibility that the nomenclature referring to the system of long vine branches originated in the period before Etruscan colonization, which was, however, still the first to which one can attribute a normal diffusion of the cultivation of vines through this area. Linguists currently ascribe terms like *rumpus, rumpotinus, rumpotinetum* to an ancient Mediterranean linguistic base, intermixed with Etruscan and then Gallic elements. Latin writers on agricultural matters used these terms to refer to the long vine branch (*tralcio lungo*), a "tree married to the vine," and "planted with trees with vines," and these terms still continue today in the romance tongues of the zone in question (for example, in the Lombard *romp*, "festoon of vines"). The very zone of the diffusion of these remnants, and of vines raised high up, seems to confirm the important role the Etruscans must have had in the propagation of this system. Thus still at the time of the Pyrrhic wars, according to what Pliny tells us, the system of cultivating "married" vines with long vine branches was common not only in the Po valley, in Tuscany, and in the Agro of Capua (where it is still to be found in our own times), but also in the area around Rome, at Ariccia, for example, a region, in fact, earlier under Etruscan rather than Greek cultural influence.

In the later Roman period, however, the *piantata* of trees with vines was commonly called not *rumpotinetum*, which was presented as a term of local technical nomenclature, but *abrustum gallicum*, that is, "planted in the gallic way," not particularly because these populations had a significant part in the diffusion of this way of cultivating vines (which they certainly did not practice in their own lands beyond the Alps), but because of the fact that precisely the Po valley, Cisalpine Gaul, where Gallic invaders replaced Etruscan colonists, was now the zone where the system of planting trees with vines had begun to impose its form most characteristically on the agricultural landscape. This was true particularly in the period following the Roman conquest of the Cisalpine region, when the new colonists, who were accustomed to consuming wine in a climatic environment that was more naturally suited to cultivating vines, must often have made use of the indigenous system of the *arbustum* to extend cultivation to low-lying wet areas, where vines cultivated in the low-lying Greek manner would produce poor results. That it was in fact an Etruscan cultural tradition, and not just particular environmental conditions, that affected the spread of the *arbustum*, seems confirmed by the fact that in western zones of northern Italy influenced by cultural radiation from the Greek colony at Marseilles, the predominant system for cultivating vines remained Greek, in fact, on a low tree (*ad alberello*) or on a dry stake (*a palo secco*), and, in the technical nomenclature of this area, one can find linguistic traces of this Greek influence up through a much later period.

It remains certain, however, that still under the Roman Empire the diffusion of the *arbustum gallicum*, whose area coincides, substantially,

Plate 1. Vines on trees in the frieze of the Casa dei Vettii at Pompei.

with that of the greatest extent of Etruscan domination, was much greater than it has been more recently. This does not mean, it is well understood, that it covered areas comparable with those planted with trees and vines at a later period; and even less that it presented those regular and elaborated forms that began to appear only after the communal revival and in the Renaissance, particularly with regard to the hydraulic system of the terrain. But otherwise, already in the Roman period, systems of cultivation of vines in festoons, *a tralcio lungo*, presented a variety of types, of which the famous frieze in the house of the Vettii at Pompei, reproduced as our Plate 1, provides an illustration.

· 4 ·

The Landscape Plan of the Roman Conquest

With the geometric plan of plots and fields in the system of fallow, with the polygonal irregularity of contours of the Mediterranean garden, and with the squared rows of trees in the *piantate* of central and northern Italy, the Greek colonization and Etruscan urban expansion formed, as we have seen, the oldest elements of the Italian agricultural landscape. But only the Roman conquest and colonization gave universal validity to the form of this landscape, with a definitive triumph of the system of fallow over that of temporary clearings. This validity now no longer resulted from the initiative of a single urban and colonial foundation, as it had been in the Greek and Etruscan age. The Roman conquest affirmed it as a systematic and general plan of colonization, where the form of the agricultural landscape became in fact a sign of the juridical condition of the population subjugated and the lands conquered, in private arrangements as in those of the dominant city.

By *form* was meant precisely, in the terminology of the Roman land surveyors, the cadastral map through which, according to the principles and methods of geodesy, this plan of systematic colonization was realized for each colonial foundation, or at any event in which the landscape of a given territory was reordered and redistributed according to the principles and methods that corresponded to the new agricultural and juridical system that the Roman conquest spread and imposed. And the Roman measurement and division of the agricultural terrain, the *limitatio*, was in fact a universally imposed form, which expressed itself in the landscape through the tracing of two fundamental lines, the *decumanus* (generally from east to west) and the *cardo* (generally from north to south), and others parallel to these at fixed intervals. The result was a regular grid imposed over the agricultural terrain, most often in the form of *centuriae* (squares with 710-meter sides, and thus an area of about 50 hectares), so that this operation was usually called a *centuriatio*; but in other cases there were plots of rectangular shape that were variously orientated (called *strigae* and *scamna*). From a codex of ancient *gromatici* (the Roman writers on geodesy), Figures 3 and 4 show the plan of the *centuriatio* of Minturno, and of Suessa in the territory of the Aurunci. As also appears from the map, not all of a community's land was necessarily divided up and assigned, and an important part of the exempted territory was generally destined for common use in pasturage, wood gathering, and so on.

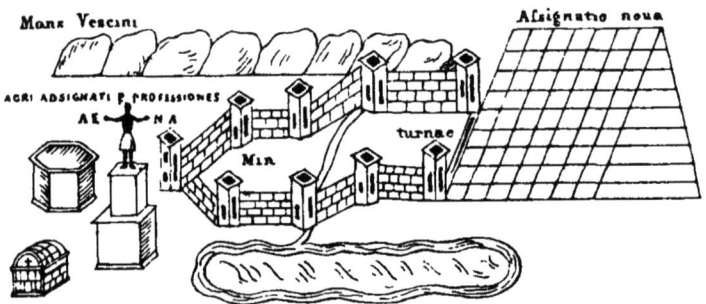

Figure 3. Old and new assignments in the *centuriatio* of Minturno, from a codex of ancient *gromatici*.

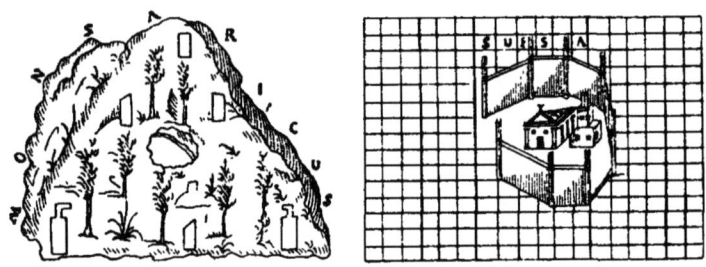

Figure 4. Assignments of lots and undivided lands in the *centuriatio* of Suessa.

· 5 ·

Roads and Aqueducts in the Roman Agricultural Landscape

Goethe rightly observed, in his *Travels in Italy*, that in ancient Rome the techniques of building assumed such importance as to give the landscape resulting from them the aspect "of a second Nature, operating for civil ends." This was true particularly for the landscape of the Roman *limitatio*, with its regular grid work that extended through many parts of the plains of Italy. And it might seem, at first glance, that the novelty of this landscape, in contrast with the Greek and Etruscan one, lay precisely here, in this greater extent, rather than in a difference of lines and forms: in this quantity, that made it a truly massive reality, "a second Nature, operating for civil ends."

Still, an absolutely new meaning, even in its forms, and not only the quantitative one of the size and extent of this type of agricultural landscape, was assumed by two constituent elements in the grid work of the Roman *limitatio* that emphasized the specific difference with Greek and Etruscan colonization. The *decumanus* and the *cardo*, the great lines, in fact, like the lesser ones that ran parallel to them, were not only ideal lines; their traces indicate the firm, stable outlines of cultivated plots, and sometimes the routes of a major and minor system of roads, whose development has no precedents as to extent in the Greek and Etruscan colonies. These routes inserted themselves organically into the network of roads of the Roman Empire, and became the decisive means for the spread and capillary penetration of its agricultural, economic, social, juridical, and administrative systems, and also of its language and culture. The quantitative extent of the Roman *limitatio* cannot be understood if one does not also understand the new form and meaning that the lines of its network assumed, both as boundaries and as a system of roads.

It remains true, however, that the Roman agricultural system and layout of the landscape did not mark a substantial progress over the Greeks and Etruscans from the point of view of the irrigation system of the agricultural terrain. Only in a much later period, as we will see, would this assume a basic significance in the Italian landscape, when the regular traces of drainage ditches and lines of trees, and a new dimensioning of fields, would come to give its modern form to the landscape of arrangements in the plains and hills. Still, to provide water for a population of colonists that became more dense from the Roman period onward, the

Figure 5. The Roman road system of the imperial age in the Italian agricultural landscape.

aqueduct became, along with the road, a basic and typical element in the Italian agricultural landscape. Thus, for example, with all the significance of "a second Nature, operating for civil ends," we are presented, in an engraving of the early nineteenth century reproduced as Plate 6, with the remains of the Roman aqueduct at Acqui, in Piedmont; while our Figure 5 shows the development attained by the Roman road network in the imperial age. Up to the present the great routes of communication often followed its traces.

. 6 .

The Roman Form of the Italian Agricultural Landscape

In ancient Italy, with an amplitude and rigor without precedent in Greek and Etruscan colonization, the Roman conquest liquidated the remains of the *Ius gentium* and the communitarian economy of the populations it dominated, which could still have coexisted with the old agricultural system of temporary clearings; and it developed and heightened the influence of the new socially productive forces that were inherent in the system of fallow and plantations of trees. In a juridical and political sense, property relationships were now made fully parallel to relationships of production in the chief state founded on a slave economy in classical antiquity.

Only the immense accumulation of human energy assured to a great state like Rome by the exploitation of servile labor can explain the massive and relentless power of "the second Nature, operating to civil ends," to which Goethe referred. From this arose also the pattern that the form of the Roman *centuriatio* impressed, along the system of great roads of republican and imperial Rome, on the network of local roads and fixed boundaries of a large part of the Italian plain, on whose landscape it made what even today remains perhaps the greatest and most lasting imprint.

In Figure 6, which is taken from a map from the Istituto geografico militare, we show, as an example, how this imprint can still be seen in our own days in the area around Cesena, where the basic lines of the agricultural landscape are still marked by the network of the Roman *limitatio*, whose sides measured about 710 meters, equal to the 2,400 feet of the Roman *centuria*. Throughout the peninsula, from Ivrea and from Cuneo to Capua, and particularly in the Po valley, from around Turin to Novara, Tortona, Pavia, Cremona, Padova, Piacenza, Modena, and Forlì, the basic lines of the *centuriatio* often determine the orientation of fields and rows of trees still today, as well as boundaries and local roads, so that even the most recent works of improvement or reorganization of the land must often insert themselves into this preestablished pattern, and adjust their forms to it.

We find ourselves here confronted with one of the most typical cases of what we could call the "law of inertia" of forms of the agricultural

Figure 6. The imprint of the Roman *limitatio* in the territory of Cesena.

landscape, which once fixed in determined ways tend to perpetuate themselves, even when the technical, productive, and social relationships that conditioned their origins have disappeared, until new and more decisive developments in these relationships come to derange them.

· 7 ·

The Lands of Common Pasturage, and the Agricultural Landscape of Pasturage in Ancient Rome

Despite the decisive progress toward realizing a complete cycle of agricultural fertility produced by the system of fallow in contrast with the system of temporary clearings, a technological limit to the development of productive forces in agriculture remained in the lack of an organic relationship between agriculture and raising animals, which was an essential condition not only for the full restoration of fertility to the soil, but also for a maximum yield from the solar energy collected in crops and chemically transformed for use by man in the form of "rich" foods such as meat, fat, and milk.

Throughout Europe, up to the introduction of artificial meadows and other types of fodder crops that spread only at the end of the eighteenth century, to provide pasturage for animals in a farming system that did not produce sufficient forage crops farmers turned to two different methods whose relative importance had a decisive part in the configuration of the agricultural landscape. Wherever—as in the system of temporary clearings, and later in the "three-field" system that dominated central and northern Europe up to the modern period—all the cultivated land and fallow of a given community was abandoned after the harvest to promiscuous pasturing of animals by all members of the community, a landscape of "open fields" prevailed, which was uninterrupted by any pattern of boundaries marked by dividing hedges, ditches, or local roads. Wherever, instead—as was more common in the Mediterranean "two-field" system with its biennial cycle of fallow and grain—other people's animals could be excluded from pasturing on the fallow and stubble of private fields, a landscape of "closed fields" tended to emerge, where boundaries were consistently fixed and had a significance unknown in the other system. But the provision of basic forage for agricultural enterprises had to be assured, in this case, by setting aside special pieces of land that were kept from cultivation for the promiscuous pasturage of animals by all members of the community, and by neighboring proprietors.

The agricultural landscape of ancient Rome, as we have already indicated, appears to have been landscape of "closed fields" (or rather plots), from which the rigor of Roman civil proprietary law excluded any promiscuous use. But this led, as we have seen, to supplementation of

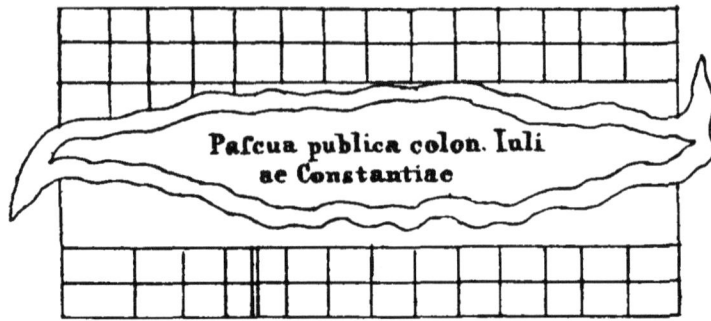

Figure 7. Lands assigned to *pascua publica* in the colony of Iulia Constantia

Figure 8. The *silva* and the *pascua publica* in a map made by geodesists.

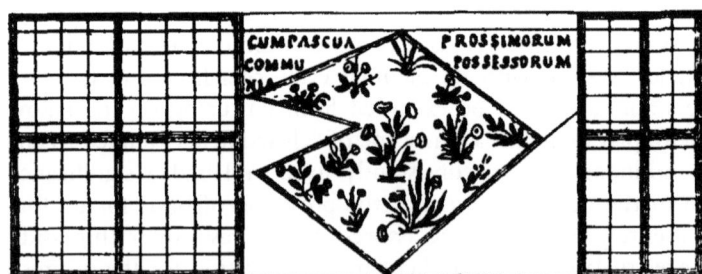

Figure 9. The lands of the *compascuo*, open to the use of nearby proprietors, in a map made by geodesists.

the basic forage of the agricultural enterprise, not only with shoots and boughs taken from trees around the fields that were widely used for feeding animals, but also through promiscuous pasturage on public lands, which were originally open to use by the whole community (Figures 7 and 8), or on the *compascuo* (Figure 9), which was open to use by the community or by nearby proprietors. Such lands were an integral part of the Roman agricultural landscape.

. 8 .

The Rustic Villa and the Landscape of the Plantation

In the most ancient system of Roman agriculture, the cultivation of trees and shrubs spread relatively late, and with insufficient importance to affect significantly the form of the agricultural landscape. When such cultivation had already developed in the age of the great tyrants of Sicily and Magna Grecia from the small closed and irregular parcels of the Mediterranean garden into great "plantations," whose layout and care were entrusted to an abundant slave labor force, this process was just beginning in Rome. Only after the Samnite and during the Punic Wars, with the deep technical, economic, and social transformations that overturned the ancient system of Roman agriculture, did the importance of an economy of plantations grow, until it affected the form of the agricultural landscape to a degree not reached in the Greek or Etruscan age. In the first half of the second century B.C., for Cato the Elder, the vineyard was at the top of the list of types of cultivation, and the vineyard and olive grove that he discusses in his *De agri cultura*, with areas of 100 and 240 *jugeri* respectively, were no longer small family-sized plots that conformed to the Mediterranean garden, but true plantations that employed an increasingly large servile labor force.

Along with the growing use of slave labor, and the ever growing expectation of a plantation economy, went the decline of the old type of directly cultivated small holdings and the growth in importance of great agricultural estates worked by slaves, the *villa rustica*, with its buildings devoted to slave quarters and workshops, its storehouses for crops (*villa fructuaria*), and eventually the pleasure house of the owner (*villa urbana*).

In Plate 2, from a Roman mosaic in the Museo del Bardo, in Tunis, is reproduced a detail of the landscape of a *villa* and its attached plantation as imagined by an artist of the imperial age. One sees the layout of buildings, and the disposition of cultivation of valuable trees and bushes, arranged in good order, around the center of the estate, as was usual, clearly for convenience in cultivating them and watching over the fruit.

The arrangement for planting in this villa seems to have been the *quincunx* (four trees in a square with one in the middle), which, along with rows, was the most common in ancient Rome. The arrangement

Plate 2. Roman *villa* and plantations in a mosaic from the Museo del Bardo.

for growing vines varied, but low growth or dry stakes, which was usual in Greek colonies, continued to prevail in the greater part of the South and Liguria; the method involving live trees that was common in Etruscan colonies, and was adopted also by the Celtic populations of the Cisalpine region, already predominated in the Po valley and the region around Capua.

. 9 .

The "Bel Paesaggio" of the Villa Urbana

In the first half of the second century B.C., in Cato the Elder's work *De agri cultura*, a sense of the agricultural landscape and preoccupation with the details of its forms still seem distant from the author's mind. We are in an age of transition, when the new dominant classes were still employed in founding their economic, political, and military power. But already in Varro, who published his *De re rustica* a little more than a century later, in 37 B.C., there is a clear awareness of a specific form of the agricultural landscape, conditioned by the requirements of cultivation ("quam sationes imponut"), a form he contrasts to that of natural landscape, given by nature itself ("quam natura dat").

Wherever man's agricultural activity begins to impose consciously elaborated forms on the agricultural landscape the way is opened to an evaluation of these forms not only in a technical and economic, but also in an aesthetic manner. Already in Varro, in fact, there is a clear preoccupation with the form of the agricultural landscape that looks not only to *utilitas* (as he writes), but also to the satisfaction of aesthetic requirements and pleasure (*venustas, voluptas, delectatio*). As for the specific form and type of plantations with trees and shrubs, it was now common to speak of *arborum et vinearum ratio*. But in a villa like the one depicted in Plate 2, for example, the order of planting, more than *ratio*, more than in geometric order, is a *kosmos*, as it was now called in the new Hellenic terminology; the order of the landscape was *beautiful* as well as being rational and useful. These aesthetic requirements were otherwise identified by Varro along with those of rationality and utility not only because the beautiful order of fields and plantations increased their productivity, but also because the "bel paesaggio" had now become an attraction for buyers of estates: it increased their monetary value (cf. Varro. *De re rustica*, I, 4 and 6).

We are at the beginnings of a process whose development clearly reflected an increasing differentiation of the classes of proprietors, whose accumulated riches now provided a margin for less primitive interests and pleasures than those of a purely economic type. The new Hellenic culture of the dominant classes undoubtedly influenced this evolution of taste, but it also corresponded to an urban development that led one to seek out in rural landscapes, in real ones as well as in fantastic ones of poetic idylls and pictorial landscapes, an alternative to the growing ten-

Plate 3. The "bel paesaggio" of country villa in a wall painting from the Casa del Centenario at Pompei.

sions of city life, and an increasing contrast between city and countryside. Thus the regular geometric forms of the "bel paesaggio" in the plantation were still conditioned by productive needs, but in the garden and pleasure villa—the *villa urbana* as it was usually called, although it was located far from the city—the "bel paesaggio" assumed an autonomous value, and even unfolded in bizarre forms that were entirely disrelated to any productive need. In Plate 3, from a wall painting from the Casa del Centenario in Pompei, which was conceived in the new taste for picturesque landscape, we see the form of one of these country villas where the detachment from productive needs assumed a particular significance. Did not Horace (*Odes*, II, 15), with poetic hyperbole that was not without foundation in reality, complain that the *regiae moles*, the princely building of pleasure villas, did not leave enough land free for the plow, and that "celibate plantains," and other ornamental plants, were taking the place of elms, which more enterprising ancestors had usefully married to the vine?

. 10 .

The Sylvan-Pastoral Landscape of the Saltus

The "bel paesaggio" of the villa urbana, even if it extended to quite extensive sectors of the agricultural landscape in the imperial age, was not the product of broadly distributed, socially productive forces that permitted ordinary mortals to enjoy its value. It remained rather the product of a huge concentration of wealth in the hands of privileged citizens into whose private treasuries flowed the greater part of the resources of all of Italy, and then all of the Roman Empire, and who meanwhile employed these resources not only in productive enterprises and works of beauty, but also in folly and decadent dissipation. Even the considerable investments that were needed to create and operate plantations presupposed this concentration of riches, so that the necessary investments, productive or voluptuous as they may have been, that allowed plantations and villas to spread through relatively extensive zones of the peninsula diverted more and more resources from other parts of the agricultural landscape. These were reduced to more meager forms, like those that supported the needs of the extensive, but impoverished, pastoral economy.

The pastoral economy, as is known, which was based chiefly on usurpations onto public lands and the large-scale employment of slave labor, also had a decisive part in the concentration of riches and formation of the great patrimonies of the new dominant class. Pliny thus recalls a Claudio Isidoro, among others, who, although he had lost much of his wealth in the civil wars, left in his will 4,111 slaves, 3,600 pairs of oxen, and 257,000 head of farm animals!

We will not expand here on the agents that already at the end of the republican age had begun to condition new growth of the pastoral economy, and a gradual decline in cultivating cereals in many parts of the peninsula. It is enough to indicate that growth of the pastoral economy brought about a notable extension of the landscape of the *saltus*—that is, according to the definition of the jurist and philologist Elio Gallo, a shapeless landscape "ubi silvae et pastiones sunt," of woods and pastures, interrupted only, if at all, by some small cultivated plots used by shepherds or guards.

A clear and lively image of the pastoral and wooded landscape of the *saltus*, which after the great clearings of the republican period gained increasingly in significance under the empire, is provided in Plate 4,

Plate 4. The landscape of the *saltus* in a mosaic from the Villa Adriana.

taken from a famous mosaic in the Villa Adriana, over which, perhaps, the new taste for pictorial landscape, whose source we have indicated, has shed an idyllic light (however severe and contained) that perhaps softens the harsher outlines of this landscape of the *saltus*.

· 11 ·

*The System of Temporary Clearings, and the
Deterioration of the Agricultural Landscape
under the Late Empire*

One cannot say that in the first centuries of the empire the progressive extension of the sylvan-pastoral landscape of the *saltus* always corresponded to a deterioration or disaggregation of the agricultural landscape. For the Apennine region between Liguria and Emilia, for instance, the *Tavola ipotecaria* of Veleia provides a documentation of how the *saltus* was expanding in this region at the time of Trajan, not at the expense of an agricultural landscape in course of degradation, but rather at the expense of the natural or seminatural landscape, on which, besides regular sylvan-pastoral use, *coloni* had began to mark their imprint with the fires of burned spots and the first glades of clearings.

In such cases as these, clearly, a certain form, if still precarious and lacking in precision, was imprinted on the natural landscape, even when one sector was given over to pasturage, which was a landscape nonetheless delimited and punctuated by the work of man. In the centuries of the late Roman Empire, despite a certain agricultural revival that appeared here and there, an increasing importance was assumed by extension of the *saltus*, which had already begun in different parts of the peninsula during the first period of the principate, by the agents of degradation of the already formed agricultural landscape. It was not by chance that the very term *saltus*, which was generally used at first to indicate a sylvan-pastoral landscape, became, in practice, synonymous with "great lordly or imperial property." The degradation of the agricultural landscape, in fact, often expressed itself through a restriction of cultivated land, and a growth in the extent of land that was uncultivated or given over to pasture.

The prevalence of the pastoral economy over the cultivation of cereals was not, to be sure, as yet so decisive in the centuries of the late empire as it would become in the age of the barbarian invasions. The changing relationship between cultivated land and pasturage at this point was not only, or so much, quantitative as it was qualitative. With the crisis of the slave labor force, and with a consequent prolongation of the period of repose of fallow land as pasture, the traditional biennial rotation between fallow and grain deteriorated on the lands of the *saltus* into a new

Plate 5. A pastoral landscape from the late imperial period from a mosaic in the central nave of San Maria Maggiore in Rome.

system of temporary clearings. This was certainly no longer that of more ancient Italy, with its beneficial clearing of virgin land that then reverted to spontaneous natural vegetation. Now more and more frequently a year of fallow and a year of grain were followed by one or more years of repose as pasture as later was practiced in the Maremmas, in the Roman and Pontine Agros, and in a good part of the South and the Islands.

This was not only a question, one must take care to add, of a deterioration of the agricultural landscape, but also of a progressive disaggregation of its precise forms. With the new orientation around great estates, arranged around their lordly and imperial *saltus* and with their emphasis on raising livestock, recognition of the right of *coloni* to pasture on all the land of the *saltus* (*jus pascendi*), and to sow on any arable land (*jus serendi*), became a necessity of production. Thus in a regime and landscape of closed fields there arose a tendency to shift to a regime of open fields, in which all the lands of the *saltus* became, in fact, open to the promiscuous pasturage of herds after the harvest.

The disaggregation and deterioration of the agricultural landscape in this period seems to correspond to a certain degradation of the forms of pictorial landscape, which can be seen in our Plate 5. It reproduces, from a mosaic in the Basilica of Santa Maria Maggiore from the period of the late empire, a pastoral landscape in which one does not fail to notice a difference in content and form from the one reproduced as Plate 4. A historian of art would certainly attribute this contrast to a difference of maturity and stylistic tradition, an association that goes well beyond the Italian agricultural environment. Be that as it may, here, as in other instances that we will consider later in this study, novelty in style, taste, and themes of depicted landscapes may refer not only to a change in tradition and cultural influence, but also, and more likely, to a true change, positive or negative as it might be, in the developmental capacities of the agricultural landscape itself. Such a change was never, to be sure, a mechanical (or photographic) reflection—it was mediated, naturally, on an artistic and cultural level, and must often be interpreted historically—but this does not mean that it lacks a significant documentary value for our study.

. 12 .

*The Barbarian Invasions and the Ruins of the
Italian Agricultural Landscape*

From the first years of the fifth century A.D., with the Gothic invasion of Alaric and the sack of Rome (410 A.D.), and continuing through the early Middle Ages with the invasions by Hungarians and Saracens of the ninth and tenth centuries, the agents of the deterioration and disaggregation of the Italian agricultural landscape, which we have seen in operation from the end of the principate to the late Roman Empire, increased frighteningly in their effects over a large part of the peninsula. From wave to wave, to be sure, the forms of the invasion, and their consequent devastation varied markedly: from pure and simple raids of predators and pirates (like, for example, the Hungarians and Saracens), which reached a peak of destructive power within a few months, to true migrations, like that of the Lombards, who finally established themselves permanently on Italian soil, or to wars of conquest, like those of the Franks, whose victorious bands succeeded in supplanting the old ruling classes of Italy.

In sum, however, the first and most evident result of the invasions was pillage, devastation, and the inexorable decline of the old centers of urban life. It would be a mistake, certainly, to adduce this decline, or in general all historical processes of this period, purely and simply to the impact of the invaders. This process had much deeper intrinsic roots in Roman society itself, which was already falling into economic decline and torn by deep social divisions. The very success of the invasions would be impossible to understand if one did not keep in mind the extent to which they were linked to the resistance and revolts of slaves and *coloni*, and to the alienation of a large part of the landowning classes through customs that had crystallized increasingly into decadent and oppressive forms. Thus the robust barbarity of peoples, already marked by the fresher forms of the *Ius gentium*, was often perceived as liberating by the oppressed and exploited classes, and in the end these new forms made a historically positive contribution to the rebirth of Italian civilization. But it is still true that the first barbarian invasions, while precipitating processes already developing in Roman society under the late empire, were colored with the dark hues always worn by foreign invasions. The cities and public works that had taken on the form "of a second nature operating for civil ends" in ancient Rome were pillaged,

Plate 6. The ruins of the Italian agricultural landscape: the Roman aqueduct at Acqui.

devastated, and pressed into inevitable decline. For long centuries, up to our own days, ruins and "dead cities" were integral and characteristic elements of the Italian agricultural landscape, which with the urban decline saw its deterioration and disaggregation accelerate and worsen.

The ruins of the "dead city," certainly, remained as a witness to an ancient urban and agricultural civilization, whose influence continued to operate in the depths of Italian society, and without the continued ferment of urban life it would be difficult to explain the unique character of the later communal revival. Even in the darkest period of the Middle Ages, and precisely within and in the closest proximity to the crumbling city walls, certain more definite and ordered forms of the agricultural landscape for which there was no longer a viable place in the open countryside often persisted. Within this closed space were often contained the most valuable groves of trees, and gardens and closed fields, while in the more distant countryside, where livestock and hunting prevailed over agriculture, the agricultural system of temporary clearings spread: open fields (to which the invaders were themselves accustomed) now dominated everywhere. A degraded and disaggregated pastoral-agricultural landscape, with fields open to hunting and pasturage, without definite forms, certain boundaries, or the relief of regular plantations of trees paralleled the landscape of ruins and "dead cities." Our Plate 6, which reproduces an engraving from the early nineteenth century of the landscape around the ruins of the Roman aqueduct at Acqui, in Piedmont, offers a suggestive image of this scene, which is still today familiar in certain parts of the peninsula.

. III .

THE EARLY MIDDLE AGES AND THE FEUDAL ERA

Plate 7. The deterioration of the agricultural landscape in Byzantine Italy: landscape and decoration in the apse of Sant' Appolinare in Classe in Ravenna.

· 13 ·

The Disaggregation of the Agricultural Landscape and Pictorial Landscape in Byzantine Italy

We have already indicated, as we will have occasion to do again as this study progresses, how in a given society the interest and taste for pictorial landscape rise and fall often in close conjunction with the interest and the taste the society itself shows for the more or less definite forms it imposes on the natural landscape, in the course and for the ends of its agricultural activity. This is not, to be sure, a mechanical reflection, that can be abstracted from the complexity of social and cultural relationships expressed by artistic intuition and taste. But it is not surprising that in a given historical environment the same social creativity that imposes regular and precise forms on the natural landscape appropriate to its productive ends also learns to compose the lines of a pictorial landscape in a coherent form, which declines and desegregates when the productive forces of society are no longer sufficient to impose well-defined and coherent forms on the real agricultural landscape.

In Italy at the time of the barbarian invasions, but particularly in Byzantine Italy where social and cultural formalism most fully developed, this relationship between the real and pictorial landscapes reveals itself with particular clarity. In content, the deterioration of the agricultural landscape and the increasing prevalence of animal husbandry over agriculture correspond to an almost exclusive interest in pastoral landscapes in Byzantine art. Still more noteworthy is the way this nexus manifests itself in the *form* of the pictorial landscape. Just as in the real landscape when the regular network of local roads and boundaries and the organized formal unity of the landscape become disaggregated, so also in the pictorial landscape the unity of the composition becomes disaggregated: it dissolves and unfolds in a repetition of its constituent elements, which, from parts of a landscape, are transformed, without any direct or real references, into decorative and ornamental motives.

In the splendid mosaic in the apse of Sant' Apollinare in Classe, at Ravenna, reproduced in Plate 7, an artist of the sixth century shows us how, in Byzantine society, these processes of disaggregation of the agricultural and pictorial landscapes accompanied one another with a formal rigor and perfection that can be recognized in the stylization of the images of rocks, plants, and animals, just as it can be in the forms of logicians and rhetoricians, and in the juridical formulations and social norms of this period. But if, in the pictorial expression, a perfect match between form and content is enough to satisfy the joy and richness of artistic intuition, this perfection cannot conceal, but rather reveals, in its reflection of contemporary reality, the decadence and meager poverty of the contents, which are those of an urban and agricultural society deteriorated into a pastoral society: disaggregated, and also crystallized into an order that suppresses all energy.

. 14 .

Castra, Curtes, Massae: *Centers of Reorganization of the Agricultural Landscape in Lombard and Byzantine Italy*

In the fury of the barbarian invasions, and with the disaggregation of the agricultural landscape that these produced, the economic and political means that had permitted diffusion of the ancient centers of the landscape's organization—the Roman *civitates* and *villae*—over a more or less vast territory were now often impeded or blocked. In the cities, to the results of pillage and devastation were added those of true economic and administrative asphyxiation, which followed their loss of hegemony over the surrounding territory. Saint Ambrose in Milan had already spoken, with reference to Emilia, of "corpses of cities half ruined" (*semirutarum urbium cadavera*); and more generally Gregory the Great, at the end of the sixth century, wrote that throughout Italy were destroyed cities, ruined castles" (*eversae urbes, castra eruta*) and a lack of tillers of the soil.

It is not surprising that urban life languished and declined in these conditions, and that the cities lost more and more the capacity to organize and dominate the deteriorating and disaggregating agricultural landscape of their old territories. It was from the *villae*, which already in the imperial age had assumed a more and more decisive function as centers of the economic and administrative life of the latifundium, that, if at all at this point, the initiative for new forms of organization of the economy, administration, and life of the countryside often departed. To be sure, even a large part of the villas had shared the fate of the cities; on them as well came down a fury of pillage and devastation. But after the first destructive impetus, the need for production, even at a lower level, regained the upper hand; and in a reduced society and nearly barren economy, the old organizational center of the latifundium appeared, often, as the location destined for revival. Here, not rarely, the new barbarian lord established himself beside, or in the place of, the old one, and elaborated new forms and ways to exact revenue in labor or kind from the uprooted and impoverished population, which he could only consume in the latifundium itself. Here he fortified himself, for defense and offense, and here he found the center for establishing his territorial lordship.

Plate 8. The *castrum*, fortified villa, a center of reorganization of the agricultural landscape of Lombard and Byzantine Italy, in a mosaic of the Mosque of the Ommiadi, at Damascus.

This new center of organization for life in the countryside, whether or not it took the place of the ancient Roman villa, also assumed typical forms that varied greatly from one region to another, just as in the lay or ecclesiastical seigneurial economy there was variation in the role of direct or indirect cultivation, or in the revenue that came from work services or was paid in kind. No less varied was the development of the territorial lordship in different parts of Italy, which asserted itself more easily in the North, while it encountered serious difficulties in Byzantine Italy. But *curtes*, or *domuscultae*, or *massae*, as they were variously called in different parts of Italy, all differed in structure and function from the classical Roman villa, and now became further differentiated as centers of organization for the life of the countryside through a variety of associated personal, juridical, economic, administrative, political, religious, and military functions. There was a variety of payments in kind, and work services, military and fiscal services, *censi livellari* or *enfiteutici*, lands held for personal service; crowds of dependents came to seek justice, each one with a different personal, hereditary, or contractual relationship with the lord.

These new centers of organization for life in the countryside did not always become centers of economic organization of an agricultural landscape. But in this disaggregated society, the common need for defense introduced a basic element of unity and cohesion, even if it was coerced and oppressive. In Plate 8, which shows a mosaic from the Mosque of the Ommiadi at Damascus, a Byzantine artist of the first years of the eighth century has left a suggestive image of one of the *castra*, fortified villas, which from the sixth century onward appeared throughout Italy as the center and stronghold for the *curtes, domuscultae,* or *massae*. To be sure, this did not as yet restore an organic unity to the agricultural landscape. But in the shade of these *castra* already matured some elements of a new rude cohesion, which the severe unity of this pictorial landscape seems to promise and express.

. 15 .

The Landscape of the Wildwood, and Hunting in the Early Middle Ages

In the centuries of the early Middle Ages while the impetus of barbarian raids and incursions continued, the emergence of fortified *castra, domuscultae,* and *massae,* although this assured to the rural population certain elementary means for defense and provided the first centers for the reorganization of a disrupted society, was not generally sufficient to promote a lasting revival of productive agriculture. There did not lack, to be sure, even from the time of the Gothic invasion, attempts to effect such a revival; nor did there lack local and partial results here and there, through clearing part of a wood, or improving and cultivating a swampy terrain. But a new invasion, and still more the general condition of the economy and society of these centuries, often made such results entirely precarious. Throughout the early Middle Ages, references to cultivated land abandoned again to woods and marshes predominate in cartularies and diplomatic codices, where the amount of uncultivated lands, woods, and marshes is always impressive.

Along with the open pastoral landscape, what clearly continued to prevail in Italy up to the year 1000 and beyond, and well into the communal period, was a landscape of woods and forests, which was the theater, in its less wild parts, of an important pastoral activity, and particularly (in the woods of oaks and beeches) of a free-range rearing of swine, that assumed a great importance in this period for assuring the population with the essential dietary resources of fats, which the decline in cultivation of olives could no longer provide.

These same pastoral activities that animated some parts of the woods with a rare human presence, and in the long run made them less wild and inhospitable, only penetrated in the early Middle Ages to a limited part of the forest landscape, which was dominated almost entirely by dark and impenetrable woods, full of threats and snares, the fearful habitat of wild beasts, and the shelter of bandits and marauders. Bears, wild boars, and above all wolves remained terrible enemies for man until the introduction of firearms. Only in the less impenetrable parts of the wild forest did armed men dare to venture for the purpose of hunting, which was not yet the preferred diversion of the upper classes, but instead an essential productive activity that, along with rearing, provided decisive resources for nourishment of the population.

Plate 9. The landscape of the wildwood and of hunting in Italy under the barbarian invasions, from a bas-relief in the Duomo of Civitacastellana.

This landscape of the wildwood in the early Middle Ages is shown by a suggestive image, in its barbaric crudity of form, that is taken from a bas-relief now found in the Duomo of Civitacastellana, and is reproduced as Plate 9. It is a scene of boar hunting, where the stylization of the few forms of trees does not lighten the darkness of the forest, and where men and beasts are shown engaged with fixed gestures in the ancient ritual of the hunt.

. 16 .

The Cultivation of Lesser Cereals, and the Medieval Agricultural Landscape of Open Fields

During the whole early Middle Ages and beyond, as we have already shown, the Italian landscape continued to be dominated, at least where there was a human presence, by activities of a sylvan-pastoral type, like hunting and rearing animals in the wild. Even when documents of this period mention rural holdings appropriated by private individuals and reduced to cultivation, there is always in their description the formula "cum cultis et incultis"—with cultivated and uncultivated land. Besides their frequent isolation among vast extents of woods and marshes, even in these cultivated lands, in fact, the uncultivated parts, thickets and meadows, woods and fens, generally made up by far the largest part.

In such conditions, while the activities of free-range rearing and hunting continued to prevail over agriculture more properly speaking, it is not surprising that the reelaboration of an organized agricultural landscape proceeded slowly and uncertainly. Thus, even where a wood was cleared or a fen made arable, what more often prevailed, between the woods and the open country, was the agricultural system of burned spots (*debbio*) and temporary clearings (*campi ed erba*), or that of fallow (*maggese*) degraded by a forced prolongation of the period of pastoral repose. In these agricultural systems, recourse to cultivation of inferior grains that required less care and were more adapted to rustic conditions than wheat, such as millet, panic grass, sorgum (or *melica*), spelt, rye, and barley, imposed itself as a technical necessity. Already in the first half of the sixth century, a letter of Cassiodorus gave instructions for the public granaries of Pavia and Tortona to distribute panic grass to the starving population; and for all the early Middle Ages and beyond, at least in central and northern Italy, inferior grains continued to have a decisive role in agriculture and in the diet of the rural population, and to this corresponded a decline in the cultivation and consumption of wheat.

To this prevalence of free-range rearing and hunting over agriculture, to the diffusion, on the few cultivated lands, of the agricultural systems of burned spots and temporary clearings, and to the growing recourse to inferior grains, corresponded necessarily, in the incipient reelaboration of an organized agricultural landscape, a general prevalence of open fields, that remained for the whole Middle Ages, and often up to the

Plate 10. The system of open fields in medieval Italy: a pig in a field of sorgum, from a miniature in the *Theatrum sanitatis*.

threshold of the contemporary age. Thus after the harvest, rights of promiscuous pasturage of flocks and herds were exercised even on the few cultivated lands in the open countryside. A concrete and significant image of this landscape of open fields, with the characteristic prevalence of inferior grains, is provided by the miniature reproduced in Plate 10, taken from a manuscript codex of the *Theatrum sanitatis*, from the Casanatense Library in Rome. In a field of sorgum—or *meliga*, as it was popularly called, a name that later passed on to be taken by Indian corn—the lack of any fence or protection left the crop exposed to damage, not only from wild beasts, but also from the voracious and devastating wild pig, who was the great enemy of the medieval cultivation of cereals.

. 17 .

*The Hilltop Town in the Pastoral-Agricultural
Landscape of the Italian Middle Ages*

Together with the *castra* that, from the sixth century onward, began to appear in the countryside to defend the *curtes, domuscultae,* and *massae* as centers of population and reorganization of the pastoral and agricultural landscape, towns also began to arise in the early Middle Ages, on the sides of steep hills, and even perched on the mountaintops.

We know that this type of settlement was the prevalent one among many Italic peoples in the age before the Roman conquest; and it was precisely the Roman conquerors who systematically destroyed these hill towns that were so difficult to assault, and drove the conquered peoples toward the plain. This systematic effort had a strategic end—besides the preoccupation of a political and military character to guarantee against offensive retribution by the conquered peoples—in the requirements of colonization itself, which was now founded on a more evolved system of fallow and plantations and required a more continuous presence of cultivators in the immediate neighborhood of the site of production.

But on the other hand, for the entire Middle Ages, the need for defense against barbarian raids and incursions, and the general lack of security in the countryside, pushed the population again to seek places of refuge in the mountains, and places to live that were less exposed to enemy attack in the hilltop towns. As well, with the general deterioration of the agricultural to a pastoral landscape, and with a return of the system of temporary clearings instead of fallow and plantations, the need for habitations in the plain, dispersed in small hamlets and scattered houses in the zones of most intensive Roman colonization, eventually declined. Now everywhere, by contrast, if only out of need for defense, centralized hamlets tended to prevail. In mountainous zones these also satisfied the requirements of nomadic types of herding and precarious cultivation, which were practiced on scattered plots of land that changed every year. Thus the concentration of population in hilltop towns on the highest sites at the center of the territory subjected to this precarious pastoral and agricultural existence became more necessary than ever, even to economize on the distances traveled. If we add to this the marshing up of many plains in valleys and along the coast, and the spread of infection from malaria, it is not difficult to see why the hilltop town became again more than ever an integral element in the Italian pastoral-agricultural landscape in the early Middle Ages.

Even in the centuries that followed, and down to our own day, wherever, in an extensive pastoral-agricultural system, the disaggregation of the agricultural landscape into open fields did not permit the formation of stable peasant communities, habitation in hilltop towns imposed itself in hilly or mountainous regions for reasons similar to those just expressed. But even when this economic rationale and the need for security gradually disappeared, hilltop towns often remained for the whole Middle Ages, and even up to our own day, as a typical element in the Italian landscape through that rural "law of inertia" that we discussed in chapter 6.

Thus when Giotto painted his Saint Francis giving away his cloak for the church in Assisi, a detail of which we reproduce in Plate 11, the first springs of the communal movement had already come to undermine, in places like Umbria and Tuscany, many of the economic and security motives that everywhere had conditioned the spread of hill towns in the early Middle Ages. But although softened by new structures that covered over the older and rougher buildings, and now slipping down toward the plain that began to be dotted with hamlets and scattered farmhouses, the town still remained perched up there on its ancient lofty site, just where the genius of Giotto knew to put it, as an element now organically inserted into the Umbrian landscape. And the first sparse plantations of trees—which following a custom that continued for a long time through the Middle Ages were arranged along terraces sloping down from the town—are evidence of how, still in the thirteenth century, the most valuable plants did not venture far beyond the limits of the protective shade cast over them by the walls of the city.

Plate 11. The hilltop town of medieval Italy in *Saint Francis Giving Away His Cloak* by Giotto.

. 18 .

The Agricultural Landscape of Closed Fields of the Italian Medieval City

Even in the most difficult centuries of the Middle Ages, it must be admitted that the rural population sought protection against the barbarian invasions not only around the *castra* that rose in the countryside, or in towns perched on the high hills and mountain slopes. The walls of ancient cities, although half ruined, offered shelter against the most fleeting sorties; and not rarely, even where the economic and administrative hegemony of the ancient urban centers had declined, they retained in Italy an importance unknown in other parts of Europe as centers of a more elaborate artisan production, as depositories of a cultural tradition or religious authority, or as consumer markets. But there was still more. In the shelter of the marshes that surrounded it, a city like Ravenna assumed a new preeminence in Byzantine Italy as a center of military, economic, and administrative defense. From the deserted islands of the lagoon, which were populated in the seventh century by fugitives from invaders on the mainland, the autonomy and then the new power of Venice arose and began to affirm itself. In Lombard Italy, a city like Pavia assumed a preeminent importance as the seat of a new kingdom and as the center of a royal *curtes*. Not infrequently in this period not only royal *curtes*, but also lay and ecclesiastical ones, had their center, or at least their economic and administrative center, in a city. Often even farms devoted to the costly cultivation of trees, shrubs, and gardens were located here, whose products would be exposed to too many risks in the open countryside, outside of the circle of the city walls, or distant from their protective shadow.

Many of the documents that have come down to us from the charters of the early Middle Ages that mention these more costly types of cultivation refer precisely to pieces of land contained within the circuit of town walls, or in the nearby suburbs. With the decline of urban life, in short, and with the disaggregation of an organized agricultural landscape in the open countryside, it was in the city (or around it) that the most precious elements of this organized landscape found, so to speak, a place of refuge, from which, in the age of rebirth of the communes, they could again radiate and spread into the countryside.

In this process, the landscape of the city itself significantly changed its aspect, and took on here and there contours of an agricultural landscape

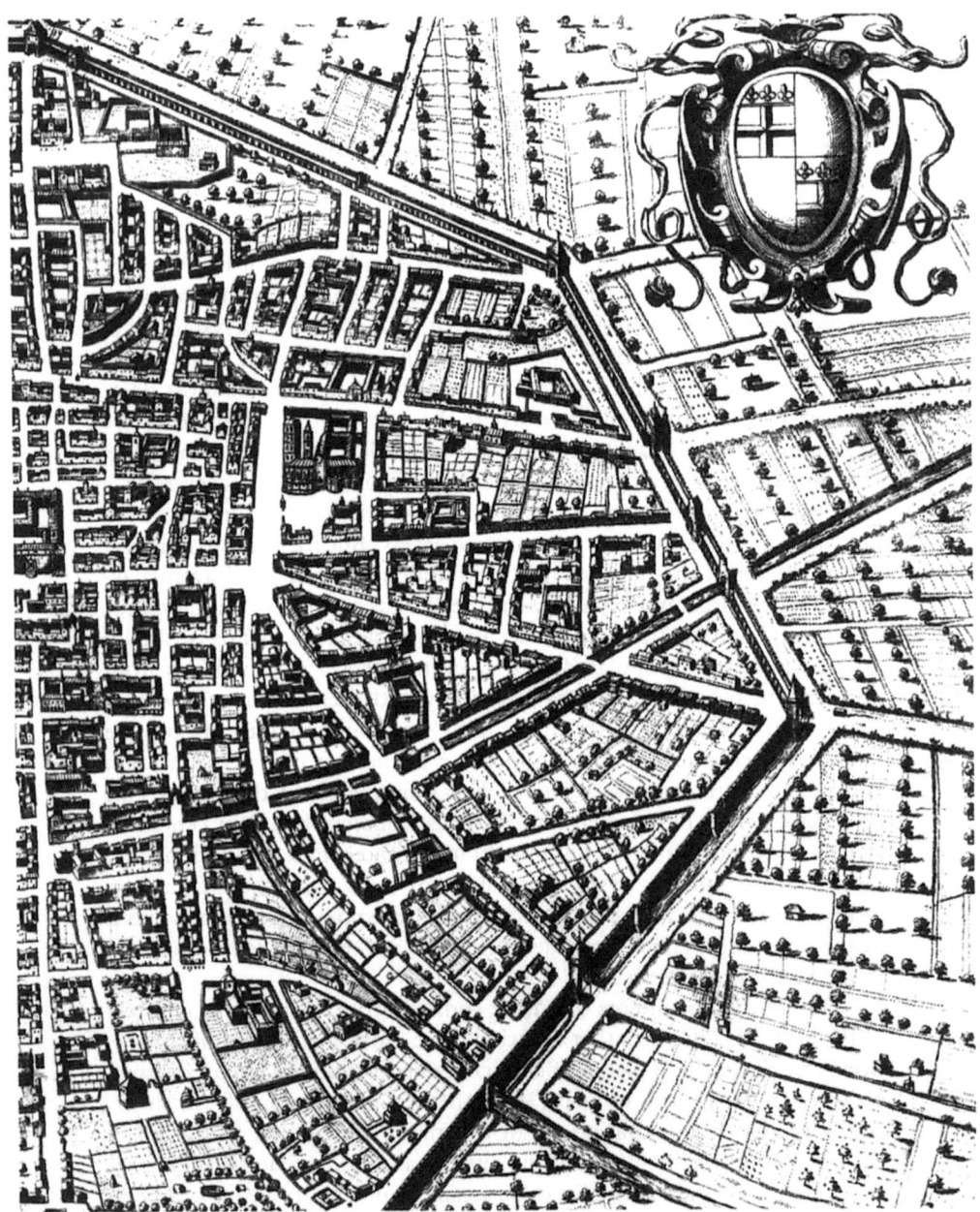

Plate 12. The agricultural landscape of closed fields within urban walls, from a sixteenth-century map of Bologna.

with closed fields containing vines, fruit trees, and kitchen gardens, and not rarely also sown fields and pastures, as at Rome itself, where, amid the half-buried columns of the ancient forums, almost up to our own days, herds grazed in the "Campo vaccino."

More often, as we have already indicated, the agricultural landscape of cities in the early Middle Ages was a landscape of closed fields, vineyards, kitchen gardens, and fruit trees, of which modern place-names of streets and quarters often still preserve a memory. With consolidation of the growing population of the cities, to be sure, this agricultural landscape became more limited in extent in the later Middle Ages. But, even at the threshold of the modern era, it had a notable importance within the girdle of walls of a great medieval city, and repeated there the forms of the suburban agricultural landscape, as will appear clearly to the reader from Plate 12, where we have reproduced a segment of a sixteenth-century map of Bologna.

· 19 ·

The Medieval Agricultural Landscape of Closed Fields: The Low-Growing Vineyard

We have already indicated how, even with the greatest deterioration of agriculture and the general disaggregation of an organized agricultural landscape, the cultivation of vines never entirely disappeared in Italy, even in the most difficult centuries of the barbarian invasions and the early Middle Ages. One can no longer speak, to be sure, of a true plantation economy; the plots of land planted in vines were reduced mostly to very modest extents, and often confined, if not within the precinct of the *curtis* or town, within its immediate neighborhood. With the condition of public security of the time, however, and confronted with the need to protect the vines from the grazing of flocks and herds, which had free access through rights of promiscuous pasturage to fields not defended by a fence, the plots of land destined for vines must generally have been, and generally were, closed fields through the whole early Middle Ages. And the land that was so laboriously enclosed was exploited to the utmost with valuable crops such as vines in specialized cultivation, in close rows, and thus grown low-down, on low trees or stakes.

All the literary and archival sources, and iconographic ones as well, demonstrate the clear prevalence, in the early Middle Ages (and often later), of this method of raising and cultivating vines, in contrast to the method of the *arboretum*, trees married with vines that were raised high, in mixed cultivation, which had been widespread in the Roman period, at least in central and northern Italy. In Figure 10, which reproduces a miniature from the *Martirologio di Adone*, illustrated in 1180 by Albert, a priest in the cathedral of Cremona, the reader will find a precise iconographic documentation of the method of cultivating vines on dry stakes, which, along with those raised on low trees, almost exclusively dominated the iconographic tradition of medieval "allegories of the months," just as it dominated the reality of the early Middle Ages.

The early medieval landscape of the vineyard, in short, even when distinguished from *topiae* or espaliers in urban gardens, with its small enclosed plots crowded with low rows of vines in specialized cultivation, rather approaches the Mediterranean garden in its contours and typology, which is not the same thing as a more modern plantation of "married vines" in central and northern Italy, or the great plantations of low-lying vines in specialized cultivation that in quite recent times have begun to prevail in many hilly zones, and on the plains of the South.

Figure 10. A vine raised low-down in the system of closed fields of the medieval city and suburb, from an illustration of the *Martirologio di Adone* of the twelfth century.

. 20 .

The Medieval Agricultural Landscape of Closed Fields: Kitchen Gardens

We have already indicated how the elements of an agricultural landscape organized in closed fields can be traced through the early Middle Ages and often much later in vineyards, orchards, and kitchen gardens, enclosed for the most part within the precinct of the city, the *curtes*, the hillside towns, or in their immediate suburbs. This is true particularly for kitchen gardens, whose cultivation, throughout the Middle Ages, must have provided people not only with important foodstuffs, but also with an essential supply of the aromatic herbs and simples that played such a large part in the medicines of that period.

And the *Theatrum sanitatis*, of which the Casanatense Library manuscript codex provides us with the richest iconographic documentation of the simples used in the Middle Ages, and from which we have taken the miniature reproduced as Plate 13, is basically a breviary of medieval medicine.

The landscape of the medieval kitchen garden is animated in this miniature by a scene showing the picking of spinach, whose cultivation was introduced and spread in Italy at the end of the early Middle Ages as a result of contacts with the Arabic world. And in the composition of this lively picture, it was not by chance that particular significance is given to the gate in the fence, which assumed an importance in the landscape of enclosed kitchen gardens that was still greater than what it must have been in the landscape of vineyards and fruit orchards.

Plate 13. A plant recently introduced—spinach—in the *hortus conclusus* of the medieval *Theatrum sanitatis*.

. 21 .

*The Arab Invasions, and the Medieval Landscape
of the "Mediterranean Garden"*

The Arab invasions that overcame Sicily between the eighth and eleventh centuries, submerging Byzantine control of the island and making bold thrusts into the heart of continental Italy, were the ones, without doubt, that exercised the greatest influence on the later development of Italian agriculture, and that also imprinted the most durable lines and shapes onto its agricultural landscape. To this Arabic influence, besides a more precise continuity of the tradition of Hellenic cultural techniques, Sicily and southern Italy owed, in great part, that agricultural preeminence that (particularly with regard to the techniques of aboriculture and horticulture) they preserved up into the fourteenth and, in certain sectors, the sixteenth centuries. It is enough to remember in the Arab contribution to the agriculture of these regions, and then to the whole of Italy, the introduction of new crops, such as rice, cotton, sugarcane—which up to the sixteenth century and beyond assumed a great importance for the agricultural economy of Sicily and other parts of the South—carob, pistachio, eggplant, spinach, and many other essential items in kitchen gardens.

As to the specific form of the agricultural language, however, the most lasting contribution of the Arab conquest came with the spread of cultivation of silk, and with the introduction into Italy of citrus fruits, or, more precisely, oranges and lemons.

In Sicily and southern Italy, which until the sixteenth century remained the most important centers of Italian silk production, the accompanying spread of cultivation of mulberry trees did not introduce, in truth, an element that was qualitatively entirely new in a landscape where aboriculture was already quite significant. More important, instead, was the role that the spread of cultivation of mulberry trees had when, beginning in the sixteenth century, the center of gravity of silk production moved toward the center and north of the peninsula, where the characteristically clipped mulberry trees would become an integral element in the *alberata* of Tuscany and the Marche and in the *piantata* of the Po valley.

For the time being, however, from the last centuries of the Middle Ages, the spread of citrus groves, with the considerable technical arrangements for planting, irrigation, and protection that they required,

Plate 14. The Arab expansion and new plants: citrus fruits in the medieval landscape of the "Mediterranean garden, in a miniature from the *Theatrum sanitatis*.

was what most typically inserted itself in the agricultural landscape of the South and of Sicily, giving a new justification and fascination to the ancient forms of the Mediterranean garden, and conditioning its further extension. For all of northern Europe, now, and even for northern Italy, Sicily and the South would be the land "where the lemon trees bloom." With their regular shapes, their shining evergreen leaves, and their exquisite fruits of gold and flame, the Mediterranean garden of orange and lemon trees would take on the aspect of paradise, and would have an important place in the rebirth of a taste for the agricultural "bel paesaggio," that was already clearly expressed in the fine miniature from the *Theatrum sanitatis*, which we reproduce in Plate 14.

. 22 .

*The Castle in the Agricultural Landscape
of Feudal Italy*

The processes of disaggregation of the agricultural landscape and the separation of the city from the countryside, which we have seen develop through the early Middle Ages with varying fortune and even contradictory tendencies, reached their culminating point between the eighth and eleventh centuries. Already in the eighth century, in the majority of Italian cities, the old Roman walls had fallen completely into ruin, and there was not enough energy to repair them. As well, the multiplication of *castra* and fortified towns in the countryside, which tended to become economically and politically autonomous centers, obscured more and more the relationship between the countryside and the cities, many of which, about the ninth century, had a rural appearance.

The new serious incursions of Hungarians and Saracens in the ninth and tenth centuries had furthered these processes, while the Carolingian conquest, for its part, created conditions for a further elaboration of the feudal political system. The base of military obligations, which assumed a particular importance in this turbulent era, remained, under the Carolingians as well, the landed estate; and royal functionaries—the counts and margraves of the Frankish nation—who in much of Italy took the place of the Lombard dukes and castellans in summoning the men at arms of a given district to an assigned place in the name of the king, and, like those, had judicial authority throughout a given territory. But the "benefice" (as it was called)—that is, the concession of royal land in exchange for service, which the counts and margraves now generally enjoyed—made them practically independent of royal power, to which they were tied only by an obligation of vassalage and the revocability of the benefice itself. As well, the immunity—that is, the exemption of a given individual or territory from intervention from the central authority that was granted to counts and margraves with more and more largesse, and also to abbeys and episcopal churches—served to integrate the fief. This was, precisely, the benefice of a vassal that was immune, a territory where there was a characteristic fusion and confusion of private property and public sovereignty, and over which the fief-holder exercised public rights that were the prerogative of royal power to a greater or lesser extent. When, with the Capitulation of Quiercy in the year 877, Charles the Bald recognized the hereditary nature of the larger fiefs in fact, the political edifice of feudalism can be thought of as formally complete.

Already much before this, however, with the crisis of the slave-based economy beginning at least in the fifth century, feudal forms of property and productive relationships had appeared in Italy. As is well known, the new mode of production that now affirmed itself more fully and spread, was founded on the feudal possession of land, and on the condition of personal servile dependence of the actual producers, who still, in contrast to what had occurred under the mode of production of the slave system, were not fully in possession of the lord. In the feudal economy, the actual producers occupied determined plots, which they cultivated with their own means of production, and returned revenue to the great feudal landlord in the form of work services, goods in kind, or money. An evolution in this direction can already be found among Roman *coloni* of the late empire, and still more in productive relationship of the age of the barbarian invasions.

The political and juridical consolidation of feudal institutions in the age of the Lombards, and still more in that of the Franks, while it contributed to a more decisive prevalence of feudal forms of property and productive relationships, was itself pregnant with important historical consequences. With regard, particularly, to the relationships between city and countryside, the concession of immunities to episcopal sees largely removed cities from the jurisdiction of the counts, to whom remained the dominion over what still today, from their name, is called the "countryside" (*contado*). This was another and decisive step in the isolation of the countryside from the decrepit cities; and gradually, through inheritance of the major and minor fiefs, as the country became fragmented into independent jurisdictions and lordships, feudal anarchy made the castle the form that, throughout Italy, dominated the landscape of the Italian countryside.

In this system, the countryside, arranged around a strong castle that dominated it from above, affirmed over the city, now economically decadent and under the control of an uncertain episcopal authority, its decisive political predominance. But this predominance lasted for a relatively brief period in Italy, and it was less absolute than in other lands of Europe. When Simone Martini, in his *History of the Blessed Agostino*, from which our Plate 15 is taken, provided us with these images of dominant feudal castles, the communal rebirth of the city had already, in historical reality, started to undermine the foundations of their exclusive dominion. But in the first phase of this rebirth, as we will see, the lesser feudal nobility would have a significant part, and the life of the communes would be strongly marked by the influence and the forms of medieval chivalry that emanated from the feudal strongholds. And

Plate 15. The dominant position of the castle in the landscape of feudal Italy, from the *History of the Blessed Agostino* by Simone Martini.

Simone Martini shows himself to be still deeply sensitive to the fascination of an ancient power, which in this painting by the great Sienese is emphasized by the strong upward projection toward the mountain tops and castles, and is associated with a space that has now become distant and abstract, as in a legend.

· 23 ·

The Revival of Plantations of Trees in the Agricultural Landscape of Feudal Italy

Beginning in the eleventh century, and often before the communal impulse came to renew the initiative of city dwellers to elaborate an organized agricultural landscape, a new awakening of agricultural effort appeared through various parts of the peninsula in the open countryside dominated by feudal towers. Even in the darkest centuries of the early Middle Ages, to be sure, concessions of land *ad runcandum, ad pastinandum*—with an agreement for clearing, or plantation—had never ceased in Italy; but now, with the threat of the Hungarians distant, in the shelter of feudal castles and consolidated territorial lordships, new and less precarious perspectives opened before these efforts. These were the lay lords who, to attract faithful armed men in their rivalries with others, were generous in granting privileges to those who would come under the protection of their castles. Or they were the abbeys and churches that, to make accumulated treasure and enormous and largely uncultivated holdings of land economically and politically valuable, employed serfs and lay brothers in the work of clearing and improvement, or that multiplied concessions to third parties. Populations less seriously damaged by the devastating incursions, who looked toward the future with less uncertainty and pressed more insistently for a revival of production, were also involved. Where enormous accumulations of riches of abbeys or churches did not permit collective works of improvement or tillage, individual initiative exercised itself most frequently in the agricultural revival by planting trees or shrubs on already tilled land, or by clearing open tracts on hillsides, rather than in the more difficult wooded or boggy zones of the plain where tillage required too much preparation. The plantations of trees involved the more rustic chestnuts or olives, which were less liable to damage from animals or pilferage in the still open fields. These were often preferred at first over vines, which continued to predominate in small closed plots.

This was the period when plantations of vines, olives, and chestnuts began to expand beyond the territories of castles, on the ancient common pastures of the *civitates*, and even around the cities. But let us not be deceived by the words *olivetum* or *terra vitata* that frequently occur in documents of the time. These were as yet only a few plants, placed haphazardly on even quite large extents of land, which did not yet suc-

Plate 16. The resumption of plantations and the persistent disaggregation of the agricultural landscape in feudal and communal Italy, in the *Garden of Olives* from the nave of San Marco.

ceed in imprinting a new or more organic form onto the agricultural landscape of open fields. And this persistent disorder of the agricultural landscape is still to be seen in the *Garden of Olives* of the Gospels, some time in the twelfth century, by the artist of a mosaic from the right side of the nave of San Marco in Venice, that we have reproduced as Plate 16. The reader will perceive in the landscape of plants that the same sad dramatic stance of the human figures is also reflected in the distorted cliffs. But the drama and unity of the composition remains, precisely, in the intimacy of man. This is a traditional, and in part conventional, pictorial landscape, on which, as on the one of historical reality, the sad force of man is reflected, without yet succeeding in imposing his active and creative forms.

· 24 ·

The Age of Improvement and the Great Clearings and Reorganization of the Agricultural Landscape in the Eleventh through Thirteenth Centuries

Between the eleventh and thirteenth centuries, as we have seen, the development of productive social forces was generally not yet sufficient to assure decisive success to individual initiative in the reelaboration of an organized agricultural landscape. The progress of repopulation of the peninsula was under way, to be sure, but at first it was relatively slow and uncertain. Technical revolutions like the one in the use of draft animals, which permitted yields to multiply many times, also spread slowly in the cultural conditions of that period. The very progress of a commercial money economy—"new men and quick profits"—which would have a decisive part in the communal rebirth, had at first a disruptive effect on the old forms rather than creating new ones.

Nonetheless, the centuries between the eleventh and thirteenth were decisive for the reelaboration of an organized agricultural landscape in Italy, thanks to the great collective works of improvement, irrigation, and tillage undertaken through new forms of social organization, which themselves became a strong productive force and had an efficacy that was unattainable and unknown in that period through individual effort. It was not that technical means or forms of organization of work changed substantially at first; these seldom went beyond what Marx called "simple cooperation." But it was now not only individuals who were more and more frequently employed in the work of improvement and tillage. The early Middle Ages had known varied forms of association, among families and kin, among neighbors, among serfs and *coloni* of the same lord. But now more and more frequently these *condomae*, these *vicinae*, these *colliberti*, these *consortes*, as they were variously called according to their origin and type of association, and finally the commune itself, or even groups of people casually brought together, appear as subjects not only of rights to use common lands, but also of undertakings to defend the water supply of a given territory through the construction and maintenance of riverbanks, ditches, and canals, or of concessions of land for improvement and tillage, as in the contracts for *abitanza* or of *castellanza* that are so characteristic of this age.

Not even these forms of association, however, would be sufficient, at this level of development of the social forces of production, to impact sensibly on the reelaboration of an organized agricultural landscape without the involvement of the socially dominant groups, who controlled the greater part of the means of production, as well as the necessary advance for any great work of improvement, through great feudal landed properties, hoarding the scarce mobile riches of that period, or the power to command that they disposed. These were, first of all, feudal lords who, to realize value from vast unpopulated territories that were economically, politically, and militarily unproductive for them, favored in every way the foundation of Cistercian abbeys, which preserved what had not been completely lost from the tradition of the more evolved techniques of the classic age, and had at their disposal large resources of wealth. They became true and proper enterprises for the transformation of land, even for third parties, and they specialized in the improvement of bogs and marshes requiring masses of lay brothers and serfs. Later, with the first signs of a revival of urban initiative, episcopal churches, and finally the communes themselves, assumed a function of the first order in the development of collective works of improvement, clearing, and internal colonization, which in the space of three centuries proceeded to reelaborate the forms of an organized agricultural landscape in many parts of Italy.

It is enough to remember that this was the period when the bases for the system of embankments of the Po and its affluents were laid down, along with the system of navigable and irrigation canals in the Po valley, and the system of collecting canals that, throughout the Po valley, formed a hydraulic system for the sowed land. These great works by themselves influenced the new forms of the Italian landscape in a decisive way, conditioning these forms, with the new means they opened to internal colonization, which was now often realized along the lines of a precise economic, political, and scenic plan.

What deep traces this plan left in the Italian landscape up to the present appear from Figure 11, which we have taken from a small-scale map of the Istituto geografico militare. In the first half of the twelfth century, the vast gravel and rocky terrain of the so-called Campagna di Verona, which surrounds the city on three sides, was still for the most part uncultivated and used only precariously by the population. But on 9 March 1185, on the proposal of Viviano degli Avvocati and other proxies of the commune, the council decided, in order to "provide people" for the commune itself, and as a defense against the people of Mantova, to excavate a ditch and found a city at the extreme margin of the Campagna. On 25 and 26 March 1186, a consul and a proxy consigned to the men

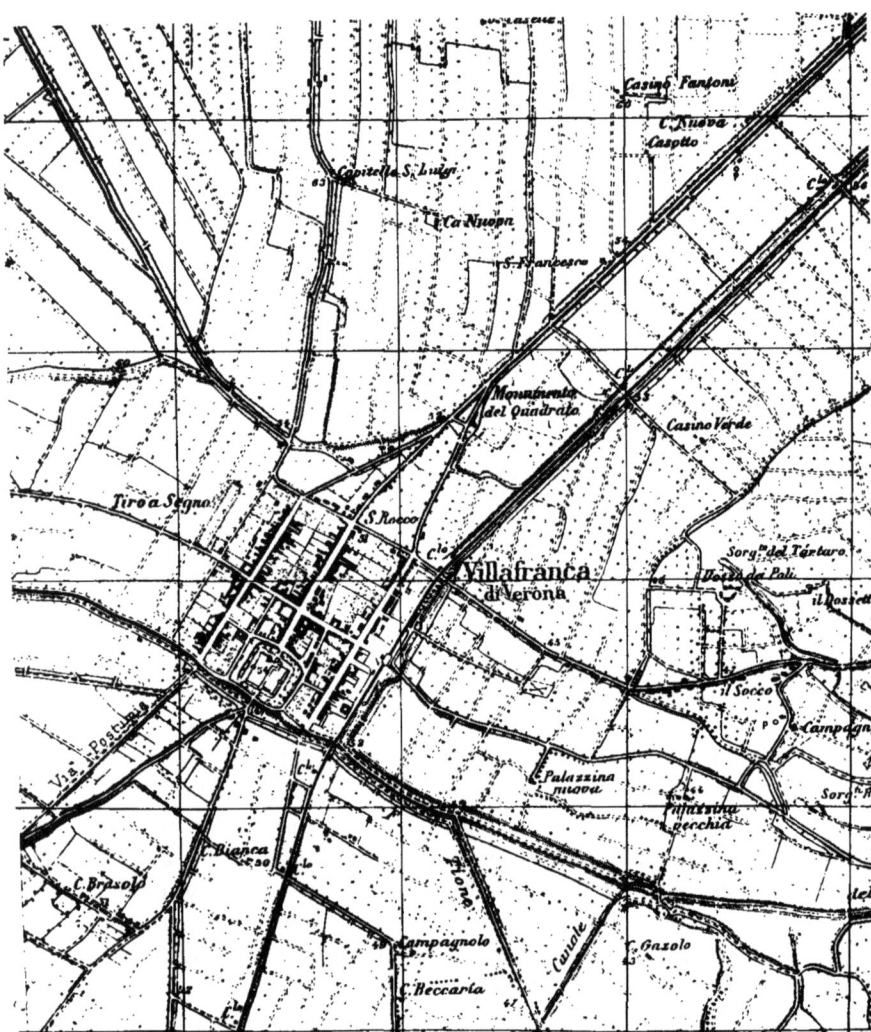

Figure 11. The traces of a plan of colonization of the twelfth century in the contemporary agricultural landscape of Villafranca Veronese.

who came out to populate this Villafranca, or "free city" (that is, exempted from any burden or engagement of the commune), and the lands assigned to it, fixed the prearranged boundaries on the spot. Each colonist was given a *manso*, a cultivatable unit according to the nomenclature already used universally in the Carolingian period, measuring 33 Veronese "fields," with one for a house and 32 "pro laborare." Six hundred fifty six additional fields were given to Villafranca Veronese as woodland or pasturage: in all 2,088 fields of land, assigned to 179 *mansi*. How the execution of this systematic plan of colonization is still today reflected in the landscape appears clearly, here as elsewhere, and despite successive changes, from the regularity of the plan that Villafranca Veronese (later made famous as the place where the armistice was signed between Napoleon III and Austria in 1859) has preserved through the centuries—since the time when, according to the prearranged scenic plan, a house was built on each *manso*, adjacent to the regular plots of uncultivated land: 32 fields that were given to each colonist "pro laborare."

. 25 .

The Landscape of Large-Scale Pasturage
in the Feudal Era

Between the eleventh and thirteenth centuries, the great Cistercian abbeys, as we have seen, were not the only agents of works to transform the land and to colonize it internally. In the context of a variety of environmental conditions, while some abbeys, like the one at Chiaravalle Milanese, specialized in the task of improving marshlands and the irrigation works that now began to spread throughout the Lombard plain, others were orientated toward the exploitation of their immense patrimonies of uncultivated lands through the techniques of large-scale pasturage, often involving transhumance. Even in this case, as in the case of improvements, the Cistercian abbeys did not limit themselves to operating on lands they already possessed. They solicited and often obtained, from emperors, kings, and great feudal lords, important concessions, even for use as pasturage, on great extents of uncultivated land, or franchises to use such lands for pasturage. Particularly generous with concessions of this kind were the feudal lords of Savoy and Monferrato. In Lombardy, as well as in Piedmont, the slopes of the Alps, where in this period sheep predominated in contrast to the cattle that have predominated in more recent periods, sometimes became almost an exclusive privilege of these abbeys and of other lesser ones. Other lay or ecclesiastical lords, and the king himself, were not slow in following the great Cistercian abbeys along the ways of organizing large-scale pasturage. Thus, in Puglia under the Norman king Roger, the ancient routes of transhumance for sheep were taken up again and reorganized from the fiscal point of view in what would now be called the "Dogana di Puglia."

Everywhere, the raising of wild pigs, which had been typical of the early Middle Ages, was still important, but now a new importance was assumed by the great flocks of transhumant sheep, and by the large-scale raising of horses, which developed and was conditioned by the preeminent place of the horse as an instrument of combat and sign of social distinction. Thus under the Swabians great herds (*razze*) of horses spread through Calabria, where Frederick II established the royal breeds. To give an idea of the importance of this enterprise it is enough to remember that, on one occasion alone, this same king removed 600 stallions from his herds in Calabria, to create a new one in Capitanata. No less important was the development and large-scale breeding of

Plate 17. The landscape of large-scale pasturage in feudal Italy, in the *Return of San Gioacchino* (school of Cimabue) in the Museo Civico of Pisa.

horses and sheep in regions like those of the Roman Agro, the Maremmas, and Sardinia: one abbey, on its foundation, received as a gift 10,000 sheep, 1,000 goats, 2,000 pigs, 500 beasts of burden, and 100 horses.

These are figures of an order of magnitude that take us back to the great breeding grounds of imperial Rome. Different, to be sure, were the historical conditions and social relationships in which this revival of large-scale breeding took place, but in the landscape, as in the pastoral way of life, the revival of large-scale pasturage did not fail to provide new perspectives and contours, ones that had seemed canceled out, in regions that were now being repeopled—if only by the sparse presence of shepherds and herds—with the forms of men who were resuming the ancient routes of transhumance. And these Roman outlines seem recognizable in the pastoral scene that we reproduce in Plate 17. It is taken from a painting of the Pisan school of the thirteenth century. A landscape rises up and gives depth to the faces and gestures of shepherds, who drive a flock before them on a long journey. The painting fixes, in this landscape and these gestures, a type that, certainly in pictorial representation, has Roman and even Hellenic and Byzantine precedents. We can find it, in reality, up to the present day in places where these same kinds of transhumant flocks impose on men and things the same glance, fixed in severe discretion.

. IV .

THE AGE OF THE COMMUNES

. 26 .

Feudal Strongholds and Villas in the Landscape of the Early Communal Age

"In June" Folgòre da San Gimignano sang at the end of the thirteenth century, illustrating his ideal summer landscape in the *Sonetti dei mesi*,

> di giugno dovvi una montagnetta
> coverta di bellissimi arboscelli,
> con trenta ville e dodici castelli,
> che sian intorno ad una cittadetta
>
> . . . .
>
> e palafreni da montare 'n sella,
> e cavalcar la sera e la mattina:
> e l'una terra e l'altra sia vicina,
> ch'un miglio sia la vostra giornatella . . .
>
> (in June there is a little mountain
> covered with beautiful trees,
> with thirty villas and twelve castles
> arranged around a small town
>
> . . . .
>
> and palfries to mount
> and ride evening and morning:
> and one place is close to another,
> so that your morning is a mile . . .).

This landscape of Folgòre da San Gimignano—which, in the new taste for a picturesque landscape awakening in Tuscan literature, seems reflected in the detail from *San Cosmo and San Damiano* of the Beato Angelico, reproduced in Plate 18—was not only an ideal landscape of this age of transition; it also reflected a reality, in which the new landscape elements of communal society (the "small town," the "thirty villas") were mixed with those of feudal society (the "twelve castles").

But it was not by chance, in the poetry of Folgòre or in the painting of Fra Angelico, that the castles and feudal strongholds were no longer distant in their menacing isolation, but nearby: "one place is close to another, so that your morning is a mile." They had lost, it seems, much of their swagger and become, along with the villa of the new bourgeois, elements of that "bel paesaggio" in paintings and poetry, for which taste

Plate 18. The "thirty villas" and the "twelve castles" around the "small city" of the early communal period, in *San Cosmo and San Damiano* by Fra Angelico.

revived in Tuscany as the new people (*gente nuova*) in the communes and the Tuscan countryside set about giving attractive and familiar forms to their fields and rural dwellings. From the eleventh to the fourteenth century, when the preeminence of the city over the countryside increased with the communal revival, we know that there was a rapid growth in the numerical importance of the lesser nobility, which for various reasons had a place of first importance, along with the merchant class (*popolo grasso*), in the politics of the early communal era. Feudal strongholds, of which there were, for instance, 52 in the zone between Florence and Fiesole in the eleventh century, increased to 130 in the twelfth century, and 205 in the thirteenth century. Thus it is not surprising, in these conditions and with this density of dwellings, that the lesser nobility of the countryside found its style of life and its economic and political interests becoming closer to the merchant class of the cities (which was already expanding its domination and property into the countryside), rather than to the great feudal lords, who still threatened from their more distant castles.

This evolution of relationships among social classes and elements of the landscape in the early communal period derived otherwise from processes that had been developing in feudal productive relationships for centuries, and that now came to decisive conclusions. Beginning in the

Carolingian period, in fact, the system of *mansi*, farms, which we have seen used as units of colonization in the founding of Villafranca, had begun to dominate everywhere in the cultivation of the great territorial lordships, while less and less importance was assumed by the Lombard *sala*, the part cultivated directly by the lord, and all cultivated land tended to be distributed in farm units, *mansi*, which corresponded to the work force of a family of *coloni*, two oxen, and a plow.

When the productive unity of the great territorial lordships (*signorie*) was thus fragmented, the way was also opened to their division and political decline. With the multiplication of subconcessions within the feudal system of even single or a few *mansi*, there was a multiplication of a proud and turbulent lesser nobility, which, by following its own economic and political interests, ended up by providing arms and leaders not only to the communal movement, but also to the resistance and antifeudal revolts of the serfs themselves. Even further, as the urban citizen communes extended their political domination over the countryside, and utilized the collective manumission of serfs as an effective weapon in the struggle against the great feudal lords and against the institutions of feudalism itself, the productive fragmentation of the old territorial lordships favored their rapid erosion through more or less forced sales of individual farms by the lords, and their acquisition by representatives of the new dominant urban classes.

The decisive blows in this crisis of property owning and productive relationships in feudal society and in the feudal landscape came, to be sure, on one hand from the resistance of the rural masses, and on the other from the contestation of the new urban citizen classes. But in this period of transition, while the final outcome of these struggles still seemed uncertain, a compromise between the "thirty villas" of the merchant class and the "twelve castles" of the small and intermediate rural nobility is what seemed to dominate the agricultural landscape of the first phase of the communal period, as it also dominated the political scene.

. 27 .

*Individual Clearings, Plantations, and Settlements
in the Agricultural Landscape
of the Early Communal Period*

Already about the middle of the thirteenth century, the elaboration of the agricultural landscape, from which Pietro de' Crescenzi, the restorer of the science of agronomy in communal Italy, would draw the sum of his experience, was beginning to develop in new conditions, in which individual initiative assumed a weight and effectiveness that it was able to obtain only sporadically in the preceding centuries.

This new effectiveness resulted certainly, chiefly, from the increase in density of population, which already by itself, and almost automatically one could say, multiplied the results of individual initiatives for agricultural change and increased the possibility of their visible impact on the form of the landscape. But the increase in population was only, of itself, an element in the general change of socially productive forces that now manifested itself more clearly in the invention of new techniques and the spread of new work experiences. Thus even from this perspective—in the increased productivity of labor—the possibilities for individual efforts to influence collectively the forms of the agricultural landscape grew and multiplied.

These individual initiatives could, otherwise, now be undertaken on a terrain that the work of past centuries had largely cleared, in both the literal and figurative senses of the word. On a terrain where collective or public initiative had already removed the trees and leveled the ground, in fact, the individual *colono* could proceed more easily not only with the cultivation of herbaceous plants, but also with planting trees and bushes whose spread became one of the most characteristic aspects of the Italian agricultural landscape in the communal period. Or likewise, on the terrain of an ancient bog, which a Cistercian abbey had drained and the public initiative of a bishop or a commune had crisscrossed with a network of *dugali* (larger common drainage ditches), even an individual proprietor would not need to fear, in proceeding to the hydraulic systematization of his plot, that with the first rain his own ditches would back up for lack of an outlet.

The work of systematization of the arable terrain, in fact, and of planting trees and shrubs, which already implied an at least elementary preparation of the soil, would become, as we will see later in our re-

Plate 19. Plantations of the communal period in the *Garden of Olives* by Duccio di Buoninsegna.

search, two of the means by which individual initiative would have the deepest effect on the agricultural landscape. In the *Sonetti dei mesi*, as we have already seen, the ideal landscape of Folgòre da San Gimignano was not composed only of towns and castles, but also of a little mountain "covered with beautiful trees"; and when (at the beginning of the fourteenth century) Duccio di Buoninsegna painted his *Garden of Olives* (reproduced here as Plate 19), his mature sensitivity to the landscape provides us with a measure of the steps that, in art as in reality, had been taken toward this ideal since the time when the artist of the mosaic in San Marco composed the image of the same garden that we reproduced in Plate 16.

Nothing is lost, in the painting by Duccio, of that human intimacy, which is also here reflected against a landscape of rocks and trees, although in a more composed and less contorted way. But the landscape now assumes a unity of its own, an autonomous significance; and the trees have become elements in a real wood, where they are not to be confused with wild trees, but instead are regularly spaced and arranged where the roughness of the rocky terrain allows.

The agronomic writings of Pietro de' Crescenzi, a contemporary of the art of Duccio, give us the assurance that we are not here, once again, witnesses of an evolution only of form and style in the pictorial landscape. Since the period of the mosaic in San Marco, a new class of communal bourgeois had assumed leadership of the progress of agriculture in the Tuscany of Duccio, as in Umbria and Emilia; and the *coloni* who planted and cared for these olives were no longer serfs. Their freedom, conquered or conceded in the struggle of the communes against feudal lords, was now inscribed in documents like that of the Bolognese *Liber Paradisi* of 1256, which did not open the doors of a terrestrial paradise to serfs, to be sure. But already, in this freedom, *coloni* and artists learned that it was no longer sufficient to project their intimacy and sad force onto the reality of the landscape: it was necessary, and possible, to place active and creative forms within it.

. 28 .

Systematization in the Plain, and the Planting of Trees Festooned with Vines

In our discussion of Plate 19, the reader will have noticed that we used the term "systematization" in the modern and technical sense of the word. By systematization one means in agronomic terminology today "the complex coordination of complementary works that serve to perfect the irrigation and drainage of the soil (which was already initiated through the major work of clearing and improvement), to the end of assuring the conservation of the soil, and of adapting it to less uncertain, more varied, and more intense production." Depending on variations in the terrain, and its elevation from the plain to hills and mountains, the work of systematization may consist of drains, embankments at the heads of fields (*cavedagne*), drainage ditches, banks, terraces, and so on. But the form these works assume generally follows a common goal of assuring the most economical hydraulic system, the most balanced use of water, and the most effective tillage of the soil, and its division into roughly horizontal fields of convenient size.

In this modern and complex sense, to be sure, a term like systematization would have been incomprehensible at the time of de' Crescenzi; but a study of his agronomic work, and of the documents of his time, is sufficient to show how the idea and practice of systematization, although of a less complex and intensive type, are what most closely characterize the agricultural revival of the communal era, not only in comparison with the feudal period, but even in comparison with classical Rome.

This new possibility of basic systematization, to be sure, was made possible by the great work of clearing and improvement of the previous period, and by the fragmentation of the territorial lordship into less insecure farm units. But in the climatic and geographical conditions that dominate a large part of Italy—with its large hilly and mountainous regions, its lands in the plain that are heavy and impermeable to water, and the excessive rain in the autumn and winter and dryness in the spring and summer—the work of clearing and improvement would have produced only meager and uncertain harvests without an accompanying effort at systematization, which assumed, from this period onward, an importance in the elaboration of the Italian agricultural landscape that was unknown elsewhere.

Plate 20. Planted vines festooned on trees and the pressing of grapes in a miniature from the *Theatrum sanitatis*.

Among these works of systematization, what had at first the most effect on the form of the agricultural landscape in the plains were those that generally accompanied the planting of trees festooned with vines—of which, again, a fine miniature taken from the already cited codex of the *Theatrum sanitatis*, reproduced in Plate 20, provides us with a concrete example that is animated by a scene of pressing grapes. Pietro de' Crescenzi, on his own part, gives us a technically correct description in his agronomic works of this new planted landscape—where vines cultivated promiscuously have spread well beyond the limits of small enclosed plots with close rows within which their specialized cultivation (on small trees or dry stakes) had most often been contained. De' Crescenzi speaks to us, with regard to the various local ways of planting, particularly in the Po valley, of vines raised on "large trees disposed in squares," and he specifies that these "are planted on the banks of ditches, or above the banks, or in fields, near large trees." Still further, he shows us that in Lombardy and the Romagna already in his time it was usual to have a field length of two hundred feet or more in these regular fields planted with trees.

As to the extent that this planted landscape attained in the communal period, interesting research carried out by Torelli[1] for the region around Mantova provides a less imprecise measure. The documents consulted by Torelli concern in all, for the thirteenth century, a surface of about 3,400 hectares of tilled land: and of these, 2,200 hectares were sowable fields without trees and pastures (*seminativi nudi e prati*), a little less than 1,000 were planted with trees and vines (*seminativi vitati*—that is, *piantate*), and 200 were specialized vineyards (*vigneti specializzati*). But while in the suburban region around Mantova specialized vineyards occupied as much as 35 percent of the surface (as compared with 31 percent sowable fields without trees and 25 percent sowable fields with trees and vines), in the zone more distant from the city that was more typically rural, such as the left side of the Mincio River, the sowable fields without trees clearly predominated, with 59 percent of the tilled terrain. But here the landscape planted with trees and vines (*piantata*), with 22 percent of the surface, already predominated over the specialized vineyards, which decreased to 2 percent of the tilled land.

---

[1] [Pietro Torelli, *Un comune cittadino in territorio ad economia agricola* (Mantova: Pubblicazioni della Reale Accademia Virgiliana, 1930).—Trans.]

· 29 ·

Individual Tillage, and Extensive Systematization on the Hillsides

In the hills like the plain, as we have already indicated, individual initiative assumed now, in the communal period, a decisive part in the development of clearings and plantations. The increased need for firewood and wood for construction, meanwhile, for an increasingly dense urban population often made deforestation itself a lucrative undertaking, and the hilly slopes closest to the towns and larger villages were becoming denuded of their mantle of trees. But, in fact, precisely on these new bare slopes began to spread the property of the new class of bourgeois owners, who here most often first bought their farms and built their villas. To these new people, quick profits provided means for monetary investment and outlays for cultivation to an extent that would have been inconceivable for private individuals of the feudal era. These farms on the suburban hills, where the urban proprietor could watch over the *coloni* more easily, thus became the chosen location for the most valuable crops, and particularly for trees and shrubs. Many communal statutes of this period oblige all proprietors to plant at least a certain number of trees or feet of vines every year, and if, in the plains, which were more adapted to the mass production of grapes, the landscape expanded with rows of trees festooned with vines, it was in the hills that the production of fruit and quality wines was most often concentrated. One should not forget that this was precisely the period when many of the most prized grape types from distant places were introduced, selected, and diffused throughout Italy. "And I want only," Cecco Angiolieri already sang in the thirteenth century,

> . . . e non vorria se non greco e vernaccia,
> che mi fa maggior noia il vin latino,
> che la mia donna, quand'ella mi caccia.
>
> (. . . and I want only Greek and Vernaccia,
> for Latin wine is more distasteful,
> than my woman, when she nags me.)

Whereas consumption of less esteemed wines was appropriate for the masses, as we see, the gourmands of the new dominant class no longer wanted to hear of "Latin wine," the product, that is, of the traditional grape types. Along with Greek and Vernaccia, wine types imported from distant lands were sought, such as Moscatello, Malvasia, and Schiava, whose cultivation was rapidly spreading precisely and particularly on the sunniest exposures of the hills where specialized low-growing vineyards continued to spread, while the mixed cultivation of vines on trees prevailed more and more clearly in the plain.

Already at the time of Pietro de' Crescenzi, the spread of deforestation and cultivation in the hills began to set before the agronomists of the Italian communes the concrete problem of hydraulic systematization in the hills, as well as that of the plains. And here now also there was need and scope for the initiative and ingenuity of individuals. Plantations of trees and bushes, like those of olives, and more particularly vines, required, if for no other reason because of the deep ditching that was needed to plant them, some kind of systematization of the type agronomists now call "extensive," in comparison with the "intensive" type that required greater care and more complex arrangements. On the more sloping terrains of Sicily, the southern coasts, Tuscany, and the Ligurian coast, there was already recourse here and there, for better cultivation of citrus fruits and sometimes vines, to arrangement of land in the hills in rough terraces (*cunzarri, lenze, fascie, ripiani*, or *terazze,* as they were variously called); and de' Crescenzi already speaks in his work on agronomy of attempts to defend against erosion from running water on the steepest slopes, through the use of supports and embankments.

It was a matter now, to be sure, of practices in a few zones of the most solicitous cultivation and the densest population, that were not yet widespread. More frequent, in the hill zones, systematizations of a more extensive type continued, like those needed in any plantation of trees or bushes, or that which appears in Plate 21 where we have reproduced a view of a seaside by Lorenzetti, the master of the Sienese school, who revived European landscape painting in the communal period. From a village situated on the coast of the sea, tilled land extends along a plain toward the interior. But now plants and trees have appeared on the inclined slope of the deforested hills through the hardworking initiative of some individual. From what one can see of the hillside in the painting, the cleared plot has been arranged in a regular form, protected against erosion from water not only by a barrier along the perimeter but also by two traverse ditches or drains (or is it possible that this plot has already been arranged in terraces?).

Plate 21. Individual clearings and an extensive systematization on a hill in the *Seaside* by Lorenzetti.

. 30 .

The Suburban Agricultural Landscape

In the painting by Lorenzetti in Plate 21, we saw a new taste for picturesque landscape strive to reveal quite minute and realistic technical details of the systematization of the agricultural landscape. The same creative activity in the agricultural landscape of the communal age, indeed the same individual genius that instilled admiration for squared fields, well-aligned rows, and astute arrangements in plain and hill expressed itself in a revolutionary renewal of the pictorial landscape, of which the *Buon Governo* (Good government) by Lorenzetti, from which we reproduce a detail in Plate 22, can be considered a remarkable example. It provides also, with its precise realism, a true and proper agricultural panorama of communal Italy.

The limits inevitably imposed by reproducing a pictorial detail makes it impossible, to be sure, to appreciate the importance of the composition from Plate 22, which in Lorenzetti's painting nonetheless now gives a fully autonomous meaning to the landscape. But in this detail of the painting, which displays important aspects of the suburban agricultural landscape, we will limit ourselves to indicating to the reader how the initiative of individuals, multiplied by *securitas*, the security that the good government of the commune guaranteed, had arrived at the point of imprinting new regular forms on the hillside landscape dominated by the city. These new forms were not, to be sure, those of the Roman *centuriatio*, or those we were able to reveal in the foundation of Villafranca Veronese, which were preordained and prefigured by public initiative. Here tillage has expanded and the rows of vines have been aligned through *individual* initiative. There is a scenic plan for each farm and vineyard of a rigor and precision that is quite unknown for any previous period. But in its complex, in its panorama, the elaboration of the agricultural landscape was left to the casual combination of these individual initiatives. And only through this casual spontaneity were the objective regularities affirmed of a new level of development of productive forces, and a new type of social relationships, which the pattern of this landscape nonetheless obeyed.

In this contrast between the perfecting of individual plans and the lack of a collective plan is expressed the whole dialectic, and also the limits of development of this landscape and of all the forms of communal society, which never succeeded, as Gramsci rightly indicated, in over-

Plate 22. The suburban agricultural landscape in communal Italy in a detail from the *Buon Governo* by Lorenzetti.

coming its individualistic, or at most corporative, development. And here nonetheless the *Buon Governo*, to be sure, succeeds in arousing respect for the underlying norm of urban discipline, which was included didactically even in every communal statute. Heedless of this discipline, the honest citizen in the foreground of the painting climbs toward the gate of the city, peacefully expecting to permit the pig he drives before him to scavenge through the streets of the city and search for food among the refuse.

Even Lorenzetti, from this point of view, may not seem very respectful of the norms of *collective* discipline; however, with regard to the norms of the agricultural technique of his time and the individual elaboration of the agricultural landscape, his painting of "good government" has a paradigmatic, if not directly didactic, value. It would have been difficult in the real landscape of Siena of his time to find all the rows of vines along the hills so astutely aligned crosswise to the direction of greatest slope, as they are in his painting, in anticipation of the modern systematization of hillsides with horizontal furrows bisecting the greatest slope of the hill (*a girapoggio* or *a tagliapoggio*, as we will have an occasion to discuss further), which is respected even where the hill had a double slope. On the contrary, the archival documentation of the time, and iconographic documentation also, shows us how—despite the disapproval of de' Crescenzi and all the best writers on agricultural subjects—it was customary instead, until a rather late period, to favor plantation and cultivation *a rittochino* (with plowing straight down the hill slope), a system that furthered erosion of the soil through the action of flowing water.

Be that as it may: in the good government of communal Italy, the individual elaboration of the agricultural landscape was necessarily paradigmatic and didactic. The *securitas* of this "good government," as well, effectively guaranteed, even through a body of rural police, the protection of closed fields against damage that might be caused by animals and thieves, and it created a habitat no longer restricted to the circuit of the city walls, but that could extend to farms and scattered houses. The plantations of trees and shrubs, and a network of side roads even from farm to farm, completed this typical and significant suburban panorama of communal Italy.

. 31 .

The Landscape of the Countryside

The complex but well-defined forms of the agricultural landscape, which individual initiative for tillage, plantation, and enclosure elaborated around the towns, did not extend in Lorenzetti's period much beyond the territorial limits of the suburbs. A further detail of his fresco of the *Buon Governo*, which we reproduce as Plate 23, shows us how the forms of the agricultural landscape trailed off at the bottom of the hill where the town was built. Beyond the cultivated area of trees and shrubs, contained here in a single enclosed plot of rectangular shape, extend shapeless sowed fields without trees, which are roughly limited by a rough road or drainage ditch that left them open or are enclosed only by a low hedge of brambles. On the stubble, however, two hunters on horseback and their dogs, engage with impunity in the pleasures of the hunt; and on the road itself, which leads along the bottom of the hill opposite the town, only the vigilance of the shepherd, it seems, is there to prevent the little flock from jumping over the easy barrier of a rudimentary hedge of thorns and invading some neighboring field. But perhaps the shepherd will lead his flock instead into the thicketed pasture that extends up toward the higher hills, which are distant, high, and bare, and wooded only in two half-hidden valleys.

But even in the more shapeless landscape of the countryside signs of new times have appeared in the *Buon Governo*. Now, even here, a convenient road climbs up toward the high hills and the wooded valley; a large farm indicates the presence of peasants close to their labors; and, at the foot of the hill opposite the city, just at the turn of the road, an isolated farmhouse documents a change in the condition of public security. And from the point of view of technology, threshing with oxen or horses, which was characteristic in less populated regions of extensive tillage, had now been replaced by an improved type of threshing that only spread rapidly after the year 1000. But still more: from the threshing floor, grain is being taken on the back of an ass to be ground at a mill powered with water. This, known to the Romans under the late empire, only gained a decisive importance in the milling technology of the communal period, thanks to a revolution in techniques for using draft animals, and in transport, that made it economical to utilize the energy of water through costly arrangements, and made it possible to bring grain in to be ground over a larger territorial radius.

Plate 23. Forms of landscape in the countryside of communal Italy from the *Buon Governo* by Lorenzetti.

Thus the new level of development of productive forces and new productive relationships had begun to affect even the landscape of the countryside, but to the extent that one became distant from the towns from which the forms of this influence spread, they appear, in historical reality, as in Lorenzetti's panorama, to be less important, less precise, and more hazy.

. 32 .

The Pastoral Landscape of the Communal Period

Even in the pastoral landscape, as we have already indicated, influences that came from the towns radiated out, in the communal period, with a new effectiveness, which is reflected by a novelty of content and form even in pictorial landscapes, from Giotto to the Sienese. Thus in a painting like the one in the Vatican gallery showing *San Gioacchino and the Shepherds*, attributed to Bartolo di Fredi or to Andrea di Bartolo, which is reproduced in Plate 24, the sensibility and style are quite different from the Giotto-like San Gioacchino from the Cappella degli Scrovegni in Padova, where the geometric synthesis of shapes contrasts in its artful simplicity with the more developed forms of the pastoral landscape of Andrea. But despite the diversity of schools, styles, and sensibilities, the reader will not miss what these pictorial landscape of the communal period have in common, in contrast with those of Roman, Byzantine, or Romanesque art. These were not only landscapes that, even in these pastoral scenes, as in the agricultural panorama of Lorenzetti, have regained an autonomous significance and unity. The civilizing and organizing influence of the city is now reflected clearly on both form and content, on both men and things, and it permeates these landscapes with a new and different spirit that was quite unknown in antiquity. This appears in Giottoesque painting in a new but severe urbanity of the faces and dress of the shepherds, in the clearly depicted sheepfold that the shepherd's goats and dogs are leaving; the animals are depicted more amiably themselves, as are the no-longer-menacing or contorted rocks and trees. In the pastoral landscape of Bartolo or Andrea, the curvature of the rocks has become sensual, almost affected; the pastoral depiction has become idyllic, and how different from that of antiquity, which we have seen in mosaics like the one shown in our Plate 4.

We are now on the threshold of the Renaissance; and the shepherds of Bartolo or Ándrea, like the "vaghe montanine pastorelle" (wandering mountain shepherds) of Sacchetti, seem transfigured by the poetic fantasy of citizens into a distant idyll, which is sometimes not lacking in irony. But now the city is near: nearer than the ancient *urbe* and *civitates* ever were. The shepherds of Bartolo or Andrea, the "pastorelle" of Sacchetti, still, to be sure, climb up to the hills with their flocks and herds to flee the summer's heat; but they pass through the "suburbs" and past villas that the cities have now deposited through the countryside, and

Plate 24. The pastoral landscape of communal Italy in *San Gioacchino and the Shepherds* in the Pinacoteca Vaticana (attributed to Bartolo di Fredi or Andrea di Bartolo).

with a few goats, they have perhaps just now left a farmhouse, which, as in Lorenzetti's painting in Plate 23, we might find at the first turn of the road along the base of the hill facing the city.

These shepherds are of a different type than the ones we encountered among the figures of the late empire, the Byzantine and feudal period, and again in the Pisan school. At earlier times, along with large wild herds, a kind of wild shepherd predominated, who not even the most lively city-bred imagination could put into an idyllic light. But now in communal Italy, with its villas and scattered farmhouses, small and medium-scale breeding was already appearing, in which the shepherd, though still going out beyond the walls of the suburb or villa, did not loose all contact with civilized life.

Even in communal Italy, however, pastoral landscapes, although more amiable ones, continued to prevail over agricultural landscapes. As well, there continued to be a promiscuous right of pasturage after the harvest over a large part of the tilled fields. Our iconographic documentation thus does not deceive us, although we should not ask it to reveal what only archival documentation should rightly provide. The art of communal Italy tells us admirably what was typical of the pastoral landscape and activity of that period, and it can give us a sense of historical development by telling us what had changed in the sensibility of the men who were involved in that historical development. But it cannot tell us why large sectors of the peninsula and islands, from Sicily to Friuli, from the Kingdom of Naples to Lazio and Sardinia, remained almost entirely distant from this development, and from the effects of the communal rebirth, or why the experience in these places was cut short in its beginnings. Nor can it tell how in these regions, and often in vast tracts of the very regions, like Tuscany, that were at the center of communal development, old ways of life and old pastoral landscapes continued, even in this period, from those of the feudal period. For this it is necessary to consult documents of other kinds that, although they express the sense of historical development less clearly, make clearer the geographical and historical limits of the communal rebirth.

. 33 .

The Landscape of the Woods and Hunting

For the changes that occurred in the landscape of forests and woods in the communal period we can make use, in part, of the same considerations that we made in the previous chapter with regard to the pastoral landscape. The process of deforestation that spread through public initiative in the eleventh and twelfth centuries continued in the following two centuries. But it no longer had the character of a general movement, and it appeared more localized in specific sectors, where particular situations conditioned its further development—thus in Venice, for example, where the widespread deforestation of the mainland for the purpose of naval construction induced the public authorities to develop a policy to protect the forest already in the fourteenth century; and also around Siena (and in other of the chief communes of Tuscany), where a special magistracy for forests was instituted in 1358. But in this and other similar cases, public initiative turned to the protection of a forest patrimony that was already in process of deterioration, rather than to provide for further deforestation. Private initiative proceeded, if anything, with unwise deforestation around the most active centers of seafaring, building, manufacturing, and agriculture, and thus threatened a serious deterioration of the slopes of hills and mountains when the local forest patrimony became exhausted, as well as a dangerous dislocation of the hydraulics of the valleys.

Where the particular conditions that spurred on private initiative were lacking, as in zones more distant from the urban centers, deforestation (in absence, generally, of any public initiative) was retarded, and it sometimes even receded as the communal period advanced. Thus the deforestation of this age was most concentrated in the immediate neighborhood of the urban centers, while between one city and another vast extents of forest remained, which are frequently mentioned in the literary and archival sources of the time. In comparison with the preceding period, to be sure, zones of deep forest began to lighten, at least in the areas of the strongest communal development. But still, a few miles from a city like Florence, the woods continued to be filled with wild beasts, wolves for instance, that are often mentioned in communal statutes, archival sources, and fourteenth-century tales; and they attacked flocks, shepherds, and even city people who imprudently ventured away from beaten paths at night.

Plate 25. The woods and hunting in communal Italy from a miniature in the *Theatrum sanitatis*.

Still at the time of the *Decameron*, and even of the "bel paesaggio" of its suburban villas, forests abounding with large game were close to parks and gardens. Hunting thus remained an important resource for the supply of meat to the population in the communal period, even if it had more and more the character of a privilege and preferred recreation for the dominant classes. The miniature from the *Theatrum sanitatis* reproduced in Plate 25—which gives much space to illustrating hunting and the quality of game—provides a concrete image of this landscape of the woods and hunting, as it appeared on the threshold of the Renaissance, in one of the zones where the deep forest seems already to have been mitigated by the closeness of city folk, who had begun to open up paths and glades, while literary and archival sources also reveal all the importance that, even in these zones, the forest of the communal period preserved for the purposes of foraging animals.

. 34 .

The Revival of Cultivation of Grain, and the
Landscape of Closed Fields in the
Communal Period and the Renaissance

With the development of plantations of trees and shrubs, and with the practice and diffusion of the first extensive systematizations of the plain and hills, the basic agricultural conditions were undoubtedly formed in communal Italy for a recovery of the system of fallow to replace temporary clearings, and of the predominance of cultivation of grain over that of inferior cereals.

These last two processes were clearly closely interconnected, just as both were connected to the progress of systematization in the plain and hills. It was rarely practical to continue a system of extensive and irregular tillage, like that on temporary clearings—even when regulated by a four- or five-year cycle—on a terrain significantly improved through some kind of systematization; while, conversely, a terrain systematized even for extensive cultivation that had returned to regular alteration of fallow and cereals could be utilized more effectively for the cultivation of wheat, which was more demanding, than for more rustic, but lower-yielding, inferior cereals.

Arguments against the system of temporary clearings had already arisen in the agronomic work of de' Crescenzi, who pronounced himself in favor of fallow, the use of dung, and the plowing under of stubble. Otherwise, beginning in the thirteenth century, consumption of wheat began to gain a clear ascendancy over inferior grains, even in the towns of central and northern Italy where wheat consumption had contracted the most during the early Middle Ages. For the rural population, however, consumption of inferior cereals continued to predominate, or at least it had a decisive importance, for the whole communal period, and often to a much later date.

This very evolution of items of consumption, which was set in motion by the productive upturn of the urban economy and the increased buying power of the urban classes, draws our attention to the limits that social conditions and the level of technical development imposed on the progress of horticulture, which were even more than those imposed on the cultivation of trees and shrubs. All the available documentation confirms that despite the return of the system of fallow and the spread of systematization on the plains and hillsides, there was a quite limited increase in yields per unit for cereals in the communal period. One can indicate at most, in general, besides an improved ability to meet the more complex requirements for cultivating wheat, a somewhat lesser uncertainty of harvests in the changing seasons; and this factor, if any, might have contributed to a certain increase in mean yields per unit.

The decisive limitation on a more rapid increase of revenue in the communal period was undoubtedly set for cereals by the as yet scarce diffusion of crop rotation with forage, and the consequent scarcity of dung. As well, with the progress of clearings, and the progressive shrinking of terrain in woods or pasture in zones nearest to populated centers, the problem of finding forage for domestic or semidomestic breeding, which was nonetheless becoming more important, became inevitably worse. This was a still more serious problem precisely in those zones where the progressive extension of a landscape of closed fields, even on sowed land, deprived cattle of the resource earlier assured to them, with open fields, by the right of promiscuous pasturage over stubble and fallow. Even for beasts of burden, which were indispensable for working the land and for the minimum of dung needed to fertilize it, there was often a lack of pasturage and hay, and it was necessary to resort to collecting branches of trees and shrubs to supplement the meager forage supply. Searching for twigs for cattle thus became not least among the tasks and concerns of peasants. When the fourteenth-century realist Umbrian poet Ser Cecco Nuccoli wanted to make a joking reference to a friend who had transformed himself from a city dweller into a peasant—"e mane e sere mange coi bevolche" (and morning and evening he eats with the beasts)—he accused him with the words: "a frasche vai mozzando col falcino" (you cut branches with a sickle), which had become a typical task of peasants, like hoeing or plowing.

Otherwise, Piero de' Crescenzi already shows us how important this need for forage was for the spread of a landscape planted with vines on trees. This provided the farm an indispensable source of wood not only for domestic and agricultural uses, but also, and particularly, of forage for cattle among the trees of the farm. Later, when artificial pastures had already begun to spread elsewhere, it became proverbial that "i Toscani tengono i loro prati sugli alberi" (Tuscans keep their meadows on the trees). Meanwhile, not only in the landscape planted with vines on trees, but also on sowed land without trees (which more and more frequently was enclosed, at least near centers of habitation), the need of twigs for cattle induced farmers to resort more and more widely to "live enclosures," which also offered greater security from the theft and damage done to stockades and palisades. A detail from the *Flight into Egypt* by Gentile da Fabriano, reproduced as Plate 26, provides luminous

Plate 26. Fields enclosed with live hedges in the Renaissance landscape, from a detail of the *Flight into Egypt* by Gentile da Fabriano.

evidence of how this landscape of fields enclosed with live hedges looked on the threshold of the Renaissance, which had a certain similarity to the French *bocage*. The landscape expressed, to be sure, the new and organic forms extending into the countryside from the circuit of the city walls. But the detail reproduced here does not let us see how limited it still was. In the full composition of Gentile da Fabriano, a mountain, on the highest slopes of which the figure of San Giuseppe is projected, indicates the extreme limit of the suburban landscape. Nearer are further open and untilled fields, which separate us from another distant city, which is surrounded, down into the plain, with another narrow circuit of vineyards and enclosed fields.

.V.

THE AGE OF THE RENAISSANCE

. 35 .

*The Origins of the Contemporary Landscape:
Enclosures, Systematization* a Rittochino
*on Hillsides, and the Landscape of Irregular
Fields* a Pìgola *in the Early Renaissance*

In Plates 21 and 22, two paintings by Lorenzetti allowed us to follow the changes that individual initiative of clearing and plantation introduced in the hill landscapes of Italy in the communal period, in zones not too distant from centers of habitation. From a natural landscape hardly touched by the initiative of any individual or group, one thus passed, in the painting in Plate 22, to a more complex picture, where the multiplication of individual initiatives had reached the point of meeting and colliding, interweaving an irregular network of fields and vineyards onto the slopes of the hills, which was marked by boundaries and hedges to separate the plots.

In the communal period, however, as appears also in the painting by Lorenzetti in Plate 22, it was still rarely that the confluence of individual initiatives for clearing and planting succeeded in covering the whole hill surface, and thus to close completely the network of irregular fields ("campi a pìgola"—that is, "a spigolo," with rectangular but not parallel sides, as these were aptly called in the region of Volterra). Only in the later communal period, and the early Renaissance, did this more elaborated hill landscape or irregular fields—of which a detail from the *Adoration of the Magi* by Gentile da Fabriano reproduced in Plate 27 gives us a clear and precise image—tend to be definitively affirmed on slopes leading down from towns and more populous villages, or on whatever sites close to centers of habitation.

The hill landscape of irregular fields had the appearance, on first sight, of the landscape of irregular open fields found in various parts of France, and particularly the *marrello* of Provence. Like this, in effect, they were born from a coincidence of clearings and plantations made without any plan, according to individual impulses and initiatives, and they corresponded to the habit of working with a hoe or a simple plow (the French *araire*), while a landscape of regular elongated fields corresponded often to the use of a plow with moldboard (the French *charrue*). But the French landscape of irregular fields differed from the Ital-

Plate 27. Hillside systematization and *campi a pigola* in Renaissance Italy, from a detail of the *Adoration of the Magi* by Gentile da Fabriano.

ian one of *campi a pigola*, which tended to be a system of closed fields; while, unlike the landscape of the Mediterranean garden—where the most valuable crops were for the most part protected by walled enclosures, or at least by rubble or ruins—in the landscape of *campi a pigola* there were less conspicuous enclosures made up of living or dried-up hedges, or simply ditches or drains.

From one end of Italy to the other, with the increase of population and the ever greater recourse to individual clearings and plantations on hillside terrains, the landscape of *campi a pigola* became, and remains up to our own day, one of the dominant and characteristic types of the Italian agricultural landscape. It often spilled over even into the plains and the treeless sown fields. But in the early Renaissance, its appearance was tied particularly to the impetus that the expansion of bourgeois property owning and individual plantations in the hills gave to the prevalence of the system of sharecropping (*mezzadria*). "Now is the time," Michelangiolo Tanaglia, an agronomist humanist, sang in Florence in the time of Lorenzo the Magnificent,

> ... or tempo pare a l'openion mia
> ne' monti aperti mai mostrarsi stracco
> di piantar viti, e del miglior qual sia
> or frequentar della pianta di Bacco:
> non s'usa far più utile postùme:
> chi l'arte sa, non serri a questa el sacco,
> che se di nulla intendo o veggo lume,
> questo piantare o diboscar terreni
> di dar più util hanno per costume.
>
> (... now is the time, in my opinion,
> on the open hills, to never tire
> of planting vines, of the best kind,
> or of attending the plant of Bacchus:
> there is no better second crop:
> who has this art [should] never close the sack,
> for I know nothing better
> than planting and clearing land
> for generally making a good profit.)

And for the whole fifteenth century, and beyond, in fact, the communal bourgeoisie did not "close the sack"; it did not economize on the investment necessary to "clear and plant land" in the hills, particularly with vines. This was no longer only to meet the needs of domestic consumption, but from a calculus of profit, because the vine is the "best second crop"—that is, the most profitable one—and because lands once planted in vines "generally make a good profit."

In this calculus of quick individual profit, and the dialectic that resulted, can perhaps be found one of the reasons why, despite the disapproval of de' Crescenzi and of Renaissance agronomists, systematizations in the hills continued to prevail, as can be seen also in our Plate 27, with the land worked *a rittochino*—that is, with furrows running straight down the line of greatest descent, rather than across it, which was the only way to achieve a minimal protection against soil erosion from running water. Thus, up to this point, private ownership of land in its new bourgeois forms was joined with the search for capitalist profit, and although it clothed Italy with a new splendid garment of vines and orchards, it also prepared dangers for future generations that threatened the very integrity of the soil.

· 36 ·

The Landscape of Enclosed Fields in the Plain and Systematization in Porche

"The best harvest in the plain is from the *magolato*," Michelangiolo Tanaglia, the agronomist Florentine humanist who we saw make himself the paladin of plantations in the hills, sang in his verse treatise *De agricultura*. In the communal period and the early Renaissance, in effect, if plantations represented the most diffused elementary form for systematization of terrain in the hills, the *magolato* (that is, systematization in *porche* or hummocks—narrow elongated rows separated by shallow drainage furrows) affirmed itself in the zones of central Italy, from Tuscany to Umbria and the Marche, and it remained until our own days the most common extensive arrangement for tilled land in the plains.

Plate 28, which reproduces a detail of the *Disposition* by Fra Angelico, shows us how, in this period and environment, the *magolato* already appeared, with its rigorously parallel hummocks that formed narrow plots in the suburban agricultural landscape of closed fields. This systematization in *porche*, which left ridges of soil, sixty to sixty-eight centimeters wide and packed down in various ways, interposed between two *solchi acquai* (drainage channels, as they are called in Tuscany), was a temporary arrangement, or more precisely periodic, and extensive: it required a relatively limited input of labor, but one that had to be repeated with every sowing. Its closely aligned water channels tended to assure drainage of rainwater from the soil where it might otherwise drown the young plants. This system is typical today in agronomically backward areas, and it can be justified only where it has not yet been possible to advance to more intensive permanent arrangements. The closely aligned water channels, in fact, protect against a lesser evil by draining away excessive water in the rainy season; but in the dry season they dry out the soil, precisely when the plants have the most need for water.

For this and other reasons, Ridolfi and other major agronomists from the nineteenth century onward made systematization in *porche* the object of justly severe criticism. But in comparison with the earlier lack of any drainage system, the diffusion of the *magolato*, so warmly recommended by Tanaglia, undoubtedly represented an important advance in the regions of central Italy during the communal period and early Renaissance. Our Plate 28 shows us why. In the smallest details of the agri-

Plate 28. The system of closed fields and systematization *a magolato* in a detail from the *Disposition* by Fra Angelico.

cultural landscape, systematization in *porche* introduced an element of geometric regularity that, aside from any other consideration, became of itself a factor of agricultural progress. The same regularity that the painting of Fra Angelico documents in the detailed internal texture of fields is what Tanaglia advised for the planting of trees, when he sang:

> Se per tramite retto e pari sesti
> fien compartiti, più grati saranno,
> e par che me' la terra omor vi presti;

> (If set in straight and even rows
> they will be more grateful,
> and you will do more honor to the earth;)

or for the external form of the fields themselves, which he wanted well squared and aligned:

> Agli orti come ai prati squadra e lista:
> a medesime plaghe è ben sien volti:
> per l'uggia a l'altre men frutto s'acquista.
>
> (Square up and align both orchards and meadows
> with the same extents and turn them carefully:
> otherwise from spite there will be less fruit.)

This is the whole agricultural landscape, summed up in its more general aspect as well as in its smallest details, with its forms reelaborated. And with these new forms, the taste for a more refined and precise "bel paesaggio" than ancient Rome itself ever knew was already prepared.

Meanwhile, it must be confessed that the diffusion of the *magolato* was connected, as Ridolfi also noted a century ago, not only to the need for hydraulic arrangements, or to regularize the smallest details of the agricultural landscape, but also to more prosaic motives, and specifically to the deficiency of dung, which was characteristic and particularly serious in the zones of central Italy because of the lack of forage crops that we have already discussed. With the arrangement in *porche*, the Tuscan peasant realized (or at least thought he realized) an economy in the use of dung, in that the fertilizer was spread only on top of the *porca*, that is, on the part of the soil that would in fact be sown, and not on the remainder of the terrain, which was occupied up to 20 or 30 percent of the surface by drainage channels.

Thus even from this point of view, a study of the forms of the agricultural landscape that were characteristic of the communal period and early Renaissance—from the plantations of trees with vines, to the live hedges around fields, to the arrangement in *porche*—leads us back to consider the decisive limits set to the development of productive agricultural forces in this age, from a technical point of view, by the lack of basic forage for the agricultural enterprise. This limitation was now made worse by new productive and social relationships, and by the progressive spread of a system of closed fields, which excluded still more terrain from the right of promiscuous pasturage.

· 37 ·

Toward a Redressed Balance of Forage: The Landscape of Enclosed Pastures and Meadows

Already in the fifteenth century, as we have seen, the decreasing proportion of uncultivated land, fallow, and stubble that was utilizable for pasturage contrasted with the growing need for livestock to improve tillage of the soil, and for a more abundant production of dung. In these conditions, the problem of redressing the balance of forage began to appear in Renaissance treatises on agriculture, just as it was posed urgently in practice on a daily basis. Already Tanaglia did not make himself the knight-errant only of new arrangements in the plains and hills; in proposing the selection of a farm site, he recommends that

> . . . con assai acqua v'abondi di prata
> che me' si possan mantener gli armenti
> per a' debiti tempi bifolcare
> e tirar rastri con ispessi denti
> o carra, o quando accade trainare;
> però di fieno e pasture provisti,
> che' la bocca le gambe dee portare.

> (. . . there be enough water in the meadow
> for me to keep the flocks
> and in due time hitch the team
> to pull rakes with strong teeth,
> or the cart, or whatever is needed;
> to get hay and provide fodder,
> for the mouth must carry the legs.)

Without hay from the meadows now, and with fewer resources for pasturage, or dung, a farm lacked even the strength of beasts of burden, which would not provide their needed droppings if they were not well nourished: "for the mouth must carry the legs," as Tuscan peasants say.

For the whole Middle Ages, the essential fodder for agricultural beasts of burden was assured, not excluding what came from natural meadows that could be mowed, chiefly by foraging. Even in the late Middle Ages, despite the progressive expansion of clearings, serious obstacles to greater extents of pasture resulted from two natural and social factors. On one hand, feudal lords tenaciously opposed reducing cultivated land to pasture so as not to be deprived of dues and other payments in kind from the harvest, which feudal custom exacted generally from land destined for the cultivation of cereals, and from kitchen vegetables and orchards of fruit, but not from land destined for production of hay. On the other hand, even where private owners had the right to enclose tilled land, not only the feudal lords, but often also the great majority of the rural population, which had an interest in exercising rights of promiscuous pasturage, tenaciously opposed the enclosure of pastures by individual owners. Thus, after the first cut of hay, which was ultimately reserved for the proprietor, meadows generally had to be left open for the customary pasturage of the whole community.

It is easy to understand how, in these conditions, the interest and possibility of extending the land destined for meadow remained quite limited. But already in the early Renaissance new agents intervened to change the conditions disfavorable to enclosure of meadows and pastures. Meanwhile, in various parts of Italy, and particularly in the Kingdom of Naples, the feudal lords themselves, spurred on by a growing demand for wool in international markets, tended to encourage sheep breeding in their fiefs. And they illegally exempted from the customary promiscuous pasturage of the local population a part of their feudal lands, which they closed off, reducing them to *difese* ("preserves," as they were called), which were reserved for their own flocks or for those of the great entrepreneurs of herding to whom they were "entrusted." From 1443 Ferdinando d'Aragona, with the pragmatic sanction *De salario*, was obliged to recognize the gravity of this abuse, making royal authorization necessary for establishing preserves, and ordering that all those which feudal lords had created without authorization to the prejudice of the general population be reopened. But this act, like the other, *De Baronibus* of 1536, did not succeed in halting the creation of preserves. The movement toward enclosure of pastures, like the English enclosures that are vividly described by Marx in the first volume of *Capital*, was otherwise, as is well known, a Europe-wide phenomenon of this period. Even in southern Italy this had an impact on the form of the agricultural landscape, in which—along with enclosed tilled fields—preserves composed of meadows and pastures, which had once been open to the common use of the general population, now became characteristic elements.

But where the development of the communes more deeply eroded the forms of feudal property, as in a good part of central and northern Italy, this process often had different aspects. Here, the initiative to enclose meadows and pastures often came from new bourgeois owners. But still it encountered serious opposition from traditional and widespread customs of common pasturage, which only gradually declined in extent and importance, and first in the new meadows that received more intensive care and were destined to be mowed. "Now," Tanaglia wrote already at the end of the fifteenth century,

> or si deen provedere e' macri prati
> di sterquilin, misto di fien con seme . . .
>
> (now one must prepare the great meadows
> with dung, mixed with seeded hay . . .)

We have now reached the point not only of a regular manuring of pastures, but also of their renewal by sowing hay grasses mixed with dung. Only one step lacked in the further passage to artificial pastures, and even this was accomplished with the diffusion of pastures of clover and alfalfa (*erba medica*). But for an artificial pasture, now, enclosure was obligatory; and with regard to a pasture of alfalfa Tanaglia aptly warns

> . . . con siepi e fosse e' pòstumi difendi
> dagli avversari, che son tutti armenti,
> che so naturalmente tel comprendi . . .
>
> (. . . with hedges and ditches protect the gleanings
> from adversaries, which are any herds,
> that, naturally, I know you understand . . .)

When one reflects that there were at least four *pòstumi* (that is, the second mowings, or gleanings) of alfalfa, it was "naturally" understood that cultivation of this type of forage would not be profitable without a permanent enclosure to exclude third parties from enjoying the *pòstumi*. And in central and northern Italy, in fact, it was particularly the diffusion of natural and artificial pastures that produced new forms of the agricultural landscape, like those that appear in Plate 29, which is taken from a detail of the *Meditation on the Passion* by Carpaccio. For open pastures and temporarily enclosed meadows, which were characteristic of the Middle Ages up to the communal period, was now more and more frequently substituted, although still slowly, the permanent enclosure of meadows and pastures with hedges, trees, stockades, and ditches, following lines that (more often than was common for tilled fields of cereals or shrubs) assumed irregular sinuous shapes, to adopt themselves, especially in the case of natural meadows, to curves of the hills.

Plate 29. Live enclosure of meadows and pastures in the Renaissance landscape, from a detail of the *Meditation on the Passion* by Carpaccio.

. 38 .

Improvements and Irrigation in the Renaissance Agricultural Landscape

We have already indicated that for the whole fifteenth century, and beyond, the spread of the cultivation of meadows remained difficult, and developed only gradually. Tanaglia himself, who pushed for a more rapid pace, was fully conscious of the limits posed by objective conditions, and he followed and accepted the scarcity of hay on Tuscan farms as a matter of fact, along with the traditional need to make use of branches from the trees festooned with vines to complete the basic forage of the estate, so that

> . . . olmi ancor con la foglia nutriranno
> gli armenti in parte . . .
>
> (. . . elms also with their leaves will nourish
> the herds in part . . .)

With regard more specifically to the permanent enclosure of meadows, fifty years after Tanaglia, Luigi Alamanni in his *Coltivazione*—the most important agronomic poem of the sixteenth century—urged the farmer, in the work to be done in the spring:

> Indi volga il pensier con l'opra insieme
> intorno ai prati, che il passato verno
> aperti, in abbandon negletti furo,
> agli armenti, ad ogni uom pastura e preda.
> Quei con fossi talor, talor circondi
> con pali e siepi, e se n'avesse il luogo,
> può di sassi compor muraglie e schermi:
> tal che il rozzo pastor, la greggia ingorda
> e col morso e col pie' non taglie, e prema
> la novella virtù, che all'erba infonde
> con soave liquor la terra, e 'l cielo.
>
> (Then turn your thoughts and actions
> to the meadows, which last spring
> were open and abandoned to the herds:
> any man's for pasturage or taking.
> These with ditches now, now surround

> with palisades and hedges, and with enough space,
> you can make walls and barriers of stone,
> which the rustic shepherd's greedy flock
> will not cut with mouth or foot, or crush
> the new life that the grass receives
> as sweet liquor from the earth and sky.)

It is easy to see, from the contrast between these neat verses of Alamanni and the rougher ones of Tanaglia, what and how many refinements the fifty years between the end of the fifteenth and the beginning of the sixteenth century brought to all forms of culture and society, as well as to the Renaissance landscape. It remains, however, that still in the first half of the sixteenth century—as appears from the verses of Alamanni themselves—the permanent enclosure of pastures remained a far from completely resolved problem. But still more: in the climatic and agricultural-geological conditions of a large part of Italy, the "sweet liquor," which according to the Florentine humanist should rain "new life" from the skies onto the meadow grass in springtime, was not enough in itself—because of the irregular distribution of rainfall during the year and, in the South, its absolute insufficiency—to assure a luxuriance of forage in the meadows. And, in fact, even in the following centuries the cultivation of artificial meadows had a decisive importance only where new widespread works of improvement and irrigation provided an appropriate preparation.

Meanwhile, between the end of the fifteenth and the first half of the sixteenth century, just at the end of this preparation, along with systematization of the course of rivers, such works rapidly began to develop, through an impulse that now generally came from the largest communes and the new *signorie*. It is enough to remember, in the State of Milan, works like those undertaken by the Sforza in the excavation of the Binasco canal to bring the waters of the Naviglio Grande canal from Milan to Pavia (in 1457), the Martesana canal to bring the waters of the Adda from Trezzo to Milan (in 1464), the irrigation of the countryside around Vigevano, and around Novara, which was extended with the excavation of the Roggia Mora, the Sforzesco canal (1482), and so on. "I was a low clod," sounded the proud inscription of Ludovico Sforza on the Sforzesco:

> Vilis gleba fui, nunc sum ditissima tellus:
> Cur? quia Sfortiadum me pia dextra colit.
> Mutata est facies, mutataque nomina . . .

> (I was a low clod, now I am the richest earth:
> Why? Because the holy right hand of the Sforza tends me.
> My face has been changed, my names changed . . .)

And in fact, the "low clod" now changed itself here into "richest earth," and with its altered appearance even the place-names in this countryside changed. But the asserted preeminence of the work of the Sforza was contested by Venice, which between the fifteenth and sixteenth centuries set to remodeling and rearranging the flow of all the rivers of its dominion on the mainland; by Tuscany, which in 1469 built the first artificial lake, and began to improve its marshes; and even by Piedmont and the popes, who also redoubled their efforts to improve and irrigate.

Throughout Italy there was a new fervor for public works, that took up again and developed with increased efficiency the plans already begun with the first great works in the Po valley in the eleventh to thirteenth century. The chief scientists and technicians of the time collaborated in the planning and execution of these works, which were more and more solidly founded on the application of scientific principles and hydraulic techniques that they elaborated through these large Italian experiments. Of first rank among them was Leonardo da Vinci, who left numerous elaborate studies, projects, designs, calculations, and topographic maps for works of improvement, irrigation, and internal navigation, from the Pontine marshes to the basins of the Arno, the Adda, and other rivers of the Po valley. According to an organic plan, like the one Leonardo imposed even on a purely picturesque landscape, now preserved in the Uffizi Gallery in Florence and reproduced in Plate 30, these great undertakings of men working together imprinted the Italian landscape anew with rational lines, and with new and fertile furrows.

Plate 30. The geometric landscape of improvement and irrigation in a drawing by Leondardo in the Uffizi Gallery.

. 39 .

*The Irrigated Meadows of Lombardy and the Po Valley
in the Age of the Renaissance*

The Renaissance effort to transform the agricultural landscape through improvements and great works of irrigation developed in quite varied conditions and environments in different parts of Italy. In the South and the territory of the popes it came up against not only technical difficulties, which this period did not yet succeed in overcoming, but also continuing administrative disorder, which continued in these regions where feudal jurisdictions frustrated the productive efforts of a bourgeoisie that lacked the experience of communal development. Even in Tuscany, insistence on a policy of internal colonization, which attempted to make repopulation precede works of hydraulic and sanitary improvement, condemned many projects to failure until the late eighteenth century. They had to be abandoned after waves of malarial infection. In the flat plains of the Po valley, by contrast, the success of great works of improvement and irrigation was often assured not only by environmental conditions more attuned to the technical abilities of the times, but also by the creation of two great and relatively centralized territorial units, Venice and Milan, which realized the powerful concentration of force and means that was necessary to carry out great public works.

Otherwise, in the Po valley, and particularly in Lombardy, the planning and execution of works of irrigation, and the systematization of landholdings necessary for their rational use, could be based on uninterrupted experience and tradition dating—if not from the times of Virgil's "claudite jam rivos, pueri, sat prata biberunt" (youths, now close the springs; the meadows have drunk enough)—at least from the eleventh century. And already in the thirteenth century there is mention in Lombard documents of the method of irrigation by *marcita*, which, by letting a sheet of water run over a meadow during the winter, prevented freezing and the end of all vegetative activity, thus making it still possible to cut grass during the season when resources of forage were most stinted.

In the Renaissance, thanks to large new works of improvement and irrigation, the Po valley, and particularly Lombardy, became more than ever the special place of irrigated meadows, and now its canals and regular fields—whose boundaries were frequently indicated by plantations of mulberry trees recently introduced into that region—began to mark the

Plate 31. An irrigated meadow of the Lombard Renaissance, in a miniature from *De sphaera* from the Biblioteca Estense in Modena.

landscape in a typical way. Plate 31 reproduces a miniature of the Lombard school of the fifteenth century, from a manuscript of the *De sphaera* preserved in the Estense Library in Modena, that provides us with a precise and lively picture of the landscape of irrigated meadows, which now expanded to an increasingly vast terrain in these regions. Here haying, of which the different operations are depicted in the miniature, had already assumed a decisive importance for the balance of forage in a new type of agricultural enterprise, where the breeding of cattle, the cheese industry, and an abundant production of dung permitted the permanent establishment of livestock, for which natural or artificial pastures, often irrigated, provided an essential supply of forage.

These developments had an enormous importance for the elaboration of new forms of the agricultural landscape, and for the future of Italian agriculture as a whole. The center of diffusion of progressive methods tended to move from Sicily, the South, and parts of central Italy toward the Po valley, where the increasing cultivation of meadows soon permitted elaboration, in practice as well as in theory, of a modern agricultural system, in which regular rotation with forage crops made it possible to end the old system of fallow and to integrate organically the techniques of breeding and cultivation. It was not by chance that still in the first half of the sixteenth century new types of cultivation radiated from the South toward the North—such as mulberry trees, rice and other—while, as we have already seen, from the thirteenth to the fifteenth century the rebirth of the science of agronomy was represented by a Bolognese (de' Crescenzi), and by Tuscans (such as Tanaglia, Alamanni, Vettori, Soderini, and Davanzati). It was not by chance either, that in the second half of the sixteenth century we find Gallo and Tarello, from Brescia, and Africo Clemente, from Padova, at the leading edge of the science of agronomy, which was now founding the modern theory of crop rotation.

· 40 ·

The Origins of the Contemporary Landscape:
The Piantata *of the Po Valley*

"On both sides," Montaigne wrote, recording in his *Journal de voyage* of 1580–81 his impressions of the landscape of the Venetian plain, from Verona to Padova and beyond, "on both sides were very fertile plains, which had, after the custom of the place, among the wheat fields, trees planted in orderly rows where grapes were hung." Already thirty years before, in the mid-sixteenth century, Leandro Alberti assured us, in his *Descrittione di tutta Italia*, that "coming down the Via Emilia, and passing through that agreeable and beautiful countryside," it seemed to be ornamented "with winding rows of trees festooned with vines," as was also true through the whole plain of Emilia up to that of the Po: "one sees artful rows of trees, on which are vines that hang down on all sides."

The works of sixteenth-century agronomists, compared with those of travelers and geographers, permit us to integrate and specify the pictures of this planted landscape along the Po (the *piantata*), which extended from the Veneto and Emilia toward Lombardy, and tended now to differentiate itself, from the *alberata* of Tuscany, Umbria, and the Marche, even if it had in common the association of cultivation of grain with that of vines, and the marriage of vines to trees.

Once again, the differentiating element is provided in the *piantata* of the Po valley, by the greater importance that permanent and intensive systematizations had there, in comparison with those that were often temporary, periodic, and thus less intensive in the *magolato* that remained characteristic of the plantations of Tuscany, Umbria, and the Marche. Thus already in the second half of the sixteenth century a writer on agricultural matters such as Gallo, in his *Giornate dell'agricoltura*, clearly reflects the new experiences of agriculture of the Po valley, and helps to specify the succession of operations that, still today, remain characteristic of improvement in this region. These were hydraulic works, which were followed by the first phase of systematization (characterized by division into great squares of the so-called *larga*—open plain—marked and drained by roads, paths, and ditches, and intended for the cultivation of cereals and pasture), and then by the second and final phase, which was characterized by division into regular fields, intensive hydraulic arrangements, and the cultivation of trees and shrubs (the *piantata*). And there was already a lively polemic in Gallo, as there was also in the Paduan Africo Clemento, against systematization in *porche*, to which writers with experience in the Po valley contrasted another, of more intensive and rational character, *a prese* or *a prace*—that is, with *porche* three or four times wider than usual, and arranged with drains and a kind of embankment (*capezzange* or *cavedagne*), which could also serve as a path, and which in the *piantata* of Emilia still today serves to collect water from irrigation ditches when placed at the heads of fields.

That these arrangements did not remain just meditations in the prescriptive literature of writers on agriculture is made evident in the first half of the sixteenth century by Folegno, among others, whose good-humored realistic works reflected above all the agricultural environment of Mantova and Brescia. When in a doggerel poem he puts in the mouth of Tognazzo, consul of Cipada, a boast of his merits as a farmer, it is not by chance that he makes him say:

> Herbida prata scio cum falce taiare seganti,
> scavedagnare argos, fossatis dingere campos.
>
> (I know how to cut grassy fields with a sharp scythe,
> to build embankments and dig ditches around fields.)

With the increasing spread of meadows, in fact, the mowing of hay became one of the principal operations of agriculture in the Mantova and Brescia regions, just as making "embankments" (*cavedagne*) and "surrounding the fields with ditches" became essential operations for the systematization of farms. Thus when Folengo wants to describe a farm gone to ruin because of the neglect of a peasant, who is love-sick, he reveals that

> vadunt ad sguazzum campi, lagus arva covertat,
> namque cavedagnis nunqual fossata cavantur.
>
> (the fields are awash, covered by a lake,
> because neither embankments nor ditches are cared for.)

The fields were flooded, that is, and covered as if by a lake, because no one was taking care of the necessary maintenance of "ditches and embankments."

The constituent elements of the modern *piantata* in the Po valley, in short, were already well advanced by the sixteenth century, with division of the surface into regularly formed fields, and boundaries marked by "embankments and ditches," along which ran rows of trees with vines. This permanent and relatively intensive arrangement was integrated with the variously tilled soil, which was arranged not in a series of *porche*, with three or four hummocks, as in central Italy, but in *prese*, *prace*, or

Plate 32. The landscape of the *piantata* in the Po valley from a sixteenth-century map of the countryside around Mantova.

colle (as they were even called), whose greater breadth might correspond to the whole field's width, much wider than what thus far had appeared in the *alberata* of Tuscany, Umbria, and the Marche.

We should not, however, through a mistaken perspective, be induced to exaggerate the degree of development of these constituent elements of the modern *piantata* of the Po, or above all the territorial extension of the landscape over which this form extended in the sixteenth century. Observations of travelers like Montaigne or Alberti referred, essentially, to the landscape visible from the main roads of communication, and the main side roads. And it was chiefly and sometimes exclusively along these—as is specifically attested by writers on agricultural matters—that one found the plantations of trees. Gallo, among others, for instance, tells us for the region around Brescia of the "great quantity of mulberry trees planted along the sides of the roads so as not to invade the fields"; and the topographic maps of the time confirm that, in the first decades of the sixteenth century, trees were frequently limited almost entirely to the margins of fields along main and minor roads. Toward the end of this same century, a map of the countryside around Mantova—from which we have reproduced a segment in Plate 32—shows us that the arrangement of fields had already made notable progress, but that the plantations of trees were still long from being complete, away from the external perimeters of farms. Even close to the city, in fact, large areas were still uncultivated, or for one reason or another not systematized. And through the whole eighteenth century, the landscape of the *piantata* of the Po valley was interrupted for long tracts between one city and another by great extents of heath, marshland, and waste, not to speak of woods and fens.

· 41 ·

The Agricultural "Bel Paesaggio" of the Italian Renaissance

In the second half of the fifteenth, and in the first decades of the sixteenth century, the evolution we have traced in the preceding chapters, which would result in the Po valley moving to the forefront of agricultural progress in Italy, was not yet fully perceptible in its effects. The "gardens of Italy" remained for observers of this period, first of all the privileged lands of the South, Tuscany, and certain hilly zones near Venice, with the subtle embroidery of their trees and shrubs. The low and gray lands of the Po valley, where now stretched misty extents of irrigated meadows, did not yet always attract attention to the technical excellence of their new agricultural system. The new taste for the "bel paesaggio" of the Renaissance did not exclude, though—through the genius of Leonardo, who transcended the vision of the Florentine and Tuscan schools—a sense of the regularity of fields, canals, and rows of trees in the Po valley. In the fascination for human and secular accomplishments that moved the men of that age to impose new creative forms on the agricultural landscape, even picturesque landscape began with him, and later with the Venetian and Emilian schools, to reconquer the full autonomy of its value. "He is not universal," Leonardo judged severely in his polemic with Botticelli and the Tuscans, who affected scorn for the autonomous value of picturesque landscapes: "he is not universal who does not love equally all things contained in the picture; like the one who does not like landscapes." And a genial landscape study, like the one we reproduced in Plate 30, reveals within a distant but precise network extended over the plain, an ideal landscape that is not only imaginary, and almost romantic at one level, but also concrete in the elaboration of fields that are regularly arranged and organized for cultivation.

One can observe that, in the particular sensuality of his ideal of picturesque landscape, Leonardo was first, and almost isolated, until the Venetian and Emilian schools, and if anything he was the precursor of innovations of the still more distant future. For the time being, a taste for a "bel paesaggio" prevailed that was more ostentatious, more embroidered and even decorated, of which Benozzo Gozzoli is the most typical example. In reality, the two currents of taste in landscape developed in a parallel manner, and not without sometimes mingling their waters, through the Renaissance and thereafter. They reflect, in the elaboration of pictorial landscape as in the historically concrete one, different techniques, practices, and social and cultural realities, which we have already begun to perceive in the confrontation between the constituent elements in the agricultural landscape of Tuscany and that of the Po valley. One can thus perceive how truly the sensibility of Leonardo seems often close to the practices and writers on agriculture who were the most active artificers of the new landscape in the Po valley. And it was Gallo, for example, who speaks to us of "landholdings beautiful to look at; convenient to cultivate" and, with his style that resembles Leondardo's, of fields "well squared one after another, not longer than forty *carezzi*, or less than thirty, or twenty-five, with ditches all around, and planted on all sides with trees." "Straighten roads, square fields, dig out ditches, haul up embankments, level meadows, make bridges, banks, canals, and sluiceways for irrigation" were the works that, for Gallo, made the plain of Brescia "all a beautiful garden." These were works, certainly, not unknown even to the practitioners of Tuscan agriculture. But what a difference of tone there is when a Tuscan, Alamanni, tells us about his conception of techniques and about the agricultural "bel paesaggio"! This was no longer that geometric "square fields"; no longer that violence done to hostile nature to tear her treasures from her, but instead a support of her forms:

> e 'n quella parte, ove natura inchina
> drizzar il passo, perché l'arte umana
> altro non è da dir ch'un dolce sprone,
> un corragger soave, un pio sostegno
> un esperto imitar, comporre accorto,
> un sollecito atar con studio e n'gegno
> la cagion natural, l'effetto e l'opra:
> e chi vuol contro andar del tutto a loro
> schernito dal vicin s'affanna indarno.

> (and in the direction, that nature inclines,
> set your step; because human art
> is no more than a gentle spur,
> a soft correction, a pious support,
> an expert imitation, careful composition,
> a solicitous aid from study and from art
> to natural reason, effect, and outcome:
> and whoever seeks, by themselves, to go
> contrary to this, labors in vain.)

And for the time being, to be sure, the more facile Tuscan style reigned in the pictorial landscape, as it did in the agricultural one, prevailing, and delivering up its splendid fruits. But it was the other, the more difficult and labored, landscape of Leonardo, that already showed the ways of the future.

. 42 .

The "Bel Paesaggio" in Tuscany

How much diversity there was in the cultural currents and tastes that inspired elaboration of the Renaissance "bel paesaggio" will not escape the reader who compares Renaissance landscapes with those (even the most elaborate ones) of the classical age. This genial care, this free invention of the most minute details, and most intimate weaving of the agricultural "bel paesaggio," could not have been the work of servile labor. It presupposed a development of socially productive forces already much superior to those of classical antiquity, but it revealed also, and above all, new relationships among men, a creative contribution foreign to the slave owner or his chained-up slaves, but possible for any tenant farmer, any peasant, or any sharecropper.

In the detail of the *Journey of the Magi* by Benozzo Gozzoli, which we reproduce in Plate 33, this individual contribution to the picture of the Tuscan agricultural "bel paesaggio" of the Renaissance appears in even greater relief precisely because of its seeming casualness. For organic insertion into such an opulent and overflowing composition, every single element of the landscape must have been elaborated by the peasant and the architect, by the woodsman and the gardener, following a sure individual taste that was already educated by the spontaneous congruency of different contrasting initiatives. This spontaneity lends its fascination not only to the fabulous composition of Benozzo, but also to the agricultural landscape of the Tuscan Renaissance itself, and it has marked, and often even today marks spontaneously by itself, all the forms of life of this region, having left an enduring spark in the taste of the population. But there was also a historical limit in this exquisite spontaneity of the Tuscan Renaissance, which in the opulence of Benozzo's composition tends otherwise to dissolve into purely decorative motifs.

Thus it might be that in composing the agricultural "bel paesaggio," as in the pictorial one, the steep slopes inclining upward in the upper-right-hand side of the illustration, needed to be freed from woods that had covered and protected them from the erosion of water. It might even be said, all aesthetic consideration apart, that some peasant had not found a farm in the plain, between the forest and the castle gardens, to employ his labors and satisfy his hunger. Thus he cleared those steep slopes and reduced them to tillage; and now his furrows plowed straight

Plate 33. The "bel paesaggio" of the Tuscan Renaissance in a detail from the *Journey of the Magi* by Benozzo Gozzoli.

down the hill are preparing a rapid deterioration of this beautiful landscape, which might prove to be fatal to him.

Of the beautiful, intricate, and opulent complexity of the Tuscan landscape, meanwhile, this work of art seems already to provide a culminating expression. We are at the limit, aesthetically and historically, of individual spontaneity (however educated and artful it might be) in the elaboration of the landscape. Beyond this limit, of now intricate complexity, individual spontaneity would no longer suffice to control and provide form: so that the forms themselves, which it had happily elaborated, would deteriorate and disaggregate; and its precocious revolutionary effectiveness would decline.

. 43 .

The "Bel Paesaggio" of the Veneto

There is no need to search in the *Adoration of the Magi* by Domenico Veneziano, which is reproduced in Plate 34, for the density of decorative and ornamental motifs that we saw in Gozzoli's *Magi*, and that reflects (exalting them to the limit of art) the more minute elaboration of the Tuscan Renaissance agricultural landscape. In the painting by Veneziano, even though the artist was not without Tuscan contacts and influence, what dominates is a tone of lordly sobriety that is fully related to the landscape's Veronese surroundings. There is nothing fabulous in this landscape or in its pictorial expression, which presents itself as a sedate agricultural and pastoral panorama of the Venetian mainland, in the early Renaissance.

In this panorama, which strikes us for its modernity, we can easily recognize elements that can still be found today in the agricultural landscape of the hills around Verona. The evolution of agricultural relationships was in many aspects different here from what was characteristic of Tuscany, and particularly of the Florentine contado. Acquisition of land by new urban landowning groups was less frequent here, but it often involved larger individual estates; individual initiative in the transformation of the agricultural landscape was less widespread and lively, but also less fragmented and disordered. As in the Po valley, also in the valleys and hills of the Venetian mainland these transformations were accomplished, not by owners of one or a few farms as often happened in the Florentine contado, but through the initiative of great proprietors, who were sometimes bourgeois and sometimes old feudal nobles.

The results, for the agricultural landscape, were less minutely elaborated holdings, to be sure, than those which appear in the Tuscan embroidery of farms, gardens, and villas; and this produced a less intense and less advanced development than what could be found around Florence. But, in contrast, the agricultural system and landscape had more coherent elements of organization. The order of the vineyards, rationally arranged in rows that followed the contours of the hill slopes, appear not only in Veneziano's painting, but—according to what is revealed by agronomic sources—also in the historical reality of this region and age. And, already before the fifteenth century, the wines of the valleys around Verona reacquired a fame that they had not entirely lost even in the darkest centuries of the Middle Ages. Between the hills and the

Plate 34. The "bel paesaggio" of the Venetian Renaissance in the *Adoration of the Magi* by Domenico Veneziano.

valleys as well, the old balance between pasturage and cultivation was not broken: and the herds of sheep, which are seen to graze in Veneziano's painting not far from the fields and vineyards, solidly supported by this balance, permitted an important production and flourishing industry of wool.

. 44 .

The "Bel Paessagio" of the Italian-Style Villa

Wherever, with his agricultural activity—as we have seen happen with the ancient Roman villa—man begins to imprint more consciously elaborated forms on the agricultural landscape, the way is opened to evaluating these forms not only technically and economically, but also aesthetically. After the shapeless pastoral and agricultural landscapes of the early Middle Ages and the feudal era, there is no doubt, in the communal period and the Renaissance, that the taste for a "bel paesaggio" was reborn and reaffirmed in close relationship with the historical and also technical, economic, social, and cultural progress that restored to men working together the ability to introduce new and precise forms into the landscape, which were organically suited to the new level of development reached by socially productive forces in agriculture. We have already revealed, indeed, how—once the limits imposed by slavery and feudalism on the old productive relationships were overcome—this elaboration and taste for a beautiful landscape touched a much wider social stratum than had ever been the case even at the culminating point of Roman civilization. But in a society divided into antagonistic classes, which was the situation in which its "democracy" developed, this division by itself tended, so to speak, to definitively reserve even the privilege of beauty to the dominant classes. The Tuscan peasant or artisan of the Renaissance could thus put into the care of this row of vines or that avenue of cypresses, into the refinement of this gate or that balustrade, all the taste and capacity for artistic creation that had developed within him. But the "bel paesaggio" that he attentively contributed to and elaborated remained essentially a beautiful landscape *for the others*, for the privileged classes, for *their* repose, *their* festivities, *their* loves; a landscape from which he was excluded, and that he could admire, only *outside* of the villa gate.

The Italian villa of the Renaissance was first of all, although with this privileged exclusiveness of the dominant classes, the same agricultural "bel paesaggio" whose elaboration we have followed from the time of the communes. From the most banal rustic villas up to splendid dwellings, such as the Medici villa of Cafaggiolo, which is shown to the reader in an anonymous painting in Plate 35, throughout the first half of the fifteenth century, the perfection of forms in the landscape was still visibly tied to needs born from the evolution of new techniques and new

Plate 35. The Medici villa of Cafaggiolo in an anonymous decorative panel in the Palazzo Riccardi in Florence.

agricultural relationships. Thus one sees in the illustration the important part still assumed by woods, farms, and rustic works of construction in the landscape of the villa of Cafaggiolo, in comparison with the relatively modest importance of the seigneurial villa and its gardens. One sees, in the cultivated fields, that same arrangement in *porche* that we have already noted as characteristic of the agricultural landscape of Tuscany, and that seems to imprint its pattern of symmetrical simplicity on all the forms of the Medici villa. But already, at Cafaggiolo, the seigneurial villa repeats, as if symbolically, the forms of an ancient feudal castle; and the ground is already laid for that evolution that would make the new princes of Tuscany come out of a family of Florentine bankers.

In the second half of the fifteenth century, and then throughout the sixteenth century, the "bel paesaggio" of the Renaissance villa expanded from Tuscany throughout Italy. It assumed a quantitative and social importance, at this point, beyond an artistic one, closely connected with the economic and political evolution pushing the ancient communal republics to join together in larger principates, founded on the power of a new nobility of former merchants and bankers, of papal nephews, and of condottieri. In Tuscany itself, the greatest flourishing of Medici villas came during the reign of Ferdinando I, when, in the last decades of the sixteenth century, the decline of manufacturing had already spurred the flow of important sums of capital from urban investment to the countryside. In the territory of Venice, out of 1,411 villas of artistic significance now in existence, only 15 go back to the fourteenth century, 84 to the fifteenth century, but 257 to the sixteenth century. "The rich nobles and citizens of Venice," Girolamo Priuli wrote, "wanted to triumph by living and giving themselves over to pleasure and enjoyment in the green countryside, and they built many palaces and spent much money." In the Venetian dominions, the flourishing of seigneurial villas reached its culminating point in the two following centuries; but already in the sixteenth century in the Papal States, the enormous concentration of wealth in the hands of popes and cardinals, who generally came from the chief families of the new nobility, led to the affirmation of a new type of princely villa, of which the Villa d'Este at Tivoli is the most famous model, a type that would be copied throughout Italy at the end of the sixteenth and in the seventeenth century. In the chief villas of this period arose the grandiose constructions and the rigid but ornate symmetries of the "Italian garden." From its agricultural and utilitarian origins, this "bel paesaggio" preserved only the neatly squared plots and flower beds, the regular rows of trees, and the sloping terraces and gardens. The ostentation and size of the constructions built by these new feudal lords, who "built many palaces and spent much money," disturbed those who had in mind a balance in the villa between utility and pleasure, which the first Tuscan and Venetian humanists, beginning with Leon Battista Alberti, had recommended and practiced. "I don't want," Tansillo warned in his poem *Il Podere*, directing his advice to the maggiordomo of the Marchese del Vasto,

> Io non vo', che le ville sien palazzi,
> che ingombrin molto; e chi vi vien, che veda
> terren, dove men s'ari, che si spazzi.
>
> (I don't want villas to be palaces
> of a cumbersome type; or to have visitors see
> land that is less plowed than swept.)

And he echoed, in these verses, Horace's distress about the *regiae moles* of Roman villas, which threatened not to leave enough land free for the plow.

. 45 .

An Agricultural Panorama of the Renaissance: Pastoral Landscapes

Between the fifteenth and the sixteenth century, pastoral systems and landscapes were elaborated and crystallized in various parts of Italy, whose types continued to dominate the peninsula and the Islands almost up to the present, even though they underwent important shifts in geographic extent between the sixteenth century and the nineteenth.

In the Po valley, as we have seen, the progressive expansion of irrigated meadows and artificial pastures generally permitted an increasing recourse to the permanent or semipermanent keeping of cattle, and brought about deep changes in both the pastoral system and landscape, and the agricultural system and landscape of this zone. Thus, Tognazzo, the consul of Cipada in Folengo's *Maccaronea* can pride himself not only, as we have already seen, as a master of the art "of raising embankments and surrounding his fields with ditches," but also as

> doctor et a stalla grassum portare ledamum,
> omnibus in seris plenas scio mungere vaccas,
> inque rotam tondam teneros calcare casettos.
> Non plasentinis malghesibus atque vacaris
> invideo lactis frescum menare botirum.
>
> (a doctor in carrying rich dung from the cow shed,
> I know how to milk the full cows every evening,
> and to pack soft cheeses into the round wheel.
> I don't envy the cowherds and dairymen of Piacenza
> in churning butter from fresh milk.)

Tognazzo was thus a doctor in the art of extracting rich dung from the stalls, a master of milking and of cheese making, with no need to envy the *vaccari* and *malghesi*—dairymen and cowherds—of Piacenza, who were already famous, like those of Parma and Lodi, for the large size of their exquisite cheeses. The techniques of breeding here had thus begun to integrate themselves organically with those of agriculture, in support of which they provided a rational completion to the fertility cycle of the soil by providing dung, and to the productive cycle through a profitable transformation of products of the fields into milk and meat.

An image of this new environment and pastoral landscape of the Po

Plate 36. The integration of the Renaissance pastoral and agricultural landscapes in the *Virgin and Child* by Giovanni Bellini.

valley is provided in Plate 36, where we have reproduced the *Virgin and Child* by Giovanni Bellini, now preserved in the National Gallery in London. In an area enclosed by a stockade, some cows and two goats are sleeping in the open, or grazing on the grass. The well—provided with a regular device for drawing up water—and the closeness to the town, besides the fence around this open space, indicate clearly that we are in a dairy farm, with a permanent or semipermanent stock of cows, which was organically connected to agriculture.

As the verses cited from Folegno remind us, though, in this revolution of agricultural systems and of breeding—which, integrated one with the other, clearly had the form of modern agriculture—the pastoral landscape did not loose its high altitude on the *malghe* and the *bergamine* (places for the summer high pasturage of livestock) of alpine and subalpine regions, which now also assumed more organized and regular forms. Thus, much more than in the communal period, the pastoral economy and life on the mountains became more closely linked with that of the plain.

At the other extreme of the peninsula, in the South and the Islands, the age of the Renaissance marked the progress of a quite different evolution of techniques, economy, and pastoral landscape. The technical know-how of the time did not permit the cultivation of artificial pastures to develop in these environmental and climatic conditions, and feudal customs also obstructed this. In contrast to the development of cattle breeding on stable farms, which grew in importance in the North, the age of the Renaissance marked the beginning of a clear prevalence of sheep breeding in the South, based on transhumance. This system integrated mountain pastures, in the summer, with what was provided in the winter months by an agricultural system of temporary clearings, which now more than ever expanded and consolidated itself from the latifundia of Sicily and the Corsi of Calabria, to the Tavoliere of Puglia and the Agro of Rome.

These were the years, in fact, when this landscape of transhumant pasturage assumed its decisive form, through a more precise organization of the "Dogana delle pecore" for the Agro (in the year 1402) and of the "Dogana di Puglia" for the Tavoliere (in 1443). But if one wants to find realistic iconographic documentation for this landscape, it must be sought in a later period—or even in the persistence of the configuration we reproduced in Plate 18—rather than in the art of the Renaissance, whose cultural interests were oriented toward less coarse subjects that were less distant from the sphere of influence of urban life.

Along with the improved agropastoral landscape of the North, what continued to provide the art of the Renaissance with its preferred subjects continued to be the pastoral landscape of Tuscany, Umbria, and the Marche, which developed, in agricultural reality as in pictorial development, in ways that were already initiated and laid out in the communal period. In the Tuscan and Umbrian schools, more than is true of those of Emilia and the Veneto, the conception of the pastoral environment was now transformed into an idyll, and sometimes in a way that seems lacking in sincerity. But here again, there is a splendid iconographic documentation that offers us not a mannered but a realistic and specific image of the pastoral landscape of Tuscany, Umbria, and the Marche. I like to recall, among the many, a lively miniature in the codex of Virgil in the Riccardiana Library in Florence, which shows a pastoral scene in a wood. In central Italy, as one recollects, the expansion of clearings and the progress of the system of fallow and of continuous cultivation developed at the expense of the system of temporary clearings. Here the more frequent enclosure of fields after the communal period continued to diminish progressively the area available for pasturage; while diffusion of artificial pastures, and the cultivation of forage crops in general, were far from reaching the extent realized in the Po valley. In these conditions, the resource of pasturage in the woods, and the use of branches for feeding livestock, assumed an increasing importance for smaller-scale breeding in central Italy. The reader can seek out a concrete image of precisely this pastoral landscape of woodlands in the miniature in the codex of Virgil, which the tyranny of space forbids us to reproduce here.

. 46 .

The Landscape of Clearings in Hills and Mountains

During the sixteenth century, the process of clearings in the hills, which we have already seen advance during the communal period, developed further, touching new regions farther from inhabited centers, and expanding more frequently into mountainous lands that had never before been broken by the plow. Conditions and agents of various kinds affected this extension of the hilly and mountainous agricultural landscape. In first place, naturally, was the geological relief of Italy, where hilly and mountainous lands occupy respectively 41 and 37 percent, and the two together 78 percent, of the arable and forested terrain. To find the necessary land to provide provisions for a growing population, it became almost inevitable, given the state of development that socially productive forces had reached in this period, to clear land in the hills and mountains. One should not forget, further, that great expanses of plain were still occupied by marshes and infested with malaria in the Renaissance; while in whole provinces—the Maremmas, the Roman and Pontine Agros, the Tavoliere, and so on—the tendency to develop large-scale breeding of sheep in the second half of the sixteenth century enlarged the territory used for pasturage, and this was cultivated only occasionally through a system of temporary clearings in a three-to-seven-year cycle.

These decades of the sixteenth century also, in many parts of Italy, saw a large flow of men and capital from the cities, where manufacturing and mercantile activities that had earlier flourished were in decline, toward the countryside. Thus, already at the end of the century, not only around the chief Tuscan towns, but around Genoa, for example, the population, earlier almost exclusively occupied in maritime activities, was now forced in increasing numbers to dedicate itself to agriculture and proceed with the deforestation, clearing, and systematization of the difficult mountainous terrain. Finally, in other cases, the extension of mountainous clearings in this period resulted, to a significant extent, from the division and sale of old communal domains, which had earlier been used by the population as commons for pasturage and wood gathering. These already had occasional clearings in so-called *prese* (preserves) but they were now more and more frequently divided up and taken over by private owners.

Plate 37. Clearings and plantations on the hillsides and mountains in a sixteenth-century panorama of Serravalle Veneta.

What is clear is that already at the end of the sixteenth century the agricultural landscape of the hills—and, above all, the greatest novelty, that of the mountains—had reached an extent and importance that it never had in previous periods, when mountainous regions remained much more exclusively zones of woods and pasturage. Plate 37, which reproduces a topographic view of the region of Serravalle Veneta at the end of the sixteenth century, shows how the agricultural landscape of the hills and mountains appeared around the ancient center of Marca Trevigiana. One can see, in the enclosed fields of the suburbs, the fairly well developed form of the *piantata* of the Veneto, with its rows of trees with vines and narrow elongated plots. But what is more interesting is the extent already reached in this period by clearings and plantations on the highest hills, and even on the mountains that appear in the background. Here, so far as one can tell, there seems to have been a system of open fields, with only a hint of permanent enclosures. This was clearly the more recent agricultural landscape, which was beginning nonetheless to elaborate and clarify its forms, and to differentiate itself from the natural landscape, not only through the signs of cultivation with furrows plowed straight downhill and sown fields without trees, but also through the traversal alignment of some plantations of trees.

. 47 .

The Deterioration of the Landscape of Hills and Mountains in the Renaissance Period

In the sixteenth century, the voices that warned of the danger of deterioration of the landscape of hillsides and mountains, which was already advanced in some zones of the peninsula, became more numerous and insistent. "The mountains," Leonardo da Vinci had warned, "are undone by rains and rivers"; and already in the age of the communes, Pietro de' Crescenzi had recommended that hilly terrain be worked across the slope, along with other measures for soil conservation, because (we quote the translation by Sansovino) otherwise the land "would be entirely carried away by the rain into the valleys, when it comes down in torrents from the slopes of the mountains."

We have seen, however, how—despite the warnings of de' Crescenzi—working the land *a rittochino* (with furrows straight down the slope) continued to prevail generally on mountainous terrains with sowable fields without trees, and it also continued in plantations of trees, so that even the elementary defense these might themselves provide against erosion caused by running water was to a large extent lost. This happened, for example, in the landscape of suburban plantations at Serravalle, shown in Plate 37, where it seems that the narrow and elongated fields of trees were, in fact, orientated down the line of greatest slope, since the working of the land customarily involved plowing with furrows that ran straight down the hill.

It is not always possible to clarify, by studying the structure of the plow and the techniques of plowing in that period, the reasons for the prevalence of plowing straight down the hill. But it is clear that from the communal period onward this system had contributed significantly to the deterioration of the landscape of hillsides in Italy. And its consequences, naturally, became quite disastrous when, with the spread of deforestation and clearings in the later communal period and Renaissance, cultivation moved still higher up the steeper hillsides and mountain slopes.

Thus, of its own accord, unplanned deforestation became, as is well known, a decisive agent in the deterioration of the mountain landscape, and a serious threat to land in the lowlands, which was also exposed to the consequences of hydraulic dislocation in river valleys that were now deprived of their mantle of vegetation. And deforestation developed during the late communal period and the Renaissance, in a disordered and lopsided manner, through individual initiative, which left untouched, depending on the varying level of profit deforestation could yield, large tracts of forest in some places, while elsewhere there remained only naked rock, denuded, not only of forest, but also of any trace of fertile soil.

In the mid-sixteenth century, Leandro Alberti, in his *Descrittione di tutta l'Italia*, shows that he was already fully aware of the seriousness of these problems, when he tells us how "with the great population increase in Italy, and insufficient low-lying and convenient lands to produce the needed foodstuffs, it has been necessary to till the high and uncultivated mountains." And while in the past from mountains covered with forests "clear water ran down amid woods and meadows in smaller amounts, and more gently," now instead "the rain . . . is not held back, and flows down precipitously, taking with it an excess of earth" from the mountains reduced to cultivation. "It empties into torrents, canals, and rivers," with an impetus and destructive capacity without precedence, "which did not happen in olden times." It was not, otherwise, by chance that from the first half of the sixteenth century, in the first verses of his *Coltivazione*, Alamanni felt the need to warn farmers against the danger of

> l'erboso ruscello, il picciol rio,
> il pietroso torrente, il fiume altero
>
> (the grassy stream, the little brook,
> the rocky torrent, the proud river)

that now more and more often

> dispregiando ogni legge, ardito cerca
> di tor dal corso suo l'antico freno
>
> (disdaining any law, boldly seeks
> to tear the old restraint from its course)

and, bursting its banks, "depreda i campi" (floods the fields).

We will have occasion to return in the course of our study to the hydraulic dislocation that the unplanned deforestation and clearings of this period worsened, even in the valleys. But until then, in Plate 38, a painting of Vasari that portrays the assault on Gavinana in 1530 by the militia of Orange and Maramaldo provides a concrete image of forms that the deterioration of the mountainous landscape assumed in the Appennines above Pistoia in the sixteenth century. The painting shows the clear connection between the most conspicuous facts of surface erosion and the distribution of clearings in the mountains. Here, as elsewhere in the Ap-

Plate 38. The deforestation of the hillside and mountainous landscape in *Gravinana Attacked by the Militia of Orange* by Vasari.

pennine region, the first phase of rapid deterioration of the mountainous landscape seems to have been already seriously advanced in the Renaissance. This would be followed by other still more serious results at the end of the eighteenth and in the nineteenth century, and in one way or another, right up to the present, it denuded many Italian slopes of their clothing of forest and reduced them to naked skeletons of stone or clay.

. 48 .

*Systematization in the Hills and Mountains
during the Italian Renaissance*

In the age of the Renaissance, an agricultural regime founded on the private ownership of land, revenue, and individual profit was more and more broadly affirmed in Italy, although to an extent and with forms that were quite different from one part of the peninsula to another. But with the incipient deterioration of the hillside and mountainous landscape, and with the results this had for the underlying plains, the historical contrast between these developments and the need to defend and conserve arable land came to assume, in its turn, a more and more clear and threatening significance.

Against the forces leading to the deterioration of the hillside and mountainous landscape—which were those, as we have seen, of unplanned and anarchic individual initiatives—the first opposition came from these forces themselves (and it could not have been otherwise in this type of regime): individual agents operating at the level of single farms and individual agricultural enterprises. Thus the age of the Renaissance was both an age of incipient deterioration of the hillside and mountainous landscape, and also an age when new arrangements were elaborated and spread that had hardly been imagined in these zones during the communal period.

One can thus refer to our Plates 16, 19, 21, 22, 23, 27, 29, 33, 37, 38, all of which show hillside or mountainous landscapes, and compare them with Plate 39, which we are about to explain. It will not escape the reader—in comparison with the shapeless *Garden of Olives* in Plate 16— what progress had been made by the time of the carefully tended trees in the *Garden of Olives* by Duccio, of the aligned rows of vines in the landscape of Lorenzetti, and the fields *a pìgola* of Gentile da Fabriano. This was an elaboration of forms in the hillside landscape that are also confirmed by literary and archival sources, and that were not embellishments of pictorial landscapes, but were rather reflected in them, with a certain time lag, as real processes developing in the agricultural actuality of the time.

Still, in all these iconographic documents, as in the majority of those from periods from before the fifteenth century, the elaboration of the agricultural landscape did not generally involve more than a few basic elements of the systematization of hills for the purpose, as we have

Plate 39. Arrangements in plantations on the hills in *Prayer in the Garden of Olives* by Barna Senese.

already indicated, "of assuring conservation of the arable land, a more balanced use of water, and more effective tillage through arrangement of fields to be roughly horizontal and of convenient size."

In the landscapes of hillside agriculture that we have examined up to this point, by contrast—and even in the most elaborate ones—while the cultivation of trees and shrubs became gradually disposed in a less disordered way along the hilly slopes, and the plantations themselves (on "banks," as in the *Marina* of Lorenzetti) provided a primitive protection of the arable soil against water runoff, there was lacking, generally, any hint of systematization of the arable land into roughly horizontal fields of convenient size, which was more easily obtainable in arrangements in the plain. Still more, even where, in accordance with the pictorial style that spread in Italy even before Giotto, mountains and hills are portrayed with natural terraces occurring along the slopes, the trees and plants generally seem to have been distributed not so much, or not only, on the plots that were roughly horizontal on parapets of slopes, but frequently instead in an entirely casual manner, or even on the steep slopes separating one terrace from another.

The painting by Barna Senese, from the first half of the fourteenth century, reproduced as Plate 39, which presents once again *Prayer in the Garden of Olives*, indicates a new step in the elaboration of the pictorial landscape of hillsides. And a reference to archival and literary documents confirms that this novelty in the pictorial landscape reflects, with a certain time lag, a true innovation in the real agricultural landscape that began to be elaborated in the early Renaissance. It is certain that this painting is one of the most ancient iconographic documents breaking with the earlier pictorial tradition and the very tradition of the masters of the school, such as Simone Martini and Duccio di Buoninsegna, where Barna Senese presents us with a hillside landscape in which the regular terraces are no longer a mere painterly gesture, but instead have become "fields tending to be horizontal and of convenient size" where trees were now regularly planted and cultivation was distributed. And it was not by chance, to be sure, that a document of this type should appear, for the first time, precisely in an environment like that of Siena in Tuscany, where, as we shall see, systematization of the hills had already gained a particular significance, and would gain still more during the Renaissance.

. 49 .

*The Origins of the Contemporary Landscape:
Systematization in Irregular Banks* (a Ciglioni) *on
Hillsides in the Age of the Renaissance*

As in the painting by Barna Senese, reproduced in Plate 39, the painting of the *Incidents in the Life of Pope San Silvestro* attributed to Battista da Vicenza, reproduced in Plate 41, is among the first in the early fifteenth century to document the more definite form that systematization on hillsides and mountains began to assume in the Renaissance.

At first sight, to be sure, it might seem that the hillside landscape portrayed in the *Life of San Silvestro* simply repeats the traditional manner of portraying mountains and hills with an arrangement of terraces sloping down from the summit. It would not be difficult, in fact, to cite at least several dozen Byzantine paintings, and others of Italian schools of the eleventh to fifteenth century, where the manner of representation is, in general, identical to that in the work attributed to Battista da Vicenza. But, as in the painting by Barna Senese, the novelty rests in the fact that in the painting by Battista da Vicenza the terraces (or, more exactly, *ciglioni* or banks) descending from the summit no longer seem to be merely a pictorial device, but have become instead largely horizontal fields of convenient size where trees are regularly planted; they are no longer casually located (as appears elsewhere) even on the steep slopes separating one terrace from another.

"The slopes of these small mountains," Boccaccio wrote portraying, in the sixth day of his *Decameron*, the six little mountains that surrounded his beautiful *Valle delle donne*, "the slopes of these small mountains descend down toward the plain, as in theaters we see the rows of seats arranged in order from top to bottom.... And these slopes, on the side where the sun reached them, were all filled with vines, olives, almonds, cherries, figs, and many other kinds of fruit trees."

The novelty was not only, as one sees, in the pictorial landscape of Battista da Vicenza, but also in the literary one of Boccaccio. And the fact that this was truly a novelty is confirmed by the admiring stupor of the three youths in the *Decameron* who, at the sight of this elaborate hillside landscape of the *Valle delle donne*, "praised it as one of the most lovely things in the world."

Some of the oldest commentators on the *Decameron* thought they recognized Fiesole in the hillside landscape of the *Valle delle donne*, and it is not therefore improbable that Boccaccio referred, in the description of the downsloping banks, or terraces, planted with fruit trees, to a still unusual example of what he might have seen elsewhere in Tuscany. But it is worth the trouble to indicate that Boccaccio makes use of the word *piagge* (slopes) to designate the arrangement of the hill, which in the Tuscan usage of this period meant simply a natural declivity gently sloping down to the plain. As in the *Valle delle donne* of Boccaccio, and the mountains of Battista da Vicenza, it is probable that in the agricultural landscape of the early Renaissance the novelty of arrangements in *ciglioni* or *terrazze* (banks or terraces) was not chiefly so much in the amount of work involved, with massive movements of earth, to break the line of the more rapid descent of the hills, as it was in a more frequent and intelligent use of the level plots or more gentle slopes, which the relief of hillsides often spontaneously offered to farmers for planting trees and plants. Some accommodation in the usual working of the land, a cautious choice in the placement of trees or a consolidation of earth or grassy sods on the downward side of slopes, would have been enough in such cases to begin the systematization in *ciglioni* (banks) that has remained up to the present. And the picture by Barna in Plate 39, which characterized the region of Siena, is enough to reveal its early appearance.

We know that in the more elaborated arrangement of banks—though a more significant shifting of soil, if there was need—a slope was rearranged into terraces that were roughly horizontal, and more or less wide, and whose sides were secured with sods, or perhaps simply by packing down the soil along the embankment. As we shall see at a later stage of our research, such embankments became very widespread, a hillside arrangement that imprinted one of the most characteristic forms on the Italian agricultural landscape. The broader diffusion occurred in a later period, but already in the Renaissance, embankment left its typical mark, on hillside landscapes around Lucca for instance. "One cannot cease to praise," Montaigne wrote in his *Journal de voyage* of 1580–81, which we have already cited, "the beauty and utility of their way of cultivating the mountains up to the summit, making these like a staircase, with great terraces that girdle the mountains all around, and reinforcing the edges with stones or other coverings when the earth is not sufficiently firm of itself. The hollows of the terraces ... are planted with grain; and the outsides toward the valley—that is, their circumference all around—is covered with vines."

. 50 .

The Origins of the Contemporary Landscape: Systematization in the Mountains through Lunettes and Grading

Systematization in banks, which already had begun to impose their characteristic form in certain zones of the hilly landscape of Italy in the Renaissance, is classified by modern agronomists as an intensive system, because of the large amounts of work involved: work that was devoted to massive movements of earth, however, rather than to elaborate regulation of water systems. From this arises the criticism directed against this type of arrangement when there was no kind of regulation of subterranean water, and the regulation of surface water was also deficient. But the diffusion of this system represented still, in this period, a great advance in agronomy in comparison with the previous lack of any permanent systematization in the hills. Where there was sufficient summer rainfall to permit the formation and consolidation of a grassy covering on the restraining embankment, in fact, such banks could last for sixty to eighty years and more.

The most adapted environment for systematization with banks was that of the typical drumlin-like hill made up of silicic or tuffaceous elements, and particularly yellow sands and pilocene tuffs. But beginning in the communal period, and more in the age of the Renaissance, with the expansion of clearings and plantations in high anticline hills and mountains—of solid rock—there was need for other types of permanent extensive arrangements, more adapted to a less amenable geological and climatic environment, where it would have been impossible to create embankments. Among these arrangements, the most obvious, for plantations of trees in rocky or pebbly soil on slopes of more than 100 percent, was with "lunettes" (*lunette*). Around each individual tree an attempt was made to hold back a little soil and water with a rudimentary shelter or semicircular retaining wall, made of thicket or rocks. From the communal period this elementary arrangement was practiced for plantations of olives particularly, in Sicily, along the Almalfi coast, in Liguria, and so on. Agronomists of the sixteenth and seventeenth century speak of it—for instance, Alamanni, when he advises that the farmer

Plate 40. Plantations on gradings in a detail from the *Penetent San Girolamo* by the pseudo Pier Francesco Fiorntino.

Plate 41. Painterly style and graded arrangements (*ciglioni*) in the *Incidents in the Life of Pope San Silvestro* by Battista da Vicenza.

> . . . circonde in giro
> a guisa di castel di sterpi e sassi
> l'arbor che sovr'un colle o 'n piaggia assiede.
>
> (. . . girdle around
> as with a castle of branches and stones
> the tree that sits on a hill or slope.)

Where the relief of the mountain or terrain permitted, lunettes, rather than around a single tree, could be around two or three plants, whose roots could thus avoid the risk of being denuded of the little earth that covered them by the runoff of water. One passed from this, by almost invisible gradations, to "graded" arrangements (*a gradoni*) using shelters of irregular form that were dug into the less rocky spots on slopes where the depth of earth might permit tillage. According to the nature of the soil, these irregular gradations could be supported downhill by their own grassy banks, shelters, or dry walls made of rocks cleared from the site.

Plate 40, which reproduces a detail of the *Penetent San Girolamo* by the pseudo Pier Francesco Fiorentino, shows us how an elemental graded arrangement (which was frequently spoken of by agronomes of the sixteenth and seventeenth centuries) presented itself in a pictorial representation toward the middle of the fifteenth century. The irregular level "plots" (*ripiani*) have plantations of trees and bushes already more regularly distributed within the plots themselves, rather than casually located here and there on the steep slopes. It is probable, otherwise, that such graded arrangements, rather than true and proper terracing, was meant by *fascia, lenza,* or other similar terms that made their appearance in documents of Liguria, Tuscany, Amalfi, Calabria, and Sicily in the eleventh and twelfth centuries.

With supporting walls, however—more frequently than with the grassy banks that appear in the painting by the pseudo Pier Francesco—gradings appeared on the highest hills and mountains of the peninsula in the age of the Renaissance; and they became, up to our own time, among the most characteristic elements of the mountainous agricultural landscape of the central and southern Appennines.

. 51 .

Systematization in the Hills in Terraces, and the "Works of Construction" of the Renaissance Period

Among techniques for intensive land utilization in the hills, which still today characterize the Italian landscape, and the system of banks, terracing also began to spread during the Renaissance. As with banks, this was a way of systematizing a terrain divided into plots. Terraces and banks, in fact, replaced the gradually descending slopes that presented themselves in nature with a succession of descending level plots. While in the arrangement of banks, however, the support for the level plots was still entrusted to packed earth or to grassy sods on their outer surface; in the arrangement of terraces, the plots were supported (as was also often the case with grading) by dry walls built with rocks retrieved on the spot when the land was cleared. But whereas, on steep slopes, the walls supporting gradings had to be adapted to accidental features of the terrain, so that the gradings emerged as irregular plots of short extent often interrupted by the natural rapid descent of the mountain, terraces instead—on less steep terrains, with slopes less than 50 degrees—descended regularly down to the valleys, in a succession of level plots that stood out on the relief of the hills as a series of concentric rings.

If embankments, as we have seen, were the characteristic intensive system for drumlin-like hills, terracing found its most adapted environment on anticline hills whose subsoil was more solid rock. Terracing began to be widely practiced on such sites during the Renaissance, as is indicated by the chief writers on agricultural matters of the period—among others, by Gallo and Africo Clemente—who speak of arrangements with *argini* or *banche* (walls or embankments) as an already current practice.

Where there was need of a more elaborate arrangement for highly valued crops, such as citrus trees, the supporting walls of the terraces were also constructed with materials brought in from elsewhere (bricks, poles, and so on), and the system of terraces developed into true and proper "constructions" (*costruzioni*, as is precisely what they were called). In the age of the Renaissance works of construction became a constituent and characteristic element in the Italian agricultural landscape in zones like the Riviera of Almalfi and the Riviera of Salò, where they were designed for the cultivation of oranges and lemons. Terraced constructions also became an integral element in the landscape of Italian

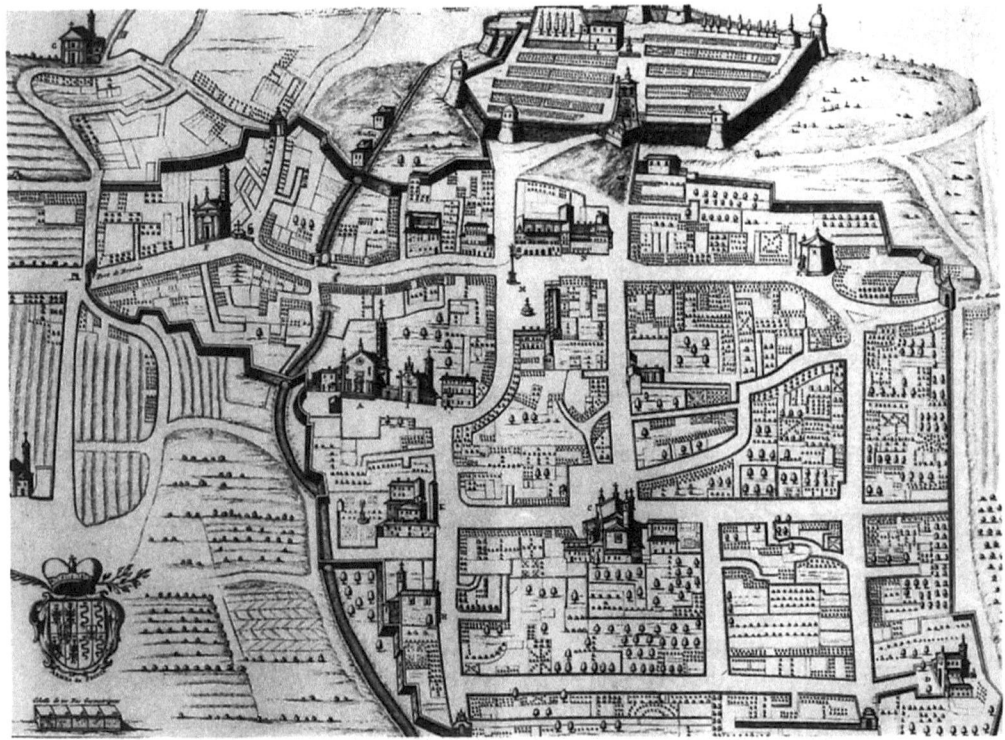

Plate 42. Works of construction in the form of terraces in a Renaissance topographic landscape of the Gonzaga villa at Castiglione dello Stiviere.

villas, as this developed and became elaborated (as we have already indicated in chapter 44) from the eleventh century onward.

Plate 42, which reproduces a handsome map of Castiglione delle Stiviere in the old Duchy of Mantova, shows us how works of terrace construction were inserted into the elaborate topographic landscape of a magnificent residence of the Gonzaga, which was destroyed in 1707. And in the following Plate 43 the reader can see the form taken by the retaining walls of terraces and works of construction in a suburban landscape.

. 52 .

*The Origins of the Contemporary Landscape:
Road Building, and the Systematization of
Hills Plowed "Crosswise"* (a Cavalcapoggio)
and "Roundabout" (a Girapoggio)

Beginning with the age of the communes, but even more in the age of the Renaissance with the extension of clearings and plantations, the development of road building in the hills assumed an increasing significance in the elaboration of new forms of the agricultural landscape. Of this development, and of the new function of roads in the hills, paintings of the Tuscan and Venetian schools provide us with a large iconographic documentation starting in the second half of the fifteenth century.

Look, for example, at Plate 43, where we have reproduced a detail from the *Martyrdom of San Giacomo* by Mantegna in the Church of the Eremitani in Padova. We have already directed the reader's attention to the interest of this painting for the purposes of our study because of its realistic representation of the detailed structure of the dry walls already widely used in this period to support the sides of banks and gradings in arrangements in the hills. But no less important is the iconographic documentation Mantegna's painting provides for the new function assumed by road building. It was precisely the access roads to towns that, snaking up along the hillsides, interrupted the continuity of the slope at different heights, and thus delimited large zones of the hillside. And it was precisely to discourage erosion of the soil down these roads that dry walls were built along their routes, with materials retrieved from clearing the land itself.

In contrast with what occurred in the arrangement of terraces and banks, however, this hillside was not divided into rectangular and roughly horizontal plots, disposed like steps. The form of the plots, like the routes of the access roads to towns, and the plantations of trees or single fields perched on slopes, was adapted instead to the lay of the terrain, without there being conspicuous shifting of soil, so that the plot usually had not only a slope down toward the valley, but also two lateral slopes going in different directions.

In the painting by Mantegna, where the landscape is inspired, as in other paintings of this master, by models provided to him by the countryside to the Euganean hills, this passive adaptation of the hillside relief

Plate 43. The road system and hillside arrangements in a detail from the *Martyrdom of San Giacomo* by Mantegna.

is marked also, and even more than by the course of the access roads or dry walls, by the rows of plantings, which are arranged across the crest of the slope. We find ourselves already confronted, in short, with the first elements of a systematization of hillsides that is designated in modern technical terminology by plowing crosswise (*a cavalcapoggio*), which involved certainly, in comparison with terracing or banks, much less labor; thus it is classified as an "extensive" hillside arrangement by modern technicians.

The control of surface water was particularly difficult in arrangements with crosswise plowing where there was no control of the subterranean water. But when, still without much shifting of the soil, the access routes to the fields—and the drainage ditches and their affluents—rather than proceeding crosswise to the slope followed the downward grade of the slope itself, the control of water was much improved. The most serious results of erosion of the arable soil could be avoided, thanks to the braking effect that circulation through twisting ditches exercises on rainwater in this arrangement, which is aptly called "roundabout" (*a girapoggio*).

The more intelligent "roundabout" arrangment—which can still today be usefully implemented, without great requirement of means, on more than two million hectares of hillside terrain in Italy—was not unknown to agronomists of the Renaissance, whom Alamanni advised with regard to drainage ditches in the hills:

> ch'ei non rovini in giù rapido, e dritto;
> ma traversando il dorso umile e piano
> con soave dolcezza in basso scenda.

> (that they do no damage below with rapid, direct descent;
> but cross over the back [of the slope] humbly and slowly
> with peaceful sweetness in their downward course.)

But until a later period, even where any systematization on hills was still often lacking more extensive and obvious arrangements like "crosswise" plowing generally continued to prevail. "Roundabout" plowing, terracing, and embankments spread only gradually and, in this period, were as yet characteristic of specific and quite restricted geographical zones.

· 53 ·

Plantations in the Hills in Central and Northern Italy, and the Landscape of Irregular Fields in the Late Renaissance

"Montepulciano"—Leandro Alberti wrote in his *Descrittione di tutta l'Italia* that we have already cited—"a noble castle, with plentiful population, situated on a pleasant hill, and producing all kinds of good fruits, and above all noble white and red wines that are renowned for their smoothness by neighboring peoples." The painting (by an unknown artist) preserved in the Palazzo Ricci in Montepulciano itself, and reproduced as Plate 44, provides a precise panorama of the agricultural landscape of this "pleasant hill" for a period not distant from the one referred to in Alberti's description. We are in one of those numerous centers, scattered through every part of Italy, from which, through the Renaissance and beyond, the fame of Italian agriculture and its most prized products radiated throughout Europe. Among these centers, to be sure, Montepulciano was one of the most renowned, and through the mouth of Francesco Redi, Baccus would later proclaim that "Montepulciano is the king of all wine." But even in less famous agricultural centers than this, foreign travelers and observers were astonished in these hillside landscapes of the late Italian Renaissance by the variety, density, and good order of the cultivation of trees and shrubs, which—an unusual spectacle in this age, even in the most flourishing zones of Mediterranean Europe, not to mention the cold north—often clothed the whole hillside relief with a delicate lacework.

The stasis, and then general economic decline of Italy beginning in the sixteenth century, did not seem at first to arrest, but rather promoted, a further diffusion of this type of agricultural landscape, of which the sixteenth-century panorama of Montepulciano provides such a characteristic image. The shrinking of markets, the crisis of artisan and industrial production, and the disasters of the great banking houses favored that "return to the land," which redirected important human energies and conspicuous investments of financial means to landholding. The urban bourgeoisie now retreated from the risks of manufacturing and foreign commerce and increased more than ever its acquisition of farms in the countryside. Around the larger and smaller centers, the zone planted with trees expanded, and dairy farms, farmhouses, and

Plate 44. The sixteenth-century landscape of irregular fields at Montepulciano in an anonymous painting in the Palazzo Ricci.

villas continued to multiply in the countryside. And with fixed investments in the purchase of land, buildings, and plantations of trees and shrubs, classes of proprietors who were becoming less active and more evidently parasitic sought to associate economic security with their own comfort and opportunity for diversion.

If we compare the sixteenth-century panorama of the "pleasant hill" of Montepulciano with those of Plates 22 and 27, it will not escape the reader that the elaboration of the landscape of irregular fields (*campi a pigola*), already advanced in the communal period and the early Renaissance, developed further in the late Renaissance, without substantially changing direction, and further extended its incidence through the complex of suburban hillsides. It was always individual (and thus casual) initiative, that joined with other not less casual individual initiatives, to delineate this irregular network of fields and gardens. The direct producers, the artificers of these irregular fields, were often peasants, liberated and made artful by the revolutionary impetus of the communal era. In this new period, while the political and cultural decadence of the age of the Counter-Reformation advanced, these continued to lavish on their fields and plantings, which were still a marvel of the world, their own treasury of labor, dexterity, and taste. But in the new historical crisis, the limits of agricultural progress founded on the casual spontaneity of individual initiatives became more and more clear. The extension of the landscape of plantings in the hills, in the form of irregular fields, as in the Mediterranean garden, was only quantitative, without any thought of substantive qualitative progress. But even from this quantitative and extensive point of view, further expansion of the specialized cultivation of trees and shrubs encountered more and more frequent technical and economic, as well as geological and climatic, obstacles. Difficulties of transport and the impoverishment of urban markets imposed well-defined limits to an extension of cultivation beyond the circuit of suburban hills. Not less important were the limits posed to an indefinite expansion of plantings—in this hilly landscape of specialized trees and bushes—by the deficient base of forage for breeding, and the consequent lack of dung to permit replenishment of the soil.

Thus in the late Renaissance, the hillside landscape of irregular fields (like that of the Mediterranean garden, which had continued to expand in the South, the Islands, and along the coasts) reached its greatest extent; and it still enchanted native and foreign observers, but it was no longer a leading feature of agricultural progress for Italy or Europe. It is enough, otherwise, to leaf through a work like Alberti's *Descrittione* to realize that, even at the moment of its culminating Renaissance expansion, the cultivated landscape of specialized trees and bushes, in the form of the Mediterranean garden or of irregular fields, remained restricted to suburban or coastal zones. It was thus still isolated amid vast extents of uncultivated land or open fields: a historical confirmation of the limits that unilateral agricultural progress revealed within itself, although it was founded on highly valued crops.

· 54 ·

The Mediterranean Landscape of Preserves, and the "Mediterranean Garden"

"Nardò in the Terra d'Otranto," Alberti wrote in his *Descrittione*, "has a beautiful, fine, and abundant territory, adorned with oranges, lemons, great groves of olives, and fine vineyards." Expressions of this kind recur in the work of the Bolognese geographer about all the inhabited centers (and they were numerous through the peninsula) surrounded by plantations of trees and bushes in the late Renaissance. A glance at Plate 45, where we have reproduced a plan of Nardò and its surrounding area in a period not much later than that of the description of Alberti, will help the reader to reduce visually to just proportions the value of such enthusiastic affirmations as Guicciardini's, who spoke of the Italy of his time as if it were all reduced to a garden.

In reality, as we indicated in the preceding chapter, at Nardò and almost everywhere in the late Renaissance, the belt of gardens and plantations of specialized trees and shrubs extended for only a narrow space around the towns. Beyond this lay the much greater domain of uncultivated land, pastures, and open fields: or of sowable fields with trees, the *piantata*, as was the case of the Campagnia of Naples, Tuscany, and the Po valley.

What we want to indicate, nonetheless, about the map of Nardò, is not so much the rapid passage from plantations to sown fields without trees, as soon as one left the city, but rather the form assumed by the landscape of plantations itself. Even in the South, to be sure, as in central and northern Italy, where the communal movement had been more intense, one can see the effects of individual initiative on the landscape, thanks to which the more well-to-do peasants or urban middle classes had made multiple purchases of land, clearings, and plantations in the suburbs. Even in the Terra d'Otranto, where universal feudal holdings still theoretically prevailed, perpetual servitude had nonetheless permitted this initiative to develop to some extent. And in some parts of our map one can see how the coincidence of individual plantations had produced characteristic forms of the Mediterranean garden in this landscape, with small irregular plots, thick plantations of trees and shrubs, and division walls. In other places, where the yoke of feudalism was less heavy than in Otranto, a more tenacious survival, or revival, of allodial property permitted such individual initiatives to develop more widely,

Plate 45. Preserves and the Mediterranean garden in a Renaissance map of Nardò.

and the landscape of the Mediterranean garden clearly predominated also in suburban and coastal territories. But still, almost everywhere in the South, and in other parts of the peninsula where feudal tenure succeeded in preserving a dominant place, the initiative for plantations was not only the individual one of peasants or members of the urban middle classes. There was also the more massive initiative of feudal and ecclesiastical lords, whose possessions almost always contained gardens and preserves (*starze*), closed and well-defended plantations destined for the pleasure of the lords and to increase their revenues.

These preserves of vines, olives, and citrus trees, which are so frequently mentioned in feudal inventories of the sixteenth and seventeenth centuries, echoed substantially, to be sure, the characteristic forms of the Mediterranean garden. But they echoed them, one might say, on a much larger scale that corresponded to the persistent economic and political dominance of the feudal and ecclesiastical lords themselves. It was no longer a matter, as in the older forms of the Mediterranean garden, of a tangle of little wooded plots divided by walls or hedges. In the preserves, and in gardens like that of the Augustinian fathers that we can see in the lower part of the plan of Nardò, or also of the bishop and the duke on the right-hand side of the plan, there were more extensive plantations that constituted an area apart, which remained as a persistent feudal or clerical domain and left a characteristic mark on the countryside. It was not now only on sown fields and pastures that feudal and ecclesiastical lords multiplied enclosures and preserves to the disadvantage of the local population. No less frequent are cases revealed in the records of the later Feudal Commission, where abusive enclosure of demesne land resulted in the formation of great plantations, or the usurpation of already flourishing plantations, whose products were awarded by the commission to the cultivators through their underlying right of *colonia perpetua* (perpetual settlement).

. 55 .

The Era of the Great Geographical Discoveries:
The Spread of Indian Corn, and the
Landscape of Agricultural Systems
with Continuous Rotation

From the end of the fifteenth century, the great geographical discoveries dislocated the traditional picture of the European economy and trade, and by different ways and means this produced indirectly deep changes in the economy and agricultural landscape of Italy. The fixed limits of this work prevent us from studying the complexity of these problems in depth. But we have already had occasion to show, for instance, the part played in the development of new forms of the agricultural landscape of Italy in this period by the return of significant human energies, and conspicuous financial means, from the city to the countryside. The return to the land was related, beyond the specific conditions and historical agents intrinsic to the society and economy of Italy, to the opening of the new grand routes of international trade, from which Italy was now excluded. Thus in pursuing our research, we will have to keep in mind certain undoubted effects of the "price revolution," which followed the influx of American gold into Europe, on the new expansion of pastoral landscapes in Italy, and on the revival of the agricultural system of temporary clearings.

Even in this more direct and immediate sense, however—and not only through the influence it had on general economic and social developments—the great geographical discoveries induced important changes in the economy and agricultural landscape of Italy. Already from their first voyages, Columbus and his companions, and then the other travelers and adventurers who discovered and conquered the New World, brought back and spread through Europe seeds of types of vegetables unknown in the old continent. Destined to a particular fortune were Indian corn, potatoes, tomatoes, tobacco, the common bean (the Old World had known only black-eyed beans of the genus *Dolichos*), not to speak of minor varieties, such as the large garden strawberry and others.

Among these products, the one that spread most rapidly in Italy, and produced the deepest changes in the agricultural system and landscape, as well as the diet of the population, was undoubtedly Indian corn, or *mais.* In the second half of the sixteenth century, and still more in the first half of the seventeenth, the cultivation of Indian corn was practiced widely in the provinces of Venice, where the great cosmographer Giambattista Ramusio, in the second edition of his work published in Venice in 1554, already provided the iconographic representation reproduced in Plate 46. From Venice, in the seventeenth century, the cultivation of Indian corn spread rapidly to the provinces of Lombardy and Emilia, and it began to appear even in the southern provinces, although it encountered more resistance in Tuscany. But toward the end of the eighteenth century, while the potato and the tomato were almost completely unknown to Italian farmers, the cultivation of Indian corn throughout almost all of Italy had replaced the traditional cultivation of other minor cereals, such as sorgum, millet, and panic grass. For the rural population of a large part of central and northern Italy, and even some provinces of the South, *polenta* made with Indian corn became a fundamental and sometimes almost exclusive type of food.

The triumphant march of Indian corn—which in Italy, unlike other lands, was used for direct consumption and not for feeding animals—encountered resistance from agricultural laborers, whose exploitation and diet were made notably worse by this invasion. Cultivation and consumption of Indian corn, in fact, often replaced not only traditional minor cereals, but even grain. Goethe, in his *Travels in Italy*, later commented on the poor nutritional results of a diet based exclusively on Indian corn for the rural populations of the northern provinces of Italy: where the triumphant march of *mais* was accompanied, almost up to our own time, by the terrible disease known as pellagra.

In entire provinces, however, from the seventeenth century onward, and still more in the eighteenth and nineteenth centuries, the fields of Indian corn expanded. And because of the requirements for its cultivation, it had a role of first importance in the passage of many provinces of central and northern Italy from agricultural systems based on a traditional biennial rotation of forage and grain to a system of continuous rotation (although it was a step backward from the agronomic point of view), where Indian corn had the function of a replenishing crop.

It is easy to understand what importance the increasing spread of continuous rotation between Indian corn and grain (or other similar patterns) had for elaboration of the modern form of the agricultural landscape: the network of tilled fields was henceforth no longer interrupted by the mottled pattern of fallow. This was a slow development even in those provinces of the Center and North where, from the sixteenth century onward, genial agronomists, like Tarello, became the pioneers of this new agricultural system for Italy and Europe as a whole. Here the greatest progress was made with the rotation of forage crops, both

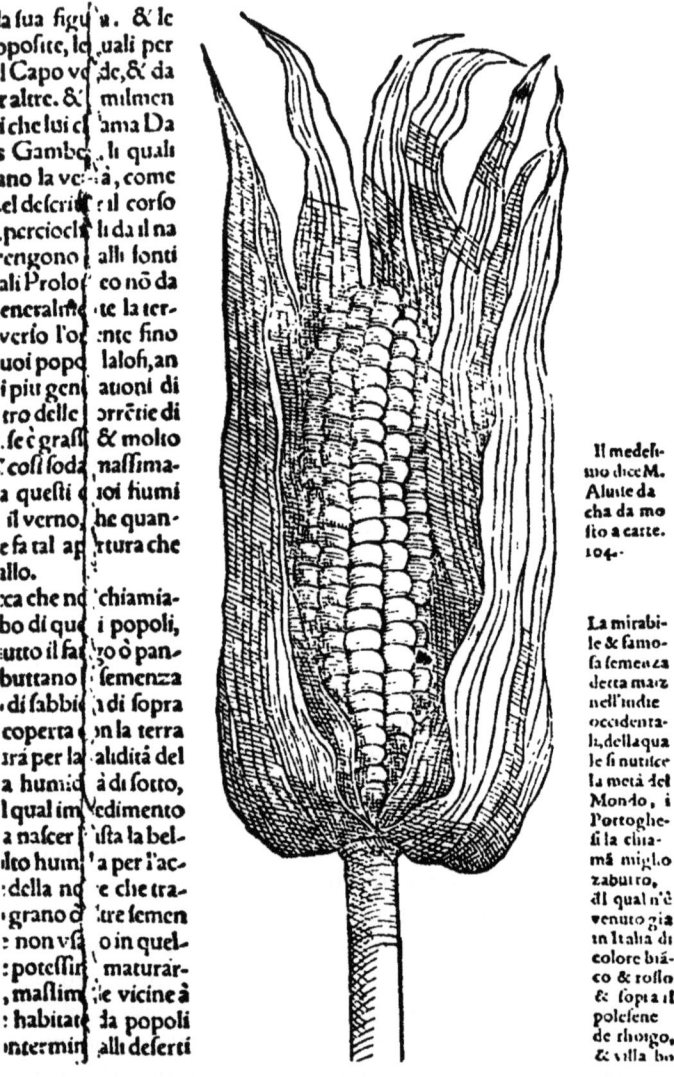

Plate 46. The spread of new crops: one of the first iconographic representations of Indian corn in the second edition of the cosmographic work of Giovanni Battista Ramusio.

Indian corn and other mowed plants. The process was accomplished in this period, and later, not without contradictions and difficulties, which appear with particular clarity from the spread of the cultivation of Indian corn. It was also characteristic of new capitalistic relationships that began to appear in the agriculture of some provinces of Italy. The history of struggles that accompanied the "march of *mais*" are truly typical illustrations of the sympathetic pages that Marx dedicated to the analysis of relative surplus value in the first volume of *Capital*. We are now at the threshold of the age when agricultural progress had the increase of profit as its single aim, and this aim was inevitably achieved through an increased exploitation of the direct producers, their chronic undernourishment, and the advance of a predatory economy that posed serious problems for the health of the population, and the integrity of the tilled soil itself.

.VI.

THE AGE OF THE COUNTER-REFORMATION AND FOREIGN DOMINATION

Plate 47. The landscape of marshes and fens in a painting by Giulio Romano.

. 56 .

*Marshlands and Improvement between the
Renaissance and Counter-Reformation:
The Landscape of Marshes, Wetlands,
and Rice Fields*

Between the age of the Renaissance and the Counter-Reformation, the experiments of great Italians, from Leonardo to Galilei to Torricelli, who founded the modern science of hydraulics, began more informed and expert improvements in river systems, which not even wars, economic and political decline, and foreign domination entirely succeeded in halting. One cannot escape the impression, however, that already in the mid-sixteenth century these initiatives were not enough to stop the agents of deterioration in mountainous landscapes, and consequent further hydraulic dislocation. In chronicles and archival documents, reports became more and more frequent of previously reclaimed lands returning to swamp, and of rivers overflowing their banks to flood and devastate the surrounding countryside, so that (as Ariosto wrote)

> guizzano i pesci agli olmi in su la cima
> ove solean volar gli augelli in prima.
>
> (fish dart around the tops of elms
> where birds once flew.)

This phenomenon had particularly serious effects in some regions of the middle and lower Po valley, in low-lying tracts of the Serchio and the Arno, in the Tuscan and Roman Maremmas, in the Pontine Agro, and in other coastal zones of the South, where flourishing agricultural centers that remained from the time of the communes and early Renaissance rapidly declined. Low-lying marshy landscapes reappeared and spread through broad zones of peninsular Italy and the Islands, and where these zones were not entirely deserted, due to malaria, they often became the theater of hunting and fishing, rather than of agriculture, of which the painting by Giulio Romano—reproduced in Plate 47—provides a lively picture. In the Maremma, in the Roman and Pontine Agros (whose landscape of fens is depicted in a painting by Poussin, in the following Plate 48), and along the coastal plains of the South, wild breeding of animals remained important, combined with cultivation of grain on temporary clearings, or on burned-over tracts; and particular attention should be given in this period to the rapid spread of rice cultivation in regions of the center and north, which up to the sixteenth century had been limited almost exclusively to Sicily and the southern half of the peninsula.

It was not by chance that this process, which from the sixteenth century onward led to the definitive transfer of the center of gravity of Italian rice production to the North, involved the landscape of marshes and fens. Between the sixteenth and eighteenth centuries, in fact, the spread of cultivation of rice in northern regions occurred exclusively through stable rice fields established in marshlands. At the end of the fifteenth century this type of cultivation was already practiced in Lombardy, and in the first half of the sixteenth century consumption of rice had spread to the region around Mantova, so that Folegno, in his mannered Latin, could sing of peasants roaring with laughter behind the back of old Tognazzo, who "omnes parebant sisi mangiasse menestram" (it seemed to everyone he was eating soup). By that time rice had already become an important export of Piedmont, and its cultivation had spread even into the Veneto and Emilia, not only in places where there were natural fens, but also in places that were flooded artificially. And the first protests were already being raised against the miasmas that emanated from these fixed rice fields and gave the domain of marshy landscapes an ever increasing extent. Nor were there delays in the call—in truth without great effect—for prohibitive legislation, to keep this miasmatic belt away at least from the chief centers of habitation.

The spread of stable rice fields, nonetheless, contributed not little to the spread of the landscape of marshes and fens; and the "march of rice," which itself became a decisive propelling force in the development of capitalism in the countryside, was not less triumphant than that of Indian corn in the northern provinces, and like this it was marked in its progress with blood and tears.

. 57 .

Agricultural Systems of Temporary Clearings,
and the New Extension of Pastoral Landscapes
between the Fifteenth and Eighteenth Centuries

Between the fifteenth and the eighteenth century, as we have seen, the relative continuity of initiatives for the conservation and hydraulic improvement of arable land gave way, in many parts of the peninsula, to a new worsening of hydraulic disorder, and a new definitive spread of fens and marshes over lands that had been reclaimed in the communal period and early Renaissance. There was an analogous development, in some parts of Italy, in the evolution and relative extent of agricultural and pastoral landscapes. Around the larger and smaller urban centers a flow of human energy and financial means toward the countryside, to be sure, permitted new clearings and plantations. This induced, in these zones, more minutely elaborated forms of agricultural landscapes, such as irregular fields and the Mediterranean garden, and thus also reduced the relative dominance of the pastoral and less elaborated agricultural landscapes characterized by temporary clearings and by fallow. Even where, in this period—as most often occurred in the Po valley—the landscape of irregular fields and Mediterranean gardens did not visibly enlarge, a further revealing expansion and elaboration in the form of the agricultural landscape was accomplished through the transformation of sowed fields into sowed fields with trees. This led to a progressive extension of the landscape of the Po that was planted with trees and with drained and irrigated artificial pastures (which were imposing their characteristic forms on the agricultural landscape), and also the development of agricultural systems of continuous rotation, which also contributed to a more modern elaboration of the agricultural landscape. Finally, one should not forget in this period the part played in the extension of more elaborated forms of this landscape, by seigneurial villas, which, as we will see further, arose in particularly conspicuous numbers in regions like Tuscany and the Veneto from the second half of the sixteenth through the eighteenth century.

In this regard, as with initiatives for conservation and hydraulic improvement, one might even consider that the process of extending and elaborating forms of the agricultural landscape from the communal period and the early Renaissance was not arrested between the late sixteenth and the early eighteenth century, but continued without visible interruptions. And in zones like Brescia, and other parts of Lombardy, in fact, despite the effect of certain regressive agents, the progressive ones, already moved by incipient capitalist development, ended up by dominating, even in this period, and they prepared the way for the surge of agricultural productivity in these zones during the second half of the eighteenth century. Even in Piedmont, the Veneto, and Emilia, although in different ways and to different degrees, the extension and elaboration of forms of the agricultural landscape did not too much regress and, here and there, even progressed. But already in zones of central Italy, and particularly in Tuscany, a further extension of the landscape of irregular fields around inhabited centers corresponded to a further degradation of the landscape over large areas; and this development appeared even more strikingly in the South and the Islands, where limited further progress in the landscape of the Mediterranean garden and of preserves marked a decisive return of pastoral landscapes and temporary clearings.

One can already see traces of this uneven development in the Renaissance period, which was already displacing centers of agricultural progress from the South and from Tuscany toward the Po valley. Within the limits we have set for our exposition, it is impossible to deepen the study of the historical, technical, economic, political, and social conditions that underlay such diversity of modes of development in the South, in Tuscany, Umbria and the Marche, and in the Po valley. We might indicate, however, one of the most important of these conditions, which has not attracted sufficient scholarly attention.

Throughout Italy, despite the persistent predominance of feudal forms of property, agricultural production in the sixteenth century began to be conditioned by trends of markets and prices, which were deeply affected in this period by the opening of the new distant routes of European trade, by the price revolution resulting from the influx of American gold, and by the rapid development of manufacturing in the Low Countries and England, not to speak of the consequences of clearing new lands, like those of the Ukraine, which began to provide important sources of grain exports for northern and western Europe.

Together, the shift in markets, prices, and costs of production acted to increase the relative importance of raising animals (particularly for their wool) over producing of grain. The developing of breeding, particularly of sheep, as we have already indicated, and the expansion of pastures and meadows, even at the expense of grain, were not only Italian developments of this period, but general European ones. They reached their height in England, where great lords reduced vast extents of culti-

Plate 48. The *Roman campagna* by Poussin.

vated land to pasturage, and excluded cultivation with enclosures, so that (in the famous expression of Thomas More from his *Utopia*) "your sheep . . . eat up and swallow down the very men themselves."

The same expression could be applied to the preserves that, like English enclosures, were abusively multiplied by the barons of the South on communal and feudal domains, so as to convert lands thus removed from customary cultivation by the local population into pasturage. From the beginning of the sixteenth century, in Rome as in Naples, voices rose up with growing concern to denounce the perils that might accrue to common subsistence from expanding pasturage and limiting cultivated land. There did not lack in the Roman Agro, for instance, legislative measures that attempted to require proprietors to cultivate at least a third of their lands. But the very repetitiveness of these provisions shows their ineffectiveness. In the territory of the Agro as in the Maremma, in the Tavoliere of Puglia as in the Corsi of Calabria—which were traditionally places of mixed agriculture and pasturage through a system of temporary clearings in a three- to eight-year cycle—a return to cultivating grain became more and more unusual, so that the landscape of pasturage in the end clearly prevailed over agriculture.

The painting of Poussin from the Roman church of San Martino ai Monti, which we reproduce as Plate 48, provides an image of this landscape in the Roman Agro as it appeared in the first half of the seventeenth century. It is a landscape of pastures, thickets, fens, and a few open fields expressive of the forms of an agricultural system of temporary clearings, where breeding clearly prevailed over agriculture. And this kind of landscape extended in this period not only through the Roman Agro, the Corsi of Calabria, the Tavoliere of Puglia, and the Maremmas, but also through vast Sicilian, Neapolitan, and Sardinian fiefs, where the ancient system of temporary clearings had often returned to prevail over systems of fallow and biennial rotation.

To get a clearer idea of the significance of this phenomenon in one sector, the table provides the goat population for some years between 1463 and 1789. As one sees, the increase was rapid in the fifteenth and sixteenth centuries, reached its height in the seventeenth century, and then began to subside. Nonetheless, the quantitative importance of breeding remained, even at the end of the eighteenth century, at twice its initial point.

Goats Counted in the Dogana of the Tavoliere of Puglia

Year	Number of Goats	Year	Number of Goats
1463	600,000	1604	5,500,000
1536	1,048,000	1697	1,792,000
1553	1,450,000	1789	1,200,000

Where there was no development of artificial pastures, the increase in number of grazing sheep and goats brought about a greater extension of pasturage, and thus a reduction in the cultivated terrain. Thus, in the Roman Agro, already in 1524 the area sown in grain was half what it had been in the previous century, and toward the end of the seventeenth century no more than a tenth of the whole area was planted, while the remaining nine-tenths was reduced to pasturage. The same was true of the Maremma of Siena, where (only in the period between 1630 and 1692, for which precise information remains) the cultivated area declined by 33 percent.

In short, throughout the South, and in vast territories of central Italy, the renewed diffusion of the agricultural system of temporary clearings, and the increase in the area of pasturage, became the agents of a significant deterioration and disaggregation of the agricultural landscape, which was often aggravated by the growth of marshlands and the endemic threat of malaria. These same agents also operated in other parts of Italy, but with less serious negative effects where an increase in breeding was accompanied (as in the more progressive zones of the Po valley) by an increase in the area used for artificial pastures, whether drained or irrigated, and by the introduction of forage in the new agricultural systems of continuous rotation. These factors of technical progress, rather than contributing to the deterioration and disaggregation of the agricultural landscape, elaborated new and superior forms for it.

These indications, necessarily brief, will nonetheless, we hope, orient the reader as to the contradictory character of processes unfolding in this period, and to the variety of directions that the development of systems and agricultural landscapes assumed in different parts of the peninsula. Certain elements in the agricultural inferiority of the South had already appeared that would become more serious in the course and as a consequence of capitalist development. And what after Italian unification became "the Southern question," the agricultural inferiority conditioned by processes already underway in this period that we have treated only summarily, would become a lasting and decisive factor in the complex limitations imposed by capitalism.

. 58 .

*The New Feudalism and the Landscape of the
Italian Villa of the Renaissance and
Counter-Reformation*

In the second half of the sixteenth century, the regressive processes that we summarized rapidly for a good part of Italy in the preceding sections began to show their negative effects more clearly, with regard both to the involution of the agricultural system and to the now visible degradation of forms of the landscape. With the new expansion of the system of temporary clearings, and of pasturage particularly, a landscape of open fields often returned to dominate even where a system of closed fields had spread beyond the more restricted circle of the suburbs in the age of the communes and the Renaissance.

This regression of agricultural systems and landscapes was only an aspect, as we have already indicated, of a more general decline of the Italian economy, which corresponded, on the level of productive relationships and political institutions, to what has been called the "refeudalization" of Italian society, and, on a cultural level, with the Counter-Reformation.

In contrast with the past, the process of "refeudalization" appears with particular significance, naturally, in those parts of the peninsula where the energy of the communes had been the greatest. Elsewhere, and particularly in the Kingdom of Naples, the communal movement had not succeeded, as we know, in decisively eroding the economic and political power of the barons, or in transforming the feudal structure of the state. In the revolt of barons in 1484, and again in that of Lautrec in 1529, the chief baronial families barricaded themselves in their fiefs, equipped armies, and dealt with the king as equals. The failure of the two revolts, however, marked the fiscal collapse of many of the most powerful feudal dynasties, the ruin of others, and the sunset of feudalism as a political phenomenon in Naples. The juridical structure of the fief and feudal productive relationships were still intact in the late sixteenth century, but fiefs often passed from the hands of the old military dynasties into the hands of bankers and rentiers, who had fewer political ambitions to exercise ancient feudal privileges, perhaps, but certainly more exacting mercantile aims.

The peasant war that took its origins from the Masaniello revolt in Naples, which also flared up through the southern countrysides in 1647–48, was the rural population's response to feudal oppression and exploitation now aggravated by the mercantile avidity of the new barons. Thus, rather than a "refeudalization," one should speak more properly in the South of a commercialization of the fief.

It was in Tuscany, by contrast, that the expression "refeudalization" can be used most effectively. Here, as we know, the communal movement had not only seriously eroded feudal property and productive relationships, but had also overturned even the feudal structure of the state. When the old aristocracy that had arisen through trade fell with the republic, the need to consolidate the principate induced the Medici to give out favors to their own faithful followers, and to seek a more secure social basis for their power. But, in a historical climate where agents of economic and social regression were already at work, this policy took a typical form: Cosimo I created fiefs, while the republic had fought and dispersed them. Tuscany was again carpeted with fifty or more lordships, where vassals were subjected to all the vexations of feudal dependents, and even the administrative role of the state was dispersed and curtailed. A third of the landed property of this region became immobilized through ecclesiastical mortmain, and the reappearance of fideicommissa and primogeniture, and then the institution of the Order of Saint Stephen, removed other large holdings from free circulation. In the seventeenth century it is estimated that entailed property accounted for no less than three-quarters of the territory of Tuscany!

The new feudalism, to be sure, in Tuscany as in other parts of the peninsula where the energy of the communal movement had been greater, and in the Kingdom of Naples, should not be confused, as to its constitution and function, with the old one. The military function had practically disappeared, and the political and administrative function was much reduced. Even in economic structure, the increasing weight of commercial capital and of commercial, monetary, and capitalist relationships had introduced deep changes over medieval feudalism. But even in regions like Tuscany, where the old feudal economic structure had been deeply eroded by the communal movement, the return of political and administrative superstructures of a feudal type clearly had negative consequences for the economic structure itself. In a historical climate of stagnation and economic, social, and political recession in the Tuscan countryside, as in that of a good part of central and northern Italy, productive relationships and agricultural contracts regained a feudal tone and coloration. Sharecropping, which in the communal period had marked the first step in a progressive evolution from serfdom to more modern types of agricultural relationships, now crystallized in a backward form, and become a factor of stagnation and of agricultural and social regression.

In the following pages we will return to the negative effect these processes of involution had on the degradation and disaggregation of agricultural landscapes. But, as we have already said for other periods, the enormous concentration of wealth in the hands of new privileged groups, of itself, influenced the forms of the landscape. And if this degradation and disaggregation expressed the general economic and social recession of Italian society, the process of "refeudalization" found its most specific expression in the agricultural landscape through the multiplication of "Italian villas," whose diffusion through the seventeenth century assumed the quantitative and social, as well as artistic, importance that we have indicated in chapter 44.

From the Renaissance through the Counter-Reformation, in fact, the multiplication of villas corresponded to the spread of the new feudalism, just as, at the end of the late Middle Ages, the multiplication of castles had corresponded to the consolidation of the old one. But the difference of forms of landscape and architectural styles expressed the considerable difference of time and historical function between the two systems. The architecture of the old castles was rough and severe in its military functionality, and the landscape they dominated, and to which they gave a first element of reorganization, was still formless. But look, by contrast, in Plate 49, at the landscape of the Medici villa depicted in Benozzo Gozzoli's *Paradise*, or even more at the princely dwelling of the Gonzaga at Castiglione dello Stiviere, illustrated in Plate 42. All military function had disappeared from the architectural structure, just as it had disappeared in the new feudal institutions, and the style of the seigneurial dwelling was already modern and civil in its luxury. Otherwise, within the precinct of the villa, the organizational elements of the communal and Renaissance agricultural landscape had become concentrated, or had taken refuge, and were exalted to a point of exaggeration. But now the increasingly wild countryside often gave way to the open and formless fields of temporary clearings and nomadic pasturage.

In the villas of the early Renaissance—like that of Cafaggiolo in Plate 34—the sober perfection of the villa landscape was still tied, as we have seen, to needs born from the evolution of new techniques and new agricultural relationships. Thus, the landscape of the villa retained, at first, an expansive and paradigmatic capacity that made it an element of organization for the whole surrounding landscape, and even the peasant farm repeated its simplest lines. But in the later Renaissance, and the Counter-Reformation, the links to utilitarian origins and productive needs increasingly lessened in the villa landscape, which was now des-

Plate 49. The Tuscan villa, in a detail of the *Paradise* by Benozzo Gozzoli.

tined almost exclusively for the indolent recreation of a parasitic landowning class. Its forms became all the more complex and elaborate as those of the surrounding countryside, in an agricultural and pastoral economy exhausted precisely by this lordly luxury, became more hazy and disorganized. In this sense, therefore, the villa could not preserve the paradigmatic function and expansive capacity in the late Renaissance and Counter-Reformation that it had in the communal period and the early Renaissance. Nor even less could it still operate—like the severe castles of the ancient feudal lords—as an element of reorganization of the agricultural landscape. On the contrary, like the *regiae moles* mentioned by Horace that removed lands from the plow, these villas, with their luxury, removed vital nutrients from the surrounding countryside, and contributed, with the exasperating elaboration of their forms, to the further disaggregation of the degraded agricultural landscape, impoverished and overtaken by woodlands, out of which they rose like splendid but sterile islands.

In Tuscany and the Papal States this phenomenon had assumed the greatest proportions already in the second half of the sixteenth century. In the Veneto, as we have already seen, out of 1,411 now extant villas of artistic interest, no more than 22 date back to the eleventh to fourteenth centuries, and 84 to the fifteenth, but then 257 to the sixteenth, 332 to the seventeenth, and 403 to the eighteenth century. However, by the eighteenth century this phenomenon assumed in the Veneto, as elsewhere, a new significance, to which we will return as our study continues.

. 59 .

Classic and Romantic Landscape in Italian Reality and Art of the Seventeenth Century

We have already indicated the contradictory character of the processes that induced elements of progressive degradation and disaggregation into the Italian agricultural landscape in the age of the Renaissance and Counter-Reformation. We are not confronted here—as in the age of the barbarian invasions—with a violent crisis that dissolved the continuity of cultural tradition and productive technologies. Despite the general economic decline and social and political regression, new and different historical conditions—it is enough to think of the spread of printing—assured to this age not only preservation, but often even notable progress in the invention of new cultural and productive techniques. Thus, as we have already noted, what strikes one in the agricultural landscape is precisely the contrast between forms elaborated to a point of exaggeration, but limited in extent, and degraded and disaggregated forms, which put their mark on a large part of peninsular Italy and the Islands: a contrast closely related, as we have seen, to the process of "refeudalization" of Italian society, and to the concentration of economic and political power in the hands of a new retrograde and parasitic court nobility.

It was not that this age lacked the general technical ability to impose more elaborated forms of agricultural landscapes on the natural landscape. What prevented the diffusion of these forms was the disrelationship between productive relationships and the social and political superstructure that retarded the development of productive forces, which new techniques might otherwise have permitted.

The degradation and disaggregation of the forms of the agricultural landscape, thus, were not accompanied in the seventeenth century, as had occurred in other periods, by a decline in techniques and tastes for pictorial landscapes, but instead their exaltation and exaggeration, which corresponded to the exaltation and exaggeration of stylistic forms in seventeenth-century literature, and to the baroque in architecture. But in pictorial landscapes as well, seventeenth-century art could not escape the reality of the degradation that had appeared in the forms of the real agricultural landscape. We are a long way, now, from the attention and minute care with which Tuscans of the fourteenth to early sixteenth century had represented the most precise details of the arrangement or boundaries of a cultivated field; distant as well from the scientific landscape of Leonardo, and from the refined landscapes of sixteenth-century Venetians. Even for seventeenth-century painters, like Albani, who wanted to continue more directly the tradition of sixteenth-century landscape painting, conventional compositions in classical taste, which became a manner and decoration, no longer had a reality of content. Minute observation and artistic attention for a countryside stamped with the work of man, and with cultivated fields, were set aside in favor of the open countryside used as a background for erudite mythological scenes, which have less power to move us.

Seventeenth-century landscape painting found its more authentic voice and most elevated inspiration when it deliberately adapted its style, form, and interpretation to the new reality of content, and liberated itself from conventional classical pastiches to express realistically, and with romantic emotion, the forms of the degraded and disturbed landscapes of seventeenth-century Italy. These were landscapes no longer dominated and refined by man, but ones that instead dominated and barbarized him; landscapes where no longer nymphs, heroes, or kindly madonnas moved, but instead beggars and evildoers, peasants and brigands, horse traders, and soldiers of fortune wandering among meandering mule tracks, leafy trees, and mysterious glades, as in the *Seaside* by Salvator Rosa, from the Uffizi Gallery, which we reproduce as Plate 50.

In this deep inspiration, this almost mournful expression, of the degraded and disturbed landscape of seventeenth-century Italy, landscape painting discovered, in its chief representatives—from Salvator Rosa to Magnasco—the value of an explicit critique, an open polemic against the new feudalism and new obscurantism, which still is not concluded today.

Plate 50. The degraded landscape of seventeenth-century Italy in the *Seaside* by Salvator Rosa.

. 60 .

*Open Fields, Farms, and Preserves in the Italian
Agricultural Landscape of the
Seventeenth and Eighteenth Centuries*

The pasturage habitually exercised by local populations on fallow lands, and even on cultivated land when the harvest was ended, also contributed to the degradation of the Italian landscape of the seventeenth century, along with the resumption of agricultural systems of temporary clearings, the new extension of woods and wastelands, thickets, and fens. The process of "refeudalization" of Italian society favored, in general, regression toward a regime of open fields, even where, in the age of the communes and the Renaissance, a regime of closed fields had expanded from the immediate area around inhabited centers to a wider suburban circuit.

In this deterioration of agricultural systems and landscapes in the Italian rural economy—and particularly among the poorest peasant populations—commons lands also assumed a renewed and growing importance (at least relatively speaking), as did lands where civic rights for sowing, pasturage, wood gathering, and so on, were exercised. This situation differed quite considerably from one part of Italy to another, according to the variety of historical circumstances that affected the regulation of land use and agricultural institutions. But everywhere, in this period, the use by local populations of commons lands, of communal and feudal domains, and of the open fields of private owners after the harvest constituted an integral element in the internal equilibrium of the peasant agricultural and pastoral economy. While the cultivation of forage crops was still just beginning to spread, this equilibrium clearly could not be maintained on peasant farms without recourse to pasturing flocks and herds on commons lands, lands of communal or feudal domains, or on the fallow and stubble left after cultivation. One can thus understand how it was precisely around the exercise of these customary rights that the most intense class struggles were fought by the peasant population during this period, and particularly against fief holders who usurped communal lands, or pretended to exempt their demesne lands from civic use for pasturage by closing them off as preserves.

But it was not only through the exercise of rights for pasturage and wood gathering that communal lands and *usi civici* assumed a particular significance in the agricultural economy of many parts of Italy. Meanwhile, in the Roman Agro, in the region of Naples, and in Sicily—where old feudal landholding and agricultural institutions preserved all their vigor—there were generally very few free allodial holdings. Here, the *jus serendi* (the right to sow, in exchange for a relatively modest part of the harvest) and the *jus coloniae* (the right of perpetual settlement) on feudal domains, more than providing an integrating factor for the peasant economy, constituted its very existence.

If the *jus coloniae* (whose exercise was generally associated with planting trees and shrubs, and thus was limited in extent by specific environmental requirements) often provided for the formation of a stable agricultural settlement, whether peasant or bourgeois, the exercise of the *jus serendi*, instead, by its very nature, was tied to the agricultural system of temporary clearings or, at least, to fallow, but it excluded continual rotation. In the seventeenth and eighteenth centuries, thus, more than ever, the agricultural landscape of a large part of the South was characterized by the absolute precariousness of peasant holdings, and by the regime of open fields, that has marked southern latifundia up to the present.

In central and northern Italy, by contrast, peasant farms had generally acquired a greater stability in the age of the communes and the Renaissance. The same new bourgeois owners who had freed serfs from bondage to aid their struggle against the feudal lords who opposed the communes felt the need to keep peasants tied to the soil with sharecropping contracts that automatically provided for the creation of stable landlord-peasant farms. Although it crystallized in backward forms during the seventeenth and in the first half of the eighteenth century, sharecropping continued to spread through a large part of central and northern Italy. Even where other relationships besides sharecropping prevailed, like long-term leases (*livelli*) and land rent to the cultivator, the farm usually remained a stable unit of cultivation and was provided with an often miserable farmhouse, which was like a symbol of this stability and was generally unknown in the South, where the precarious cultivators on the changing strips of feudal latifundia continued to live grouped together in large villages.

As already in the age of the communes and the Renaissance, and even with the general degradation of the agricultural landscape in the seventeenth and eighteenth centuries, the farm and farmhouse were what in central and northern Italy (more often than in the South) interrupted the disordered monotony of open fields, and imposed less precarious forms on the landscape, although also new uncertainties. But even here, one should not forget that—as soon as one left the most immediate neighborhood of the inhabited centers—these farms were islands scattered in the great sea of woods and pastures, wastelands and fens, which was expanding its dominion over even more vast extents.

Even in parts of central and northern Italy, however, recourse to pasturage on stubble, fallow, and communal lands remained essential for the internal balance of the farm during the seventeenth and at least the first part of the eighteenth century. Here *jus serendi* generally did not apply to demesne lands of the fief; but the right to settle (*far presa*) on communal domains retained and reacquired a notable importance, especially in zones of the hills and mountains. This involved clearing a tract of land, often by burning it over, and cultivating it until its fertility was exhausted; after which it was abandoned again to common use for pasturage, or to thicket or woodland.

According to local custom, and the varying density of population, the right to settle on commons lands fell to the first occupier, or it was regulated by a periodic distribution of plots. It introduced, however, forms into the landscape that in their very precariousness often repeated that of irregular fields. In Figure 12, a sketch preserved in the archive of Castel del Piano (in the province of Grosseto), published by Imberciadori, shows how this landscape of clearings was integrated into a farm in Tuscany at the beginning of the eighteenth century.

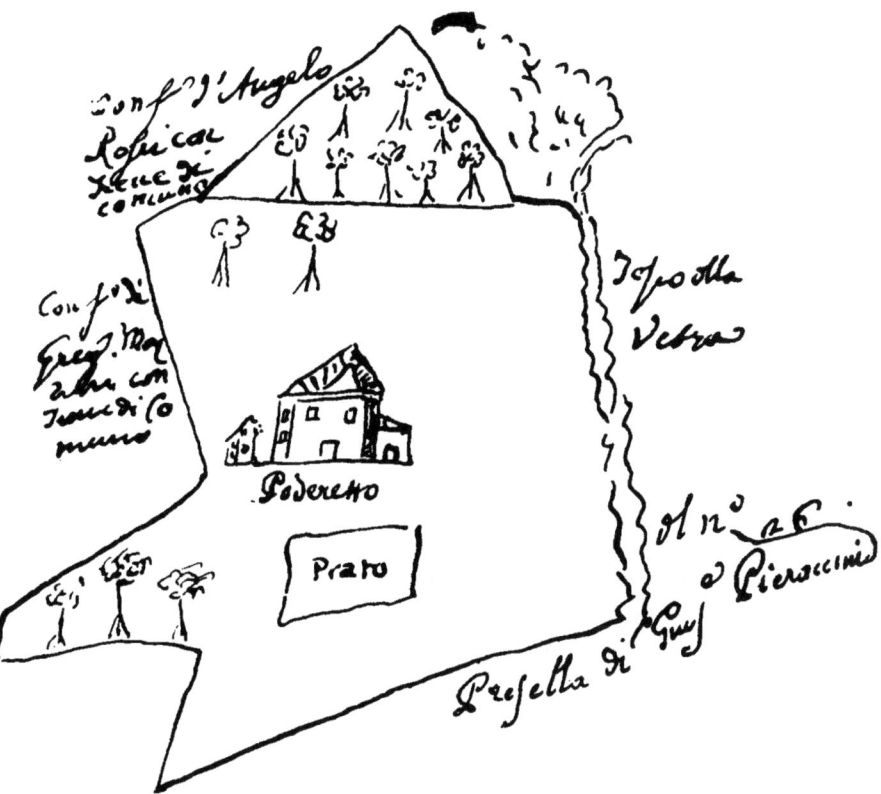

Figure 12. A farm and its clearings in the eighteenth-century landscape of Tuscany, from a sketch preserved in the Archive of Castel del Piano (Province of Grosseto).

· 61 ·

The Landscape of Industrial Crops and Agricultural Systems of Continuous Rotation in the Seventeenth and Eighteenth Centuries

We have indicated the contradictory character of the processes that, from the late Renaissance, introduced elements of progressive degradation into the Italian landscape through the reduction in area of cultivated land, and the new expansion of pastures, wastelands, woods, and fens.

This degradation and disaggregation in the forms of the agricultural landscape reached their climax in the second half of the seventeenth century: but it did not result from a single process that marked everywhere a retreat in techniques and agricultural systems, as occurred in other periods. The general economic and financial marasmus, the "refeudalization" and reactionary superstructure, and the misgovernment by foreigners did not always succeed in reversing the direction of processes in evolution, which even despite these obstacles succeeded in affirming themselves with an elemental energy, animated as they were by the historical agents that, even in Italy, were already preparing the advent of the new productive relationships of capitalism.

One might consider, for example, the revival of systems of temporary clearings and pasturage that contributed so much, as we have seen, to the degradation and disaggregation of agricultural landscapes in this period. Considered from the standpoint of this most visible feature, to be sure, the revival represented an essentially regressive agent, which returned entire sectors of the Italian agricultural economy to a level that historical development had surpassed centuries and sometimes millennia earlier. But in its historical significance, clearly, no one would want to identify the system of temporary clearings of this period in the Tuscan Maremma, for instance, with the system that the Etruscan expansion had replaced in this region. And it was not only a question, one must be careful to point out, of the evident superiority of the agricultural practice and technical means in operation during the seventeenth century, but also of the deep difference of relationships of production (and exchange) that now conditioned the revival. If in the pre-Roman age, or in the late Middle Ages, the spread of a system of temporary clearings had been associated with a natural or budding economy, where commerce had only a marginal part, in the fifteenth to eighteenth century, in contrast, the revival of older systems was closely linked, as we have seen, to the trend of relative prices for grain and wool, not only in local markets, but in the European market. And this new connection between agriculture and the mercantile economy, far from being a regressive development, was destined to become the decisive element in its historical progress. The increasing number of preserves, on the other hand, that excluded entire populations from age-old rights of cultivation and fattened the flocks that "eat men," although many zones in the Italian countryside were depopulated in this period, revealed the cruel aspects of the process of primitive accumulation that Marx thought a necessary prologue to the drama of modern capitalism. This was destined to revolutionize productive relationships in the Italian countryside and, through these, the techniques and forms of the agricultural landscape.

With regard to the effect of the revival of the system of temporary clearings, however, this revolution remained for the moment essentially negative. But in other sectors, already in the seventeenth and eighteenth centuries, the mercantile development of agriculture and first consequences of primitive accumulation began to have positive effects on the elaboration of forms in the landscape.

Consider, for example, the results of the spread of a few industrial raw materials, like hemp, of which a good picture in the seventeenth-century landscape of Emilia is provided by a painting of Guercino, preserved in the Gallery of Cento, that we reproduce in Plate 51. In the late Middle Ages, as we know, the cultivation of hemp spread throughout Italy, and it was integrated into the industrial production of textiles alongside the more ancient linen. Still at the time of the communes the two types of cultivation were often associated, not only in the same geographical regions, but even on the same farms, and this production was still devoted mostly to satisfying family and local needs. But already in the Renaissance, and even more in the period we are examining, a mercantile development of agriculture that the economic marasmus did not entirely interrupt brought about an increasing regional specialization of hemp production. Like other industrial crops, this textile fiber was commonly produced in peasant households throughout Italy to fill family needs, but it tended to concentrate in a few geographical zones, like Emilia, where it assumed a decisively mercantile character and provided a fundamental commodity in the currents of interregional and international trade. The same occurred for the production of linen, which also became concentrated in determined geographical zones, for dye-producing plants, and for other items.

Plate 51. Industrial crops and agricultural systems of continuous rotation: the harvesting and retting of hemp in a painting by Guercino.

This process of incipient regional specialization of products generally brought about important changes in the internal arrangements of individual farms, and in the prevailing agricultural system of a given region. Thus in the region of Bologna, from the time of the communes, the cultivation of hemp was practiced in an agricultural system of continuous rotation, which enlarged through the development of hemp production based on just this process of regional specialization. In other regions of the Po valley, as for example the zone around Lodi and Parma, which from the time of the communes was famous for its production of particularly prized cheeses, the same process of regional specialization influenced further progress in the cultivation of forage crops in this period and in implanting the agricultural system of continuous rotation, whose diffusion contributed also, as we have seen, to the rapid march of Indian corn.

The same mercantile development of agriculture, in short, that on one hand conditioned the revival of pasturage in a system of temporary clearings and the degradation and disaggregation of the agricultural landscape, encouraged on the other hand other processes that, from this period, introduced new and superior organizational elements into the landscape, and precisely those characterized by agricultural systems employing continuous rotation. And this contradictory development had all the more significance in that these contrasting processes developed with a differential effect in different parts of peninsular Italy and the Islands, so that, still within the prevailing generally regressive trend, there were clearly sectors where elements of a reorganization of the agricultural landscape began in this period to assume a certain importance to counterbalance the elements of disaggregation.

. 62 .

Origins of the Contemporary Landscape: The Southern Landscape of the "Mediterranean Garden"

It was in the Maremmas, in the Roman and Pontine Agros, in Sardinia, and in the region of Naples and in Sicily, in the sixteenth and seventeenth centuries, as we have already indicated, that the agents of degradation and disaggregation of the agricultural landscape generally prevailed more clearly over those working toward reorganization. It was not only a question of the particular seriousness of the reappearance of agricultural systems of temporary clearings as a result of the new expansion of pasturage. The same process of mercantile development of agriculture that produced an increasing regional specialization, would determine in fact, during this period in the South, both a restriction of cultivating grain and an increasingly serious crisis of some types of highly prized crops that had given a relative agricultural preeminence to these regions. It is enough to remember that these were the centuries when, under the pressure of foreign competition, the cultivation of cotton began to decrease, and sugarcane practically disappeared, which earlier had made significant contributions to the agricultural prosperity of Sicily and Calabria. The production of rice and silk, which up to the sixteenth century had been almost a monopoly of Sicily and the zone of Naples, now saw its center of gravity made toward the provinces of northern Italy, where there were more favorable environmental conditions, and where these crops ultimately assumed an importance that had not been reached even in the period of their greatest prosperity in the South.

We find ourselves, again, confronted with the maturation of the elements of agricultural inferiority in the South, which, as we have already indicated, would be further aggravated through the process and in the context of the capitalist development of Italian agriculture. That this was not a question, as some have wanted to show, of a kind of natural inferiority, but only a relative inferiority—tied to a particular level of technical development and conditioned by a particular system of productive relationships—is confirmed not only by the earlier flourishing and preeminence of southern agriculture, but also by the energy with which, although in difficult conditions, the landscape of the Mediterranean garden survived and expanded in these regions during the six-

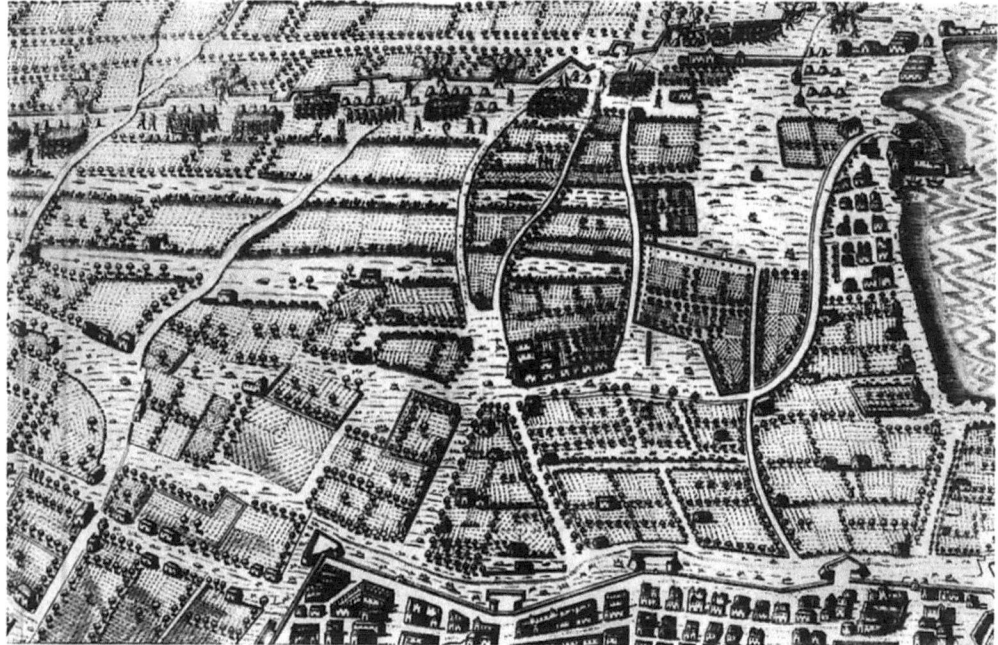

Plate 52. The belt of plantations of trees around Palermo in a map of the city at the beginning of the eighteenth century.

teenth and seventeenth centuries, with its cultivation of particularly valuable and prized trees and shrubs.

Plate 52, which reproduces a map of Palermo and its suburbs, shows us how the landscape of the Mediterranean garden, at the beginning of the eighteenth century, was already beginning to be elaborated in its modern form through a wide belt around one of the major centers of Sicily. In various parts of the South, in fact, even in the seventeenth and eighteenth centuries, plantations of trees and shrubs continued to expand through broader expanses of territory, and this gave rise to currents of trade that assumed an increasing importance in the foreign commerce of these provinces. Far from producing a degradation and disaggregation of the agricultural landscape, here mercantile development and the regional specialization of crops, such as citrus fruits, almonds, and so on, produced instead elements of its maturation and reorganization.

It was not that, even in these centuries, the expansion of plantations of trees and shrubs conditioned an expansion only of the landscape of the Mediterranean garden. It was often more likely, as one advanced through the eighteenth century, that there was a greater parcellation on large tracts planted with vines, almonds, citrus, and so on, where enterprising feudal elements or large bourgeois renters already practiced cultivation of a capitalist type. But in the period in question when the landscape of the Mediterranean garden was expanding, significant importance was also assumed by plantations of trees and shrubs carried out through the initiative of direct cultivators, or small and medium bourgeois owners, who controlled a not insignificant part of the free allodial land. This was an initiative that appeared not only on allodial lands, but also—thanks to the *jus coloniae*—on feudal and communal domains.

What is certain is that between the seventeenth and the eighteenth century, the landscape of the Mediterranean garden—even if it was far from reaching the extent it has reached in the South in our own days—continued to enlarge gradually, and that it already assumed (especially close to inhabited centers) forms that were not much different from the current ones, with small plots and their enclosures, among which ran the intricate web of suburban roads bordered with white walls topped with the shining green of orange branches. And from the slopes of Vesuvius to the Cape of Sorrento, from the slopes of Etna to the Conca d'Oro, a trip along these roads and a glance at the dates inscribed on the gates is enough to convince the reader of the part that plantations of trees in the seventeenth and eighteenth centuries had in the elaboration of this form of landscape.

. 63 .

The Alberata *of Tuscany, Umbria, and the Marche, and Systematization of Fields with Trees in the Seventeenth and Eighteenth Centuries*

In the region of Tuscany, Umbria, and the Marche, as in the South and the Islands, the agents that most effectively contested the tendencies toward degradation and disaggregation of the agricultural landscape in the seventeenth and eighteenth centuries were represented by the spread of plantations of trees and shrubs. In contrast to what happened in the Po valley, one cannot find a sensible diffusion of artificial pastures in these regions; and rarely, also, forage crops and Indian corn were joined together in rotation to facilitate the transition between an agricultural system of fallow and a system with continuous rotation. From this point of view, thus, in central Italy, the agents of degradation and disaggregation of the agricultural landscape continued, through the whole first half of the eighteenth century, to find the way relatively open to their negative effects, not only in regions like that of the Maremmas, where an agricultural system of temporary clearings returned to dominate, but also in the more varied sectors of central Italy, where with the spread of sharecropping contracts an agricultural system of mixed cultivation had come to predominate.

In general, however, despite the effectiveness of the agents of degradation, the construction of new farms marked some progress even in the seventeenth and eighteenth centuries. We have indicated how the construction of new farms, with their various outbuildings, of itself interrupted the disordered monotony of a degraded landscape by introducing new elements and new centers of organization. But on new farms as well as old, a lack of artificial pastures left the problem of essential forage generally unresolved. And as in earlier times, there was an effort to compensate for this, according to the Tuscan proverb, by "tenendo i prati sugli alberi" (tending meadows on the trees)—that is, by expanding not the cultivation of specialized vineyards but rather of vines festooned on trees, which with their branches augmented the meager farm resources of forage.

To this spread of the landscape of the *piantata* (or rather *alberata*, as it was called here) corresponded, besides the need for firewood and feed for cattle, a decisive importance now assumed by the cultivation of mulberry trees for raising silk worms in these regions. Rows of mulberries, like the rows of other trees festooned with vines, thus more and more commonly marked typical tracts of the landscape of Tuscany, Umbria, and the Marche.

As one sees, the part that plantations of trees and shrubs played in the reorganization of the landscape was different, and assumed different forms, from what we have seen in the South and Islands, where more often, even in this period, the forms of the Mediterranean garden continued to develop. In the hills, by contrast, the landscape of irregular fields continued to prevail, as did hillside arrangements with irregular banks or with plowing across or around the hill, which we have already illustrated for the Renaissance period, whereas in the plain there was a notable expansion of land planted partly with trees (*alberata*—as this arrangement was now conventionally called to distinguish it from the *piantata* of the Po valley).

In fact, the *alberata* of Tuscany, Umbria, and the Marche progressively became more differentiated from the *piantata* of the Po valley, above all for the greater density of planting trees and shrubs. In central Italy, to be sure, climatic and environmental conditions did not generally permit the absolute prevalence of cultivating trees that was characteristic of the more southern landscape of the Mediterranean garden. But in contrast with northern Italy, regions like Tuscany, Umbria, and the Marche still enjoyed a Mediterranean climate, where the cultivation of vines (and other types of shrubs) found a natural environment and tended to assume an importance of the first order (even to supply traditional needs of the peasant population of these regions), which it did not have in the harsher climate of the Po valley. In this period, however, in Tuscany, Umbria, and the Marche, the surface devoted to mixed cultivation tended to subdivide itself into relatively narrow fields (of fifteen to thirty meters) that were not very elongated, and where the plantations of trees and shrubs were arranged in close rows. This arrangement clearly permitted planting a considerably larger number of trees and vines on a given surface than in the larger fields of the Po valley, where the rows of trees and vines were spaced some thirty to eighty meters apart.

Plate 53, where we reproduce a seventeenth-century map of Fermo taken from Blaeu Mortier, shows us how, in the second half of the seventeenth century, the landscape of the *alberata* in Tuscany, Umbria, and the Marche was elaborated around this important town of the Marche. As appears from our plate, the rows of trees were placed along the middle of the narrow and more or less elongated fields, which were separated from one another by ditches or drains. But elsewhere, even in the Marche and Umbria, and particularly in Tuscany, according to the precepts of Davanzati in his *Coltivazione toscana delle viti*, the rows were

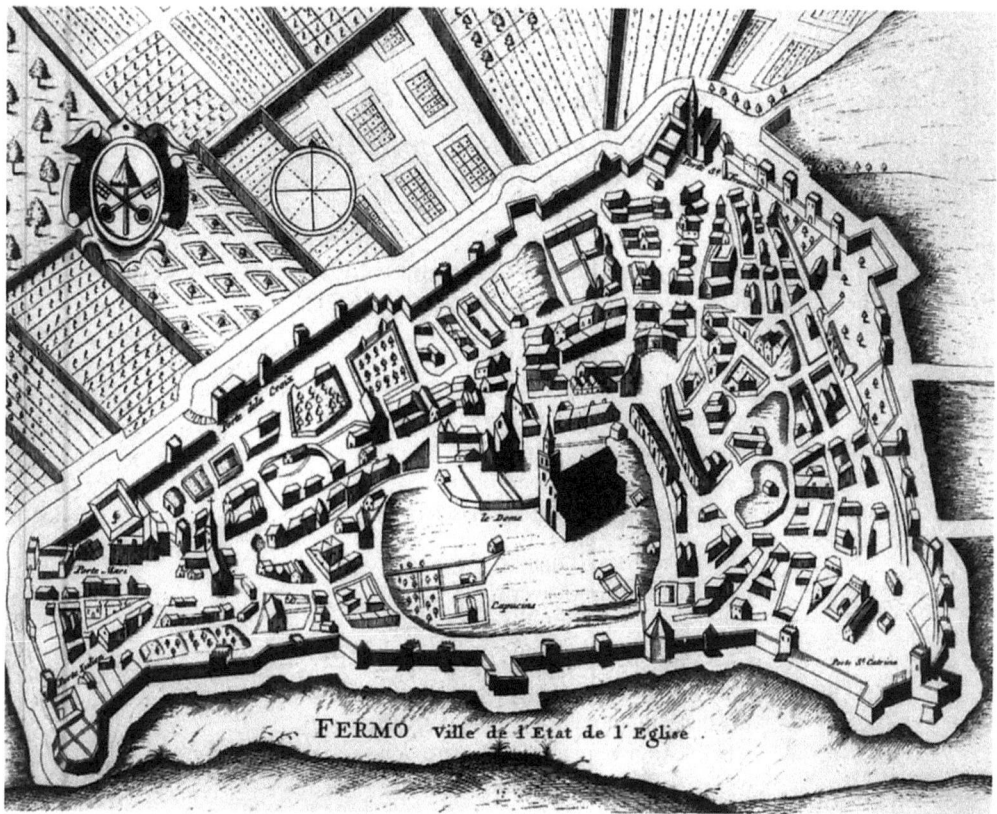

Plate 53. The *alberata* of Tuscany, Umbria, and the Marche from an seventeenth-century map of Fermo.

much closer together, two by two, along the middle of fields and along both sides of the boundary ditches, or sometimes only along the two sides. In this last case, we are confronted with the characteristic arrangement *a prode*, or *a rivale*, which from the end of the eighteenth century up to our own day had been characteristic in the plains of the whole region of Tuscany, Umbria, and the Marche.

. 64 .

The Piantata *of the Po Valley in the Seventeenth and Eighteenth Centuries*

"But remember," wrote at the end of the eighteenth century the author of the *Ecloghe rusticali*, published in Treviso in 1794,

> . . . ma ti ricorda ve' di piantar raro,
> e in buon ordin, che in questo è posta ogn'arte:
> le viti amano il sol spendente, e chiaro,
> quindi d'ombra soverchia non sien sparte,
> e certi folti ombrosi pergolati
> ai giardini ed agli orti sien lasciati.
>
> Pensa e ripensa, il meglio e' non si trova
> che in larghi campi aver lunghi filari:
> del vincin campo il lavorio lor giova,
> e tocche son dai caldi rai solari:
> si lasci che la vite in alto mova
> ma dar la mano alla vicina impari,
> e bei festoni veggansi intrecciare
> che per molt'uva si vedran curvare.
>
> (. . . but remember to plant thinly,
> and in good order, all art consists in this:
> vines like the bright clear light of the sun,
> thus keep them clear of overhanging shade,
> and save thick shady arbors
> for gardens and orchards.
>
> Think and rethink, is it not best to have
> long vineshoots in wide fields,
> they will profit from the field below,
> and the sun's hot rays will reach them:
> give the vines space to climb up high
> and stretch out to their neighbors facing them,
> and you will see beautifully laced festoons
> produce plentiful grapes.)

The warning expressed in these verses corresponds exactly, as we have already seen, to the needs of vines cultivated in the heavier wet soil and harsher climate of the Po valley. And by "thinking and rethinking," in fact a millennial experience, agriculturists perfected on these lands the typical arrangement of the plain in the *piantata* of the Po between the sixteenth and eighteenth century, which permitted them to have "long vineshoots in wide fields."

We have already indicated, in chapters 40 and 63, how beginning in the sixteenth century, and more in the centuries that followed, the *piantata* of the Po increasingly differentiated itself from the *alberata* of Tuscany, Umbria, and the Marche, not only in the greater length and breadth of the fields, but also in the increasing importance assumed by permanent and intensive hydraulic arrangements, which were much more time-consuming than the temporary extensive arrangements in *porche* that often still characterized the *alberata* of central Italy.

The tracts (*prese* [or *prace*]), drained by ditches or embankments, into which the great fields of the *piantata* of the Po were often already divided, did not yet in the seventeenth and eighteenth centuries involve the great movements of earth that would be required from the early nineteenth century up to our own days to give fields the right curvature of earth, and proper drainage. Meanwhile, rather than this ulterior qualitative perfection, the progress of systematization in the Po valley concentrated on their quantitative extension.

One sees, for example in Plate 54, an eighteenth-century map of Mantova, taken from the *Voyage en Italie* of de Beaurain, that can be compared with another from the late sixteenth century, which we reproduced in Plate 32. In this comparison the reader will not miss how, in contrast with the sixteenth century, the landscape of enclosures of the *piantata* now covered practically the whole terrain, at least nearest to the city. Fields that in the sixteenth century still seem not to have been cultivated, or were sowed without trees, or barely edged on one side, if at all, by a row of trees festooned with vines, now presented themselves, unless they were adapted to accidental features of the terrain, in regular forms with boundaries marked on every side by a row of trees, and a road, a ditch, or an embankment. There were still, to be sure, some gaps that remained to be filled, even in this suburban zone. Some rows of trees are missing on sides of fields that were too large, or were already divided in practice, and would be further subdivided. But in general, in the immediate zone of the city, the landscape of the *piantata* of the Po seems to have been almost completely elaborated. And literary and archival sources confirm that from the sixteenth century to the end of the eighteenth, throughout the plains of the north, this landscape in fact spread to new zones where it impressed its characteristic forms.

"A rich and fertile . . . land, planted everywhere with fine trees and cut through by many canals," President de Brosses had written in his

Plate 54. The extension of the eighteenth-century *piantata* of the Po valley around Mantova from a map in the *Voyage en Italie* of de Beaurain.

Lettres familières ecrits d'Italie in 1739 about the landscape he had observed in his travels between Milan and Mantova. And elsewhere, his description of the characteristic landscape of the *piantata* of the Po is even more precise. At Cremona, from the height of a tower, "the land that one discerned seemed a forest, so covered it was with trees"; in the region of Ferrara "the whole country is thickly planted with trees, so that from a height one seems to see a forested plain, formed by the tops of trees." In the second half of the eighteenth century, another famous French traveler, de La Lande, repeated and clarified the same observations.[1] "The vines," he wrote, for instance, with regard to the territory of Piacenza, "are most abundant; one sees them growing up the trunks of elms and extending along the roads, like garlands from one tree to another, with a pleasing symmetry. This land is a vast plain, where all the

[1] [Joseph Jérôme Le Français de La Lande, *Voyage en Italie* (Paris, 1786).—Trans.]

estates are closed in with hedges and trees, which makes the terrain seem covered by them, without being entirely wooded. Our armies were inconvenienced by this in the last Italian wars."

This observation was repeated by Goethe, for the territory of Cento, where also, from a tower, he seemed to perceive a sea always in movement, made up of the crowns of poplars, which (as in the present) were planted along the banks of ditches. Seen from above and in perspective, in fact, the rows of poplars, elms, maples, and so on, which surrounded the planted fields of the Po, presented themselves to the observer as an open forest, and with their undulating crowns inevitably recalled to the minds of French travelers similar aspects of the landscape of the Norman or Breton *bocage*. But it is still more important to mention how already in the eighteenth century this landscape of the *piantata* of the Po had reached such an extent that it seriously hindered visibility for maneuvers of the French army in the wars of Italy. And de La Lande himself, like other travelers in the eighteenth century, repeats descriptions like the ones cited earlier not only for the plains of Emilia and the Veneto, where the landscape of the *piantata* of the Po was a more ancient tradition, and not only for Lombardy, where it had become more widely diffused from the sixteenth century onward, but also for Piedmont (between Alessandria and Tortona, for example, and between Cuneo and Turin), where agricultural progress and the landscape of the *piantata* had spread more recently.

Still at the end of the eighteenth century, however—as we have already indicated, and it should not be forgotten—the landscape of the *piantata* of the Po valley was far from having reached its later extent. Between one city and another, for long tracts, it was still interrupted not only by sowed fields without trees, and open meadows, but also by heath, woods, and fens. From the sixteenth century onward, however, and particularly during the course of the eighteenth century, there was notable progress. In the Po valley (more than in the plains of central Italy) the diffusion of the *piantata* became in this period a decisive factor of resistance to the agents of disaggregation of the agricultural landscape and, even more, was an essential factor in its reorganization. And this was not only because the landscape of the *piantata* extended through the Po valley in this period with a rapidity well beyond what occurred in the plains of central Italy. Not only this quantitative fact was important. Even more was the fact that the seventeenth- and eighteenth-century spread of plantations was closely connected with a general progressive evolution of agricultural systems along the Po, which was difficult to match in central Italy. This involved not only the rapid spread of the cultivation of mulberry trees and rice fields, but also a further spread of drained and irrigated artificial pastures, and the introduction of forage crops, Indian corn, hemp, linen, and other industrial plants in continuous rotation, which involved improved methods of working the land and new hydraulic arrangements. Not all the effects of these improvements, which became more rapid and widespread during the eighteenth century, were as yet clear at the time. But by the last decades of the century they had brought the lands of the Po valley decisively to the lead of agricultural progress in Italy, and had also prepared conditions for a new energetic elaboration of the forms of the modern agricultural landscape.

. 65 .

Ecclesiastical Mortmain, and the Disordered Italian Landscape of the Age of Enlightenment

In the seventeenth and eighteenth centuries, as we have seen, historical agents operating toward reorganization and new and superior forms had already begun to counteract the degradation and disintegration of agricultural landscapes, although with a different effectiveness from one region of the peninsula to another.

But still at the beginning of the eighteenth century, which in the end called itself the "century of Enlightenment," these agents had hardly begun to be felt in most of Italy. Indeed, precisely in the second half of the seventeenth century, the economic, social, and political marasmus of the peninsula seems to have reached its climax. In the region of Naples, and in Sicily, the peasant uprising of 1647–48 had not succeeded, for lack of support and guidance from the urban classes, despite its energy, in shaking the foundations of the feudal order or of foreign domination. But even in the provinces of the Center and North, where a new productive energy began to show itself here and there in the countryside, increasingly harsh feudal conditions kept it within narrow limits, through a thousand restrictions and vexations. Throughout Italy these restrictions removed from circulation and more modern productive investment not only large extents of feudal land, but also the immense sterile patrimony of ecclesiastical mortmain, which had particularly serious consequences for the degradation of the agricultural landscape.

The monstrous concentration of landed property in the hands of avaricious and unproductive groups was favored, as is well known, by the obscurantism of the Counter-Reformation, in which the "refeudalization" of all social relationships had developed in Italy. In the various Italian states of the early eighteenth century it is calculated that about a third of all landed property had fallen under the control of ecclesiastical mortmain, and it was thus not only excluded from circulation but also abandoned to an avaricious and clumsy administration devoted to unproductive ends. Any possibility for agricultural improvement, and any desire to invest capital in the countryside, was blocked by this landed patrimony, which could only increase in size (and it continually grew); it never diminished. The economic and social damage done by this type of ecclesiastical property was further aggravated by the idleness, ignorance, and peculation of the innumerable legions of priests, nuns, and

Plate 55. In the dawn of the age of Enlightenment: *Stormy Landscape with Fleeing Friars* by Magnasco.

friars, who were exempt from any productive work, and, along with the few wealthy prelates, vegetated like parasites on this immense sterile patrimony.

It is not surprising that in these conditions, even before attacking feudal property, it was precisely against ecclesiastical mortmain, in the first decades of the "age of Enlightenment," that the polemics of those in Italy who undertook to open new ways for investment of capital in whatever kind of work of agricultural improvement in the countryside focused, and even more the polemics of those with concrete proposals to further the economic, social, political, and cultural modernization of the nation. The first concrete success in liquidating ecclesiastical mortmain came only in the second half of the century, but this had already become a decisive focal point for those in the new class preparing itself to conquer landed property and the direction of agricultural production. But meanwhile, at the end of the seventeenth century, in the persistent climate of clerical obscurantism, a third of the agricultural land of Italy was still abandoned to ecclesiastical mortmain. The landscape—where it was not entirely waste—seemed disaggregated and ruined, like the one depicted with polemic strength and vigor by Magnasco in the painting we reproduce in Plate 55. It was a stormy landscape that, at the dawning of the "age of Enlightenment," vividly highlights the flight of a group of friars.

.VII.

THE AGE OF ENLIGHTENED DESPOTISM AND REFORMS

. 66 .

*The Landscape of the Eighteenth-Century Villa,
and the Italian Mode of Development of Capitalism
in the Countryside*

Among the Italian states, the Republic of Venice was undoubtedly the one that was best able to resist the interference of the church and foreign powers in the age of the Counter-Reformation. Thus in the seventeenth and eighteenth centuries, when the decline of commerce of the Serene Republic, somewhat later than that of Florence, produced even here an increased flow of capital into the countryside for investment in landed property and agricultural activities, these encountered relatively less resistance from the extension of ecclesiastical mortmain than was the case of other states of the peninsula.

But the predominance of feudal landed property still remained decisive in the territory of Venice during the seventeenth and first half of the eighteenth century, and this was concentrated in the hands of an aristocracy that preserved its economic and political power intact. Only in the second half of the eighteenth century, in fact, did initiatives appear more clearly in the Venetian countryside of that new middle-class group which Goethe found, for instance, at Verona on the mainland at the end of the eighteenth century to be "quite active and always involved in affairs." Meanwhile it was still the initiative of the aristocracy that at first assumed the most evident role in the renewed interest of the propertied classes for their landed property. Already in the seventeenth century, as we saw in chapter 58, no less than 332 large seigneurial villas were built in the Venetian countryside, a number almost equal to that of all the villas built in the five centuries between the eleventh and the sixteenth! But in the eighteenth century this record figure was again surpassed: no less than 403 new great villas arose in the Venetian countryside, and we could add to these another 137 that were built largely in the early nineteenth century when this development seems, however, to have weakened.

We are confronted here, clearly, with a process that assumed, already in the seventeenth but much more in the eighteenth century, not only an artistic but also an economic and productive significance, and that had from this point onward a profound influence on the systems and forms of the agricultural landscape. Up to the present day, in fact, the Venetian landscape remains strongly imprinted with the forms of the eighteenth-century villa, just as the Tuscan countryside was earlier imprinted by the forms of the Medicean villa. But much more frequently than what occurred with the Tuscan or Roman villa in the second half of the sixteenth century, the great seigneurial villas of the Veneto, from the second half of the seventeenth through the eighteenth and into the nineteenth century, were no longer places of idleness and diversion. Instead they became the centers of true and proper lordly agricultural businesses, where investments of capital went not only to luxurious buildings or elaborately intricate gardens, but also, and in increasing amounts, to true works of transformation and agricultural colonization, the development of uncultivated lands, plantation of useful trees and shrubs, hydraulic projects, and the implanting of new farms.

It was not that the Venetian aristocracy of the eighteenth century was about to abandon its placid festivities, which precisely in this century even still more impressed foreign observers, for the complex tasks of running agricultural enterprises. As in the past these were mostly entrusted to administrators or intermediaries, who from the actual administration of these enterprises were often able to draw substantial profits that made them able to participate in the new agricultural middle class. But it remains in fact that capital now flowed to seigneurial villas not only for the purposes of beautification and display, but also for the ends of true and proper productive investment, which in the eighteenth century often flowed from foreign commerce and trade toward landed property and agricultural activity.

This line of capitalist development of seigneurial estates had a particular importance in the dominions of Venice—where in the end it assumed a decisive importance in the capitalist development of agriculture—but found a broad application also in other regions of the peninsula. In Piedmont, in Lombardy, in Liguria, in Sicily, as in the Veneto, the second half of the seventeenth and the whole eighteenth century were a flourishing age of great seigneurial villas, which, while also serving the idleness and diversions of the urban aristocracy, began even here to assume a notable importance as centers of capitalist investment in the landed economy, and as centers of reorganization of the agricultural landscape into great seigneurial business enterprises.

The painting by Gaspare Van Wittel, preserved in the communal gallery of Prato, which we reproduce as Plate 56, provides a characteristic detail of this landscape of seigneurial estates as they appeared more and more frequently in the first half of the eighteenth century. The scene shows a lordly villa and close-by vineyard (which would eventually be rented out), local people working and storing up agricultural products,

Plate 56. A villa and seigneurial enterprise in the mid-eighteenth century, in *The Harvest* by Gaspare van Wittel.

and old and new farms, where now a greater number of peasant houses began to arise.

Even where—for instance, in Tuscany—a larger number of luxurious seigneurial villas had been built in the different economic and social circumstances of earlier centuries, it is not difficult to follow a similar development of function, from the end of the seventeenth through the eighteenth century, in the sense of a transformation from places of idleness, hunting, and diversion into centers of capitalist investment in the landed economy, and of reorganization of production and agricultural landscapes. Precisely Tuscany became the characteristic place in Italy of the *fattoria*, the center of the complex organization of the farms that made up a great seigneurial estate and were generally contiguous to the seigneurial villa. The great names of the Tuscan landed aristocracy, which still today dominate the agricultural economy of this region, replicate, almost without exception, those who were the owners of these same great lordly estates already at the end of the seventeenth century.

The capitalist evolution of the seigneurial estate was not, however, necessarily tied in the eighteenth century to the construction or transformation of the lordly residence. In the region of Naples, for instance, the eighteenth century was when, if anything, many feudal lords abandoned their castles or let them go to ruin, to pass the greater part of the year in the capital, or in the major cities. And here, often, or in the immediate neighborhood of the city, they built villas where they could pass the summer. But despite this, in many provinces of Naples, in Puglia for instance, there are not signs of the same capitalist evolution of seigneurial estates that we have found in other parts of the peninsula. Here, clearly, this evolution assumed different forms from those in the estates of the Venetian aristocracy and in Tuscan *fattorie*.

What made these different forms similar, however, was the gradual penetration of new relationships of capitalist production into older feudal relationships, which from this penetration were deeply changed in content and historical meaning, even if many of the old forms remained. We find ourselves confronted with the first steps of what Lenin properly called—comparing it with another, more developed one, which he called "American"—the "Prussian" mode of development of capitalism in agriculture. And it was not by chance that Lenin himself, elsewhere, also called the Prussian mode of development an "Italian" mode. For a large part of Italy, in fact, this mode of development clearly predominated in the eighteenth century and beyond. That is, far from overthrowing old relationships and old feudal arrangements, and thus permitting capitalist society to expose the land to the full energy of socially productive forces, new forms of exploitation and capitalist oppression were grafted onto the old trunk of feudal exploitation and oppression, so that the energy of productive forces remained stunted, and the economic, democratic, and cultural evolution of society itself was impeded.

From the first half of the eighteenth century onward, one can say that there is not one significant fact in the agricultural, economic, social, political, or cultural history of Italy that can be put into proper perspective without considering the particularities of this mode, into which the development of capitalism in the countryside was now launched. Meanwhile, to be sure, even a feudal aristocracy in process of becoming more bourgeois—which, in the Italian mode of development of capitalism in the countryside, was often the protagonist of new initiatives in agriculture—might be forced to assume positions that might seem progressive, in the jurisdictional struggle against the most hated ecclesiastical abuses, as well as in the physiocratic struggle against outdated feudal restrictions. But already in the age of enlightened despotism—whose policies this aristocracy-becoming-bourgeois clearly supported—the limits and contradictions of the Prussian mode of development appeared in Italy; and more were discovered through the new tasks that the French Revolution and then the national Risorgimento urgently posed to Italian society. But new ways had to be tried, and new classes would have to appear on the political and social scene, before these problems would be confronted, not to say resolved, in the nineteenth century. And right up to the present, as we have indicated, the influence marked from the first years of the eighteenth century by the Prussian (or Italian) mode of capitalist development in the countryside is unfortunately still visible in the landscape as well as in the agricultural relationships of a large part of Italy.

. 67 .

*The Landscape of Farms in the Po Valley,
and the Crisis of Sharecropping in the
Second Half of the Eighteenth Century*

In the Po valley, as in Tuscany and other parts of central Italy during the eighteenth century, a class of large and medium-sized renters gained increasing importance in the capitalist evolution of seigneurial estates, and they had a role of increasing importance in this evolution. Up to the middle of the century, however, even in the lands of the Po, as in other parts of the peninsula, where the communal movement had been most important, the prevalent form of cultivation on seigneurial estates remained that of sharecropping—or the *tipo colonico parziario* generally —which old and new proprietors, from the time of the freeing of the serfs, had most often adopted to keep their former serfs securely tied to a determined piece of land. As in Tuscany—in contrast to what generally occurred in Naples and Sicily, where the peasant was more rarely settled on a given plot—seigneurial estates in the Po valley were also generally divided into a certain number of farms, which (like the ancient *mansi* of the Carolingian period) corresponded in size to the work capacity of a peasant family, and they were also continuous entities for cultivation.

Thus with the seigneurial villa and its annexed buildings for preparing and preserving products, and with its pieces of rented-out land, the agricultural landscape of the Po was dominated substantially, as in Tuscany, by the forms of peasant farms, of which Plate 57, where we have reproduced the splendid fresco by Tiepolo in the Villa Valmarana on Monte Berico, provides us, precisely from the middle of the eighteenth century, an exact image of the Po valley's characteristic lines of trees, and a peasant house, before which the peasants are enjoying a brief rest from work.

In these conditions, the role of the larger and medium-sized renters reduced itself, at first—as is still true today in the Italian South—to being intermediaries between the peasants and the lord proprietors of the estate: for whom, in substance, they did no more than anticipate revenue, freeing him from the cares of administration and the uncertainty of the harvests. Already, to be sure, in this most ancient function, these larger and medium-sized renters played an important part in the capitalistic evolution of the seigneurial estate, where, in the Po valley

Plate 57. The farm and the landscape of the *piantata* in Tiepolo's *Rest of the Peasant*.

and Tuscany, the generalization of this type of rental already in the first half of the eighteenth century became one of the most significant symptoms. Through the intervention of these men, profit, and no longer revenue of a feudal type, began to be the determining factor in the whole undertaking and the regulator of the agricultural economy as a whole. Still, this regulator, even if it already weighed heavily on the peasant economy by imposing new burdens, acted only, so to speak, *post factus*, after the productive process had ended, in the division of the harvest and other burdens, rather than in the productive process itself, which continued to unfold largely according to ancient relationships and norms.

We will see further (in chapters 73 and 74) how, in the Po valley, this picture rapidly expanded, beginning in the second half of the eighteenth century, to the point of permitting larger and medium-sized renters to intervene much more decisively in the processes of agricultural production. It is certain, however, that throughout Italy, although to a varying extent, large and medium-sized renters of the old and new type made up the largest and most active part of the agricultural middle class, which, along with groups within the landowning aristocracy that were becoming more bourgeois, had a decisive role in the politics of the age of reforms.

What we now hasten to indicate, however, was the crisis of the stratum and system of sharecropping, which corresponded, in the second half of the eighteenth century, and thereafter, to this ascent of the stratum of larger and middle-sized renters. In the very capitalistic function of these last, in fact, an element of heightened pressure and increased exploitation of the peasants had already appeared and did not fail to make its presence felt wherever the large renter inserted himself as an intermediary between landlord and sharecropper. In the second half of the eighteenth century, the moments of rest, which Tiepolo so splendidly illustrated in his fresco, would become more and more brief for sharecroppers of the Po valley, as well as for those on Tuscan estates, subjected as they were to the double burden of ancient feudal exploitation, over which was now imposed a new form of capitalistic exploitation. Only in the Po valley, however, did this process lead to a deep crisis of the sharecropping system, a rapid proletarianization of the former sharecroppers, and the triumph of large capitalistic agricultural enterprises, which up to our own days remained the predominant and typical enterprises of this region.

Throughout central and northern Italy, in fact, there multiplied during the eighteenth century signs and cries of alarm for the impoverishment and increasing proletarianization of sharecroppers. But in the Po valley, this process took on, for reasons that we will have occasion to return, more precise forms, which the studies of Prato and Pugliese have

Share of the Harvest Due to the Proprietor

Year	Grain	Marraschi[a]	Rione[b]
1286	1/4	1/4	—
1628–71	2/5	1/3	1/2
1749–60	1/2	1/3	1/2
1809	1/2	2/5	2/3

[a] Winter wheat. [b] Un-husked rice.

put into clear relief.[1] One sees, for example, from the table, how between the thirteenth and the nineteenth century there was an increase, among sharecroppers of the region of Vercelli, in the share of the harvest that went to the proprietor. It was in the second half of the eighteenth century that the division of the part of the harvest made up by *risone* (unhusked rice), which was beginning to assume a dominant part in the agricultural economy of the region of Vercelli, worsened most rapidly for sharecroppers; and one understands easily through this development how the proletarianization of sharecroppers must have proceeded with a quicker rhythm, thus creating conditions ripe for the dominance of the new system of capitalist cultivation.

It was not only in the region of Vercelli, as we shall see, that the crisis of sharecropping and the triumph of the new system of cultivation were closely linked to the rapid spread of new crops and new agricultural systems. Throughout the Po valley, in fact, the affirmation of the new type of capitalist agricultural enterprise was accompanied by a true and proper agronomic revolution that, in the second half of the eighteenth and first half of the nineteenth century, brought a new and deep transformation to the agricultural landscape of this region.

[1] [Giuseppe Prato, *La vita economica in Piemonte a mezzo il secolo XVIII* (Turin: STN, 1908); Salvatore Pugliese, *Due secoli di vita agricola. Produzione e valore dei terreni, contratti agrari, salari e prezzi nel Vercellese nei secoli XVIII e XIX* (Turin: F. Bocca, 1908); Salvatore Pugliese, *Condizioni economiche e finanziarie della Lombardia nella prima metà del secolo XVIII* (Turin: F. Bocca, 1924).—Trans.]

· 68 ·

The Age of Reforms in Italy, and the Agricultural Landscape of Closed Fields in the Second Half of the Eighteenth Century

In the second half of the eighteenth century, the capitalist development of the seigneurial estate, along with the initiative of the new bourgeois strata toward development of a true and proper revolution of agricultural systems and methods for cultivating estates, solidified, as we have indicated, in forms that differed considerably from one region of peninsular Italy and the Islands to another. Still, in every part of Italy, there were signs in this period of certain common needs in the elaboration of forms of the agricultural landscape. And these were clearly related to the new importance that capital and capitalist investments assumed in agriculture and in the transformation of its systems.

We have already seen how, during the age of the barbarian invasions and throughout the Middle Ages—not without some influence from the laws and customs of the barbarians, but above all because of the general prevalence of pastoral-agricultural systems of temporary clearings—a system of open fields had finally become the dominant one even in Italy. Characteristic of this regime was a conception of property rights over the soil that was quite different from the traditional quiritarian ones. If over cultivated land and its harvest the right of the proprietor continued, in fact, to be recognized (however, with many limitations), this right seemed to decrease, so to speak, as soon as the harvest was finished, or when the land, in whatever way, was no longer the object of active cultivation. On stubble or on fallow and uncultivated land, the local population's rights of communal pasturage were exercised more or less freely, competing with that of the proprietor, who was thus denied the possibility of closing off his own lands, or making a preserve when these were uncultivated, or after the harvest was completed.

It is easy to understand how the system of open fields was closely connected to the prevalence of quite primitive agricultural systems that—like those of temporary clearings or fallow—did not introduce significant improvements or stable investments of capital into the landed economy. At this relatively retarded level of development of socially productive forces, an open field system was essential, lacking cultivation of artificial pastures, for the balance of forage in agricultural enterprises, and it assured at least a minimum of subsistence to the poorest populations of the countryside. These obstinately defended their right to lead their tiny flocks to pasture on uncultivated land, stubble, fallow, and even to cultivate precariously little pieces of open land on communal or feudal domains.

In contrast with what happened in the lands of central and eastern Europe—which were organized in the Middle Ages, at least from the Carolingian period, in the three-field system—open fields in Italy became, as we have seen, a decisive element in the degradation and disaggregation of the forms of the agricultural landscape. Thus in the age of the communes, and then in the Renaissance, the process of reelaboration of these forms, and of reorganization of the landscape itself, proceeded hand in hand with a vigorous reaction against the predominance of the system of open fields that can be followed through the communal statutes of this period. On lands closest to inhabited areas, where plantations of trees and bushes were starting to spread, and on lands used for pasture, the tradition of the old quiritarian property—with the right of the proprietor to enclose his own plots permanently as a preserve, and to exclude any outsider from pasturage or other communal use—tended to reaffirm itself.

We have seen how, in the age of the communes through the Renaissance, there was a varied development of this return to a system of closed fields in various parts of the peninsula and the Islands. It was greater in Italy of the communes (in the form of irregular fields or *campi a pigola*, the *piantata*, irrigated fields, and closed drained fields), but it was generally less in southern Italy and the Islands, where it often played itself out in the forms of the Mediterranean gardens, preserves, and pastures closed off by the barons. But even where, as in certain parts of Tuscany and the Po valley, the system of closed fields had appeared in the Renaissance, vast extents of territory, and certainly the greater part of peninsular Italy and the Islands, remained subject to the system of open fields, which still remained the predominant one.

Beginning in the second half of the sixteenth century, however, and even more in the seventeenth, the process of refeudalization, and the general return of pasturage and pastoral-agricultural systems of temporary clearings naturally revived this predominance, reimposing it even on lands that had been conquered by a system of closed fields in previous centuries. It was necessary to wait for the first decades of the eighteenth century for more numerous voices among jurists, writers on agricultural questions, practitioners, and statesmen to reappear who called for a limitation of communal usages and the right of proprietors to enclose their lands. This need, in fact, became, in the eighteenth century, along with the issues of mortmain and free trade in grain, one of the basic points of

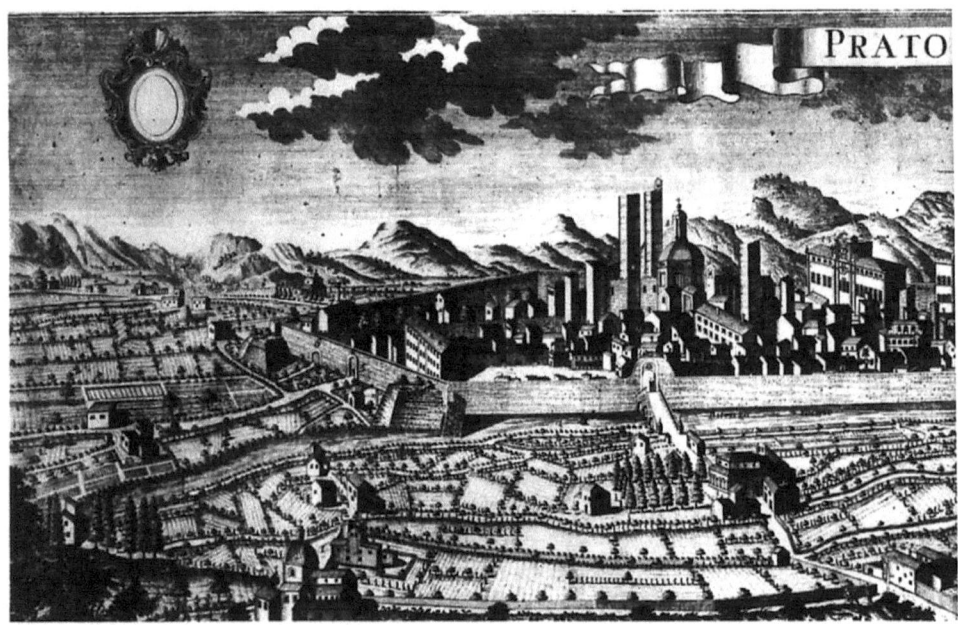

Plate 58. The expansion of the landscape of closed fields in an eighteenth-century view of Prato.

contact between the aspirations of the landed aristocracy that was becoming more bourgeois and the new middling agricultural bourgeoisie.

It is easy to understand, in fact, how the regime of open fields, which was compatible with a landed economy based on agricultural systems of temporary clearings and on insignificant and precarious investments of capital, should come to be considered an increasingly serious obstacle when plantations of trees and shrubs, drained and irrigated artificial meadows, works of hydraulic improvement in plains and hills, and stable investments in rural works of construction assumed an increasing importance, in lordly estates as well as in the affairs of enriched peasants or bourgeois renters. These now appeared with large sums of capital to transform the agricultural economy of entire regions. Particularly serious seemed the obstacle posed to agricultural progress by the system of open fields when there was a desire to implement new agricultural systems with continuous rotation, and the cultivation of artificial meadows. Full disposition of crops by the proprietor or renter strongly opposed, precisely, the traditions of communal usage and the laws that were generally in force.

It is not surprising, therefore, that the most intense class struggles and passionate polemics developed in this period around the rights of

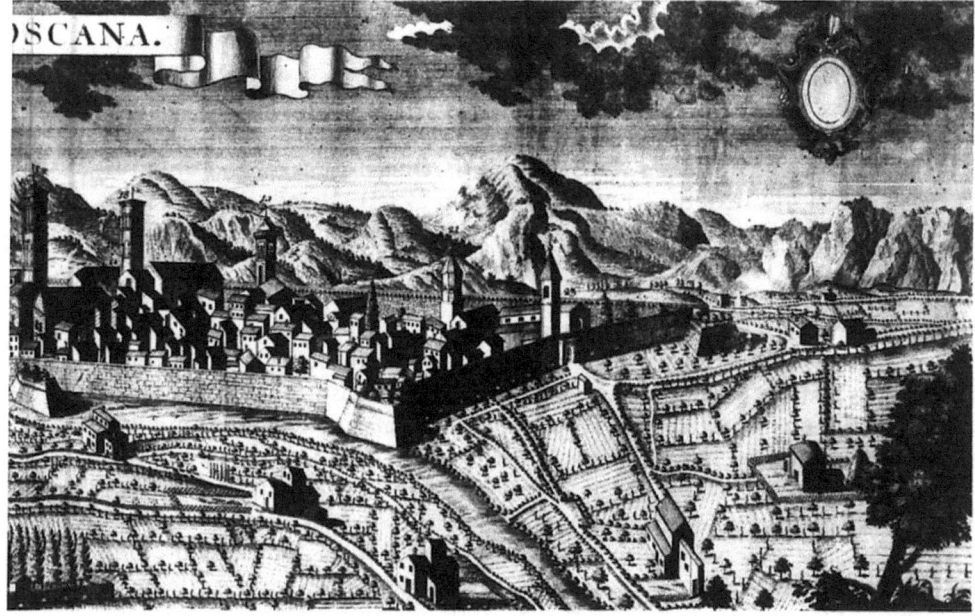

enclosures and preserves, in which the aristocracy that was becoming more bourgeois and the new middling agricultural bourgeoisie attacked not only the established interests of the more backward groups, and the great landholders with their transhumant flocks, but also the situation of the poorest populations in the countryside, who saw their most vital interests threatened. It is impossible to understand the nature of class under "enlightened despotism," or the historical significance of the "age of reforms" in Italy, if one does not keep before one this variety of interests that came together, or became opposed, around the problem of enclosures. This problem was only partially confronted and resolved, and the compromises and limitations involved were typical of Italian reformers of this period.

What contributed to defeating any uncertainty of "enlightened despotism" on this issue, however, in the last decades of the eighteenth century, was the evolution of prices and profits, that were once again becoming less disfavorable (after centuries) to cultivation of grain, as compared with large-scale wild breeding. Thus measures were taken in various parts of Italy to limit the rights of pasturage, to facilitate the enclosure of land, and to remove impediments to freedom of cultivation and rotation. I will mention here only the measures taken by Pietro Leopoldo in Tuscany, who restored the right of free cultivation and rotation for all types of land (in the years 1775 and 1780), and the right to cut down woods and make clearings (in the years 1776 and 1780); and those measures that abolished the rights of pasturage in the province of Pisa and the mountain of Pistoia (in 1776), and those that permitted redemption of pasturage rights in the Maremma of Siena, and even required enclosure of the lands redeemed (in 1778). Already in 1765 Venice proceeded likewise to limit the *pensionatico* and other pasturage rights in its possessions; in 1786, this measure was confirmed and extended, and rights of pasturage were entirely abolished in some cases, and limited in the future in others, to favor the enclosure of land. More tenacious was the resistance of the old regime of open fields in the Papal States and the South. In the region of Naples, it was not until 1792 that freedom was granted from rights of pasturage and enclosure of land was freely permitted; but this provision remained a dead letter, and one had to await the legislation of the Napoleonic period here that undermined the feudal system for the regime of closed fields to find juridical support.

It must otherwise be admitted that the progress of enclosure advanced, in the various Italian states, not so much through the effectiveness of legislation as through the strength of new capitalist interests and the impulse arising from them to transform agriculture. Thus, in the last decades of the eighteenth century, the system of closed fields—whose further spread and elaboration is shown concretely in a view of the suburbs of Prato reproduced in Plate 58—came to prevail in much of Tuscany and other parts of central Italy, where farms with mixed cultivation predominated, and in the Po valley. However (aside from the Alps and Appennines), a system of open fields continued to predominate in the Maremmas, in the Roman and Pontine Agros, and in most parts of the South and the Islands.

. 69 .

*Capitalism in the Countryside: Deforestation,
Clearings, and Erosion of the Mountainous Landscape
in the Age of Reforms*

"But the miserly peasant," sang a Venetian humanist agronomist, Lorenzi, in 1778, in a fine little poem with the significant title *Della coltivazione dei monti*,

> . . . ma l'avaro villano, e a farsi grande
> per nuovi spazi di campagna intento,
> sempre più desîose l'ali spande,
> sì come vela, che si spieghi al vento.
> Gl'intatti boschi assale; e da le ghiande
> scuote la quercia un dì cara a l'armento,
> cara al pastor che, su la terra ingombra
> mentre 'l gregge pascea, sedeva a l'ombra.
>
> Ivi l'erbe cresceano, util pastura
> agli agnelli, ai giovenchi, ai tardi buoi . . .

> (. . . but the miserly peasant, intent
> on expanding to new space in the countryside,
> with growing desire spreads his wings,
> like sails unfurled to the wind.
> He assails the untouched woods, and shakes
> the acorns from the oak, once dear to herds
> of the shepherd, who sat in the shade
> on the uncleared land, while his charges grazed.
>
> There grasses grew, a useful pasturage
> for sheep, heifers, and slow oxen . . .)

Now, concluded Lorenzi, after having lamented the disastrous consequences of unwise deforestation and clearings, now

> . . . più non resta al bue prato, né selva
> al gregge, o tana a più romita belva.
>
> Dunque o questo, o nessuno a tanti mali
> rimedio avanza, che a l'antico onore
> tornar de' paschi le pendici, quali

le mirò l'avol tuo fatto pastore.
Rivedi i lichi, e i più pendenti, e frali
di miglior suol rivesti, e tratti fuore
gli erranti sassi, al falciator prepara
ciò, che mal si vendemmia, e peggio s'ara.

(. . . there is no more pasture for the oxen, or woods
for the herds, or dens for the more solitary beasts.

Thus advance this, or no remedy for such ills:
return the old honor to the pastures on the slopes
your shepherd ancestors so admired.
Restore the lichens, and the slopes, and
reclothe them with better soil, and clear away
the fallen rocks; prepare for the scythe
what should not be harvested, and is worse for it.)

What might be given back to the woods, the meadows, the pastures on the mountain slopes, eroded by unplanned deforestation and clearings, was in short, already in the last decades of the eighteenth century, a problem that began to appear urgently to the eyes of the most informed agronomes. There did not lack, as well, in the various Italian states of the mid-eighteenth century, legislation devoted to protecting the forest patrimony, which deteriorated more and more rapidly, in the preserves around Naples: for instance, through laws of 1749, 1756, and 1762. But the very insistence of such measures shows their ineffectiveness, and this is confirmed by all contemporary sources, while the sources themselves reveal how effective, instead—to the end of a rapid expansion of deforestation and clearings—was the abandonment of the protective legislation in the vast forested zones of Tuscany that was decreed in 1776 and in 1780 through the reforms of Pietro Leopoldo.

Certainly the liberating provisions of Pietro Leopoldo responded much more exactly to the "spirit of the times"—or, more concretely, to the new needs imposed by the increasing penetration of capitalistic interests into the rural economy—than the laws protecting the forest patrimony that were enacted with little effect in other Italian states. From the mid-eighteenth century onward, to be sure, to the irresistible spread of deforestation and clearings was also added significantly the land hunger of the "miserly peasant" who, as Lorenzi wrote, became more "intent on expanding to new space in the countryside" and joined the attack himself to clear and cultivate "the untouched woods." But this was not only, or so much, a question as it had been in the past of the poorest part of the rural population clearing a little piece of woodland to seek to obtain the measure of grain necessary for direct consumption.

Population growth was not sufficient by itself to explain the magnitude of deforestation; the growth resulted instead from the new opportunities presented by the cultivation of grain. This now offered, in fact, more attractive profits both to the seigneurial estate and the better-off part of the peasant population, which was developing its economy in a commercial and capitalist direction, so that also from this point of view the search for capitalistic profit motivated and decisively regulated the pace of clearings.

This pace undoubtedly accelerated in the second half of the eighteenth century, but it still does not explain the full extent of deforestation, which far exceeded in this period that of clearings and new lands reduced to tillage. In its intervention in the process of agricultural production, in fact, capital—as we have seen with regard to the predominantly intermediary function of large and medium sized renters—still encountered no small resistance. Older, instead, was the effect of commercial and usurious capital on the exploitation of the forests. It was not a question here of a more decisive intervention into the processes of production. It was enough instead for commercial and usurious capital to intervene *post factum*, when the cycle of production was complete, to buy up the right to cut the woods of an indebted landlord, or a financially strained community. Clearly, in this way, deforestation could proceed with a rhythm independent of clearings, and much more rapidly.

Beginning in the mid-seventeenth century, in fact—a period when one can speak, if anything, of a reduction rather than extension of land under cultivation—there is evidence in many parts of Italy of an accelerated pace of deforestation, which shows the increased intervention of commercial and usurious capital in the exploitation of forest resources often in the past abandoned exclusively to local use. In the Veneto, for instance, whose forest patrimony has a historical and statistical documentation of particular richness and precision, we know that the annual levies of wood increased before 1662, and then began to decline. Similar figures can be found for other Italian states.

In no other field, one might say, did the intervention of capital into the landed economy produce negative effects so quickly and seriously as in this, and in deforestation well before unplanned clearings. In the eighteenth-century age of reform, and still more in the Napoleonic period and the first half of the nineteenth century, the spread of clearings worsened the consequences of deforestation, whose frantic pace was now set by capitalist profit, sadly accompanied by the desperate land hunger of impoverished peasants. What disasters deforestation and clearings on the steeper mountain slopes were preparing for many parts of Italy is shown with persuasive effectiveness by a late eighteenth-century popular print, reproduced as Plate 59. This, in the first panel,

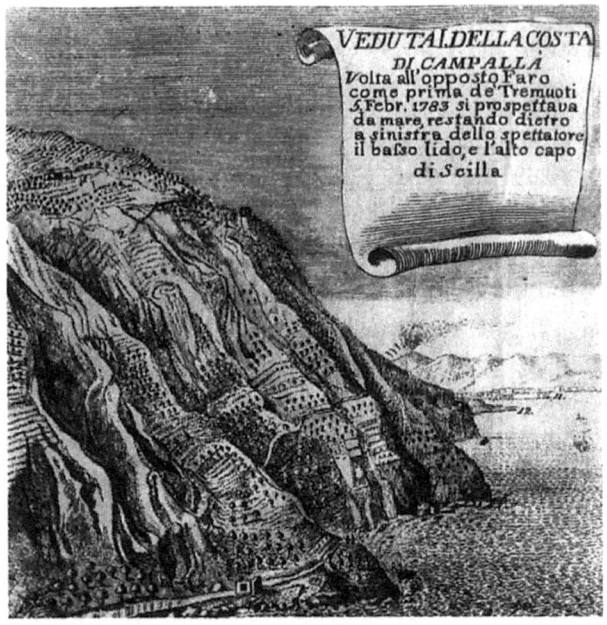

Plate 59. Deforestations, clearings, and erosion of the hillside and mountainous landscape in a popular print of the late eighteenth century: the Calabrian coast of the Straits of Messina deforested, reduced to cultivation, and eroded in an enormous landslide.

shows a cliff on the Calabrian side of the Straits of Messina, which was already deforested and under cultivation in 1782; and, in the second, its definitive erosion after the enormous landslide of 24 March 1790, which, seven years after the earthquake of 1783, provoked a true and proper tidal wave, "giving," as the legend of the print reads "fuel for the imagination of popular writers on Calabrian earthquakes."

· 70 ·

The Landscape of Landfills: Colmate di Piano *in Tuscany during the Second Half of the Eighteenth Century*

"With admirable labor," already about the middle of the eighteenth century, Spolverini, a Veronese humanist agronomist, wrote in his little poem *La coltivazione del riso,*

> con mirabil lavor natura cinse
> d'altissime foreste e boschi annosi
> (insuperabil siepe) i monti e l'alpi
> per difender i colti aperti piani.
>
> (with admirable labor nature girds
> the high forests and ancient woods
> —impassable hedges—of the mountains and Alps
> to defend the fruitful open plains.)

But now man, with deforestation and unplanned clearings that had brought clearing fires and plowshares to the tops of the highest slopes, had overturned the old order of nature, and unleashed

> ... diluvii d'acque
> a inondar le campagne, a render vane
> de' pii cultori e le speranze e l'opre.
>
> (... deluges of water
> to flood the countryside, and frustrate
> the hopes and labors of pious farmers.)

Consequently, Spolverini concluded, after giving a lively description of a disastrous flood of the Adige,

> l'arte rustica langue, ed osa a pena
> di commetter al suol gli usati semi,
> e le terre impiagar col ferro acuto,
> sol per giusto timor che d'anno in anno
> a rapirli non scenda o turbo, o fiume.
>
> (the rustic art declines, and hardly dares
> commit the usual seeds to the soil,
> or cut the earth with sharp steel,
> out of just fear that year by year
> they will be ravished by whirlwind or river.)

Already Spolverini called, as had Lorenzi, for a policy of reforestation of the mountains to protect against disasters provoked by deforestation and clearings on the slopes, and for building dikes in the valleys. And he did not omit the need for a true and proper policy of soil conservation directed against the unleashed torrents and rivers, which more and more dangerously continued, so that

> ad un tempo medesmo intere balze,
> e antichissime selve e rupi e sassi
> e dure zolle giù rotando e ghiaie
> con orribil fragor a poco a poco
> i monti a trasportar nel solco fondo.
>
> (all together whole cliffs,
> and ancient woods, and banks, and rocks,
> and sods overturned, and gravel,
> with horrible crashes little by little
> transport the mountains down the riverbed.)

And also—added Lorenzi with a different voice of his own, turning to the farmer—

> se i gran fiumi, che innalzando vanno
> l'antico letto con le torbid'onde
> te di paura impallidir non fanno
> vincendo omai le conosciute sponde,
> il tuo dolor ti mova, e 'l proprio danno:
> e mirando il color de l'acque bionde,
> pensa, che in preda a lor, benché non pare,
> vanno a second i tuoi poderi al mare.
>
> (if the great rivers, that raise their
> beds with turbulent waters
> don't make you pale with fear
> [by] overflowing their accustomed banks,
> your sorrow moves you, and your own loss:
> and looking on the color of the muddy water,
> think, that in their clutches, though it seems not,
> your farms are going to the sea.)

Neither Lorenzi nor Spolverini tells us, however, of the method of *colmate di piano* (landfills in the plain), which had nonetheless been adopted for centuries in various parts of Italy to utilize, for the improvement of low-lying, flooded, and marshy lands, the very impetus of these floods, which otherwise were the agents of further erosion of the soil in the mountains and valleys, and were so well described by Lorenzi when he tells us of the "muddy waters" taking the earth from farms to the sea.

The technique of landfills in the plain—which Leonardo da Vinci and then the Italian school of hydraulics had developed in theory and practice—consisted, as we know, of systematically diverting the muddy waters of the flood over low lands that were flooded or easily flooded, so that sedimentation of the earth carried by the water might make these progressively rise in level. Through hydraulic improvement, and this gradual process, low-lying lands could be protected from the danger of flooding and marshing-up, and even more, the stratum of alluvial soil produced in the landfill offered conditions particularly suited to agricultural cultivation.

Before pumping stations using thermal (and then electrical) energy permitted artificial settling of water on low-lying marshlands, *colmate* were practically the only method for hydraulic improvement of land, wherever the relief of the terrain did not permit sedimentation of water using drainage canals. This was the case, most often, in the lowlands of the Tuscan interior. And it was thus that in the eighteenth century and first half of the nineteenth century, Tuscany became the chosen place for improvement by means of *colmate*, which were given a new impulse by the reforming efforts of Pietro Leopoldo.

In the new initiative for improvement through *colmate*, which extended, for that matter, also to other parts of Italy, public and private interests were furthered in Tuscany by agents that were not motivated only by the peculiarities of the geographical terrain. One should not forget as well the particularly serious consequences of "refeudalization" in seventeenth- and eighteenth-century Tuscany, after the flourishing communal and Renaissance periods, for the general degradation of the agricultural landscape, and particularly for the spread of marshlands. It is enough to remember, without speaking of the Maremmas, that the marsh of Fucecchio expanded in this period through the Valdinievole, and arrived as far as Pescia; that the Chiana River overflowed from Arezzo as far as Cortona, and that marshes had expanded in the regions of Pisa and Volterra. We have already seen also how even, and precisely, the Leopoldine reforms contributed through the almost complete freedom given to deforestation and clearings, which further aggravated the hydraulic disorder of the hills. Thus, in the economic and social revival of the late eighteenth century, the problem of improvement in the plain, and the reclamation from marshes of land that had once been under cultivation, was presented here all the more urgently.

From the beginning of the eighteenth century, otherwise, capital, which had flowed with the decline of manufacturing and trade from the cities toward the countryside, began to be employed in works of improvement, despite the resistance of feudal elements and ecclesiastics, who when eventually obliged to contribute to the expense of these works saw taxation as a threat to land held in mortmain. Thus already

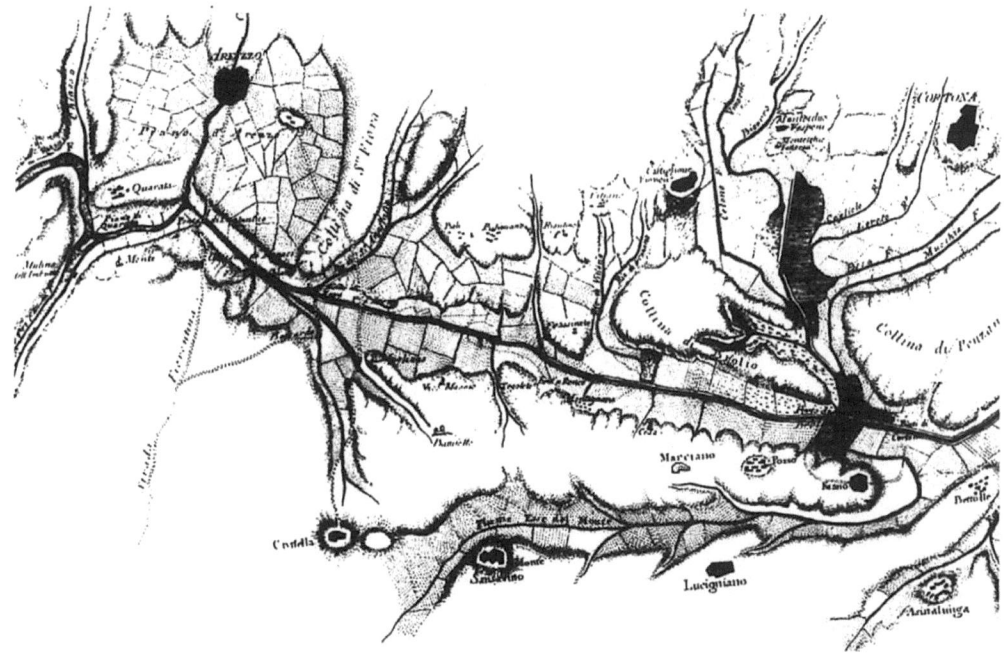

Plate 60. Eighteenth-century improvements with *colmate* in the Valdichiana in a map from the *Memorie idraulico-storiche* by Fossombroni.

in 1736 improvement in the Valdichiana, begun in 1702, had restored to regular cultivation no less than 46,128 stiora of land (about 2,500 hectares) along the main channel of the lower Chiana; and in the mid-eighteenth century there is frequent mention in contemporary sources of marshy lands recently improved around Pisa, Volterra, and so on.

It is not easy to provide precise statistics for the total extent of lands restored to cultivation through hydraulic improvement in the Leopoldine period, which extended successively from the Valdichiana to the Valdinievole, to the marshes of Fucecchio and Bientina, and to the Maremmas. It is true that some of the Leopoldine projects were destined to partial failure for technical, economic, and social reasons. But the fact remains that their cumulative effect was to return almost the whole Valdichiana to cultivation, as well as thousands of hectares in other zones. This was a success of undoubted importance for that period, and one that promised further and more important results in the decades that followed.

An image of the new forms that improvement with *colmate* induced into the landscape of the valleys of Tuscany in the eighteenth century is provided by Plate 60, taken from a volume by Fossombroni, one of the most famous Tuscan masters of the art of hydraulics in this period, that refers to the improvement of the Valdichiana, which Fossombroni himself superintended. One sees, in Fossombroni's fine map, the form of the great fields now filled with harvest, which still reflect the former basins of *colmate*. Some of these also, where the work of improvement was not yet completed, are interspersed, with darker hatching, among other cultivated fields.

· 71 ·

The Origins of the Contemporary Landscape: Systematization in the Hills in Banks and Terraces

On lands increasingly recovered by landfills from marshes and fens in the second half of the eighteenth century, even in Tuscany, the plantation of trees and shrubs, so characteristic of the agricultural landscape of this region, resumed only slowly. During the process of improvement, the lands gradually recovered for cultivation were at first generally used for pasturage or meadow; only when the improvement was completed did plowing begin of the great fields without trees that were destined for the production of grain and that generally reiterated the divisions of the former basins of the *colmate*. Meanwhile, the only trees were perhaps poplars planted along the edge of embankments and canals. Many years would pass before a regular division into farms, through the construction of peasant houses and the planting of trees and shrubs, would make the landscape of these recently improved lands blend, almost, into the surrounding countryside. This was already marked with the traditional forms of the *alberata* of Tuscany, Umbria, and the Marche, or with the forms of more or less ordered mixed cultivation where trees and shrubs had an important part.

In the hill country of Tuscany, and for that matter of all central Italy from the time of the communes and the Renaissance through the eighteenth century and thereafter, every act of clearing, improvement, and systematization was tied traditionally to the initiative of planting trees and shrubs; in any activity this was generally an essential part. Clearings on the hills and mountains, which spread in Tuscany and elsewhere in the age of the Leopoldine reforms, were accompanied by a new rapid spread of plantations of trees and shrubs in the hills of central Italy. The increase of population—which for all of Italy passed from 11 million in the first years of the eighteenth century to 16 million in 1770, and reached almost 20 million at the end of the century—was not accompanied in this period by a development of manufacturing comparable with that of England or France, and it thus obliged the new generations to seek their subsistence more than ever in agriculture. In a land like Italy, which was prevalently mountainous and hilly and where the monopoly of lordly and clerical lands remained almost absolute, a resort to clearings and plantations on even the highest slopes was often obligatory for new generations of peasants.

In these circumstances it is understandable that the problem of systematizations in the hills and mountains began to assume a greater and more general importance in the eighteenth century than it had previously. We have already indicated that, even with the agricultural flourishing of the communal period and the Renaissance, cultivation and plantation *a rittochino* (with plowing straight down the hill) had remained—despite the warnings of the best writers on agricultural subjects—absolutely predominant on hills throughout Italy; and that, as well, the only hydraulic systematization in the hills was generally what came from rudimentary ditches dug for the purpose of planting vines and olives.

There did not lack, to be sure—as was indicated in chapters 48–51, examples of more elaborate arrangements in hills and mountains—*a lunette, a gradoni, a terrazze, a ciglioni, a girapoggio* (with lunettes, gradations, terraces, banks, and plowing around the hill)—and even true and proper hillside works of construction were known and practiced from the time of the communes. We have seen, even, how systematization in banks began to appear in the landscape of the region around Lucca in the Renaissance period, and that works of construction for the valuable cultivation of citrus fruits began to mark the landscape of the coast and subalpine lakes. But these were still isolated instances, or restricted zones favored by particular environmental circumstances. Even in these, as well, the most elaborated arrangements for the hills were far from having reached a full development.

Only during the eighteenth century, and particularly in the second half of the century and during the nineteenth, did the need for such elaborate arrangements on hills and mountains become more general, so that their realization produced deeper changes in large parts of the Italian landscape. It was not a matter only of confronting the dangers, which were now clear, of serious erosion of arable soil in the hills. In the second half of the eighteenth century, the planting of new zones of trees and shrubs would often have been entirely impossible, or futile, on the steeper slopes that now often had to be cleared, if there had not been recourse to complex works of terracing, banks, or other methods, to make normal cultivation and harvesting possible by carving out little planted plots on the hilly or mountainous surface.

"And you, whoever you are," advised Lorenzi, revealing these new conditions, in his little poem that we have already cited on *La coltivazione dei monti*,

> e tu, chiunque sei, cui de' maggiori
> die' la fortuna i campi a i colli in cima,
> né più sceglier gli puoi pingui, o migliori

di que' che a gli avi tuoi toccaro in prima,
impara l'arte, e mostrala a' cultori.

(and you, whoever you are, whose fathers
left you the fortune of fields on tops of hills,
don't seek more fertile ones, or better
than what your elders had at first,
[but] learn the art, and show it to the farmers.)

And to "learn the art" of more elaborate systematization for hills and mountains, in truth, and to "show it to the farmers," the best Italian agronomists of the eighteenth century dedicated themselves, and not least Lorenzi himself, who was—like Landeschi, Lastri, Testaferrata, and other Tuscan masters of hillside improvement—not only a theorist, but also a famous improver himself in the Venetian hills. "His fathers left him on their death," Ippolito Pindemonte later wrote, remembering his work with emotion,

gli lasciar su la morte i padri suoi
balze dirotte, diroccate piagge

. . . .

ma poco andò, che mobili ondeggiaro
campi dorati, in delicati clivi
s'addolcir le più aspre erte, e impararo
a per novelle vie correre i rivi:
del rovo invece e dello spin, regnaro
gelsi babilonesi e greci ulivi,
s'apriro i lochi più riposti e cupi
e l'uva s'innostrò sovra le rupi.

(his fathers left him on their death
steep cliffs, derelict slopes

. . . .

but in a little time, [these] changed to
waving golden fields, delicate inclines
softened the harsh slopes, and the brooks
learned to run in new paths:
instead of brambles and thorns, reigned
Babylonian mulberry trees and Grecian olives,
the most hidden dark places opened up
and grapes ripened over the rocks.)

In fact, throughout the eighteenth century, the peasants of Tuscany and the Veneto, and the agronomists who elaborated and illuminated their experience, were in the forefront of all Italy for the spread and perfection of systematization in the hills and mountains. "He is wise," Lorenzi wrote further, already providing an elaborate theory for the management of surface and subterranean water in these arrangements,

> o saggio lui, che di frequenti mura
> quasi panche alternate il suol distingue!
> il declive s'allenta, e fa pianura;
> l'acqua più non depreda il terren pingue:
> passa l'umor secreto, e ne l'arsura
> cola, e la sete de le piante estingue:
> il sasso in fronte la difence, e poco
> temon di ria stagion pruina, o foco.

> (he is wise, who with successive walls
> almost benches, arranges the soil!
> the slope is tamed, and becomes a plain;
> water no longer devastates the rich soil:
> the secret humor passes through, and not dryness,
> and quenches the thirst of the plants:
> stones in front protect it, and [there] is
> little efflorescence, or runoff.)

Lorenzi also tells us how already in his time gunpowder, even, was commonly utilized in terracing the Venetian hills and mountains, and he advises using a plentiful work force in these hillside arrangements, which could be easily obtained at low wages during the winter months.

In the Veneto these practices were already common, as is confirmed by a variety of sources; and in other parts of peninsular Italy and the Islands as well, at the end of the eighteenth century, terraces—or, on the highest slopes, systematization with lunettes or grading—appeared more and more noticeably in the agricultural landscape. Thus in Liguria, the characteristic landscape of *fasce* (ridges along the sides of hills) began to affirm its predominance not only along the coast, but also in the valleys extending into the hinterland, as at Polcevera. In Lombardy, the elaborate landscape of works of construction continued to develop on the shores of the subalpine lakes. In Tuscany, as we shall see, the system of banks gained a particular importance in this period. Finally, in the South and the Islands, only with the erosion of the feudal system in the nineteenth century did terraces and grading assume the importance they still have today. In the late eighteenth century, by contrast, they were found in quite limited zones where true and proper works of construction were used for the cultivation of valuable citrus fruits.

The region, however, where systematization in the hills to break the slope had assumed a decisive importance in the agricultural landscape

Figure 13. Systematization *a ciglioni* (in banks) in the *Tableau de l'agriculture toscane* by Sismondi.

already at the end of the eighteenth century, was undoubtedly Tuscany. Even here, from the beginning of the century, there was fairly widespread systematization with terraces, grading, and lunettes with dry walls, which already in the middle of the century were diffused a bit everywhere in the Tuscan hills, and were adopted not only for the most valuable crops, like olives and vines, but even for chestnut trees. From this time, an agricultural landscape with terraces and dry walls became characteristic of the hills of Florence and of Chianti. But more than terracing what became characteristic of the hillside landscape of Tuscany in this period was arrangement in banks, that from the region of Lucca spread rapidly around Pescia, San Miniato, the Val d'Elsa, and a bit everywhere on the typical Tuscan drumlin-like hills with silicic or tuffaceous elements, and especially on pilocene yellow sands, or wherever rainfall permitted the formation of grassy embankments.

Figure 13—taken from the original edition of the *Tableau de l'agriculture toscane* by Sismondi—provides an image of the forms that this landscape and systematization in banks took in Tuscany at the end of the eighteenth century. The example of agronomists and improvers such as Landeschi, Lastri, and others of their school, who also elaborated the first elements of the new arrangement of hills graded to a single slope, contributed to the spread and perfecting of these forms in Tuscany and elsewhere. Outside of Tuscany, arrangement in banks spread rather slowly at first; still at the beginning of the nineteenth century, a great master of agronomy, Filippo Re, traveled through large tracts of the central and northern Appennines without finding a trace of them.

. 72 .

Hillsides Plowed a Tagliapoggio *in the Second Half of the Eighteenth Century*

We have already indicated the part that Landeschi, along with Lastri and other Tuscan agronomists and improvers of his school, had in the diffusion and perfection of hillside systematization in banks, and in the elaboration of a theory adapted to the needs of effective management of surface and subterranean water in such arrangements. This same theory inspired, also, the tenacious battle that Landeschi and his school subsequently fought against systematization and plantation that worked the land *a rittochino* (with plowing straight down the hill), which contributed so much to erosion of the soil in the hills, in favor of systematization of the land *a traverso* (crosswise to the slope). "It seems impossible," Chiarenti, a disciple of Landeschi, wrote decades later, "that as recently as the last century there was only one Tuscan who clearly understood [the necessity of systematization *a traverso*] and masterfully demonstrated it to the others."

It is certain, in actuality, that well before Landeschi, Pietro de' Crescenzi, and even Latin agronomists, had deplored the disastrous effects of working the land *a rittochino*. But, nonetheless, the examples and teachings of Landeschi and his school were the first to defeat this millennial tradition, and spread in Tuscany, and then throughout Italy, the example of systematization and plantations that worked the land *a traverso*, which—except for banks and terraces, which were a particular type for hills with multidirectional slopes—contributed so much to imprint the landscape of the Italian hills with the forms that characterize it still today.

If Landeschi should be considered the tenacious and effective propagandist, rather than the inventor, of systems for working the land crosswise, he and his school undoubtedly deserve the credit for having introduced and perfected important innovations in such arrangements, which made their application much more effective in all cases, and they certainly were the most frequent, where the profile of a hill had a double or multiple slopes. "Desiring to make these fruitful," Landeschi taught in this regard, "they must be divided into two or three parts, to arrange the line of banks and the cultivation on each one; keeping firm, in making such divisions, to the aim of depriving water of any ability to wash away soil."

Plate 61. Systematization A: *a rittochino*; B: *a cavalcapoggio*; C: *a tagliapoggio* on a hill with a united two-directional slope; or D: divided into *ciglioni* (banks) from a plate in the *Dizionario di agricoltura* by Gera.

We have already indicated, in chapter 52, how—even with systematization and working the land crosswise—arrangements *a cavalcapoggio* (plowing across the summit of the hill), which were already fairly widespread in the Renaissance period, were inconvenient when there were two lateral slopes in different directions, as well as when there was a downhill slope: since, if not along the downhill slope, at least along the two lateral slopes, working the land *a rittochino* was damaging with respect both to the difficulty of working the land and also to the poor management of runoff water.

With the innovation of Landeschi—which gave birth to a new type of hillside arrangement, called *a tagliapoggio*—these inconveniences of the *cavalcapoggio* system were attenuated or avoided by dividing the slope in as many sectors as were "contained between any two lateral secondary ridges extending down to the bottom" from the summit, so that within each sector the banks or drainage ditches would be approximately level and would join at the intersection with the next sector, thus avoiding the problems of hills with multidirectional slopes.

Plate 61, taken from Gera's *Nuovo dizionario di agricoltura*, provides at the letters A and B a theoretical schema of arrangements for working the land *a rittochino* and *a cavalcapoggio*. On plot C, however, we are shown how—through the innovations of Landeschi and his disciples—the two different lateral slopes are eliminated, which had still remained in the *cavalcapoggio* arrangement. The arrangement *a tagliapoggio*, with its characteristic division of the hillside into sectors having the same slopes, was applied by Landeschi and his disciples particularly to systematization in banks, and generally where there were multiple slopes. But not less important was the application of the same principles to hillside arrangements with a one-directional slope, and a system of encircling ditches that followed, in their paths, the line of the hill. This arrangement *a girapoggio* (around the hill) was particularly adapted to loam and clay slopes, were it was not possible or convenient to use horizontal ditches or rectangular plots, and its use spread, during the course of the nineteenth century, from the Crete of Siena and the Biancane of Volterra, where it was first practiced, to the Calanchi of Emilia, and a bit everywhere in Italy where this basic arrangement of the surface could provide a first line of defense against erosion from water and degradation of the arable soil: on at least two million hectares in the hills and mountains.

In Tuscany itself, as we will see, from the first decades of the nineteenth century, new and more elaborate types of hillside arrangements began to compete with those *a tagliapoggio* and *a girapoggio*, which Landeschi and his school had done so much to elaborate and spread. But these new types were inspired largely by the same theory and practice of managing water that was first elaborated by Landeschi, Lastri, and their school; and arrangements in banks, or *a tagliapoggio*, continued up to the present in Tuscany and other regions of the peninsula as characteristic forms of the Italian hillside landscape.

. VIII .

THE AGE OF THE RISORGIMENTO

. 73 .

The Po Valley Landscape of Irrigated Meadows, and Cultivation with Continuous Rotation in the Eighteenth and Nineteenth Centuries

Throughout Tuscany, and to a lesser extent in Umbria and the Marche, the age of reforms marked a period of great impetus for plantations of trees and shrubs, which expanded and thickened rapidly on the hills and plains. Already about 1760, Targioni-Tozzetti called attention to the fact that in his time rows of olive trees, for instance, were planted generally twenty braccia apart, while at the beginning of the eighteenth century they seemed to have been rather more thinly planted in the Tuscan agricultural landscape. De La Lande, another witness, calculated that not less than 100,000 new olive trees were planted in Tuscany in the period of reforms; and even more revealing were certain changes introduced into the agricultural landscape of central Italy through a rapid increase in plantings of vines, mulberries, and fruit trees.

Through improvements and systematization in the plain and hillsides, through the building of numerous peasant houses on old and new farms, but above all through the increasing density of plantations of trees and shrubs grown in mixed cultivation with other crops, it was in this period that large zones of the Tuscan agricultural landscape began to take on an appearance not much different from the current one. And in these plantations, in fact, were invested an important part of the capital that now flowed from the cities toward the countryside. We have already indicated, in chapter 64, how this Tuscan phenomenon was echoed to some extent in the Po valley. If in central Italy there was an expansion of the landscape of the *alberata* of Tuscany, Umbria, and the Marche through the eighteenth century, in the plains of northern Italy certainly not less noticeable was the progress of the *piantata* of the Po valley. One must clarify, however, that especially in the second half of the eighteenth and then first half of the nineteenth century, the lines of development and meanings of the two processes were profoundly different, and in the end they completely diverged.

One remembers meanwhile, to begin with—given the breadth of its fields and the much greater distance between rows of trees—that the *piantata* of the Po had much less density than the *alberata* of Tuscany, Umbria, and the Marche. And this fact had all the more importance in that, as we have seen, during the course of the eighteenth century there was a further development of the tendency in the *alberata* to increase the density of trees and shrubs, with rows planted closer together, whereas in the *piantata*, if anything, the tendency was the opposite.

For the same surface area, therefore, an extension of the *piantata* of the Po valley in the eighteenth century involved a deployment of trees and shrubs that was much less significant than was required for a corresponding extension of the *alberata* in Tuscany, Umbria, and the Marche. But one should also consider that when we speak of an "extension of the landscape of the *piantata* of the Po during the eighteenth century," we refer explicitly also, and above all, to the perfection of its forms through the elaboration of hydraulic arrangements and in the regularity and alignment of the trees. A comparison between a sixteenth- and an eighteenth-century map of the region around Mantova (in Plates 32 and 54) shows how in this zone, for instance, a landscape of meadows and fields, more or less irregularly and incompletely planted with trees along the margins of the fields, had been elaborated over the course of two centuries to assume the characteristic forms of the *piantata*. It would be easy to put together the documentation of an analogous development for other parts of the Po valley: and it is clear that, in cases like this one, an extension of the planted landscape involved essentially the development of hydraulic arrangements rather than further planting of trees, which were needed only to complement a few rows along the margins of the fields.

The extension of the landscape of the *piantata* of the Po during the eighteenth century did not always correspond as well to a similar increase in the surface area planted with both trees and vines. Often, instead, precisely the opposite was the case: while the *piantata*, with the employment of capital and labor that the elaboration of its forms involved, became the normal type for sowable land with trees in the Po valley, other lands that had been sowed but more or less irregularly planted with trees and vines, or that had been made into specialized vineyards, were now made into irrigated meadows or rice fields, so that the cumulative surface with trees and bushes diminished rather than increased.

One might consider, for example, in the following table (taken from a study by Pugliese) the situation in the region of Cremona. As one sees, the surface in vineyards notably diminished between the sixteenth and eighteenth centuries, while an increasing importance was assumed by the extension of irrigated meadows and pastures, in which the region of Cremona came to specialize. As in the region of Lodi, where between the sixteenth and eighteenth centuries there was a similar evolution, we are confronted with one of the most characteristic consequences of the

Proportions of Types of Cultivation in the Region of Cremona
(percentages of the total territory)

	Sixteenth Century	Eighteenth Century
Sowable land (with trees)	29.5	19.1
Sowable and irrigated meadows; rotated and dry pastures	20.9	34.3
Total sowable land (without trees)	50.4	53.4
Rice fields	—	1.9
Vineyards and fields with vines	42.4	36.3

penetration of capitalistic and commercial relationships into agriculture, which expressed itself in a regional specialization of cultivation and in investment of capital into works of irrigation and land improvement.

One might consider also, in this second table, the types of cultivation around Novara. Here, as one sees, the percentage of the surface under cultivation increased considerably: ample extents of woods, pastures, and heath (of which the percentage decreased from 36.3 to 28.8 percent) were cleared between the sixteenth and eighteenth century. But even if the percentage of sowable land without trees decreased from 52.4 to 50.1 percent, that of vineyards and fields with vines increased only a little, from 6.6 to 7.2 percent. Even here, the fundamental fact that emerges from the cadastral data is a notable regional specialization of cultivation, and an increase of cultivation of rice, which occupied an insignificant terrain in the sixteenth century but in the eighteenth century occupied 8.9 percent of the total surface.

Proportions of Types of Cultivation in the Region of Novara
(in percentages of the surface)

	Sixteenth Century	Eighteenth Century
Sowable land (with trees)	39.4	38.2
Sowable and irrigated meadows; rotated and dry pastures	13.0	11.9
Total sowable land (without trees)	52.4	50.1
Rice fields	—	8.9
Vineyards and fields with vines	6.6	7.2
Woods	18.9	13.7
Pastures and heaths	17.4	15.1

It is true that the rice fields of Novara and Vercelli, like the meadows of Cremona and Lodi, were for the most part wide long fields, next to irrigation channels edged with poplars, mulberry trees, and so on, which repeated for travelers certain of the most characteristic forms of the planted region of Emilia or the Veneto. Even here, with the prospect of a sea with waves made up of the tops of trees that reminded French travelers of the Norman or Breton *bocage*, there was the same preoccupation with wide fields and rows of trees that were distanced from one another, so that the crops would not be put into shade, and there was the same regularity in the network of irrigation channels, ditches, and embankments. But in zones like that of the Lombard or Piedmontese plain near the Po, where there was resurgent springwater, the landscape was dominated by a more minute network of drainage ditches and irrigation channels. With the progress of rice fields and irrigated meadows, as well, the cultivation of quality vines encountered increasing difficulty in the planted zones, and a progressive regional specialization of crops favored its abandonment. Thus the garlands of vines festooned among the regular lines of poplars, mulberry trees, and so on became more rare. The *piantata* preserved, even here to be sure, the old function of providing the rural population with faggots and wood to burn and build with—and more often than previously its mulberry trees offered peasants the new resource of raising silk worms—but in the new agricultural system, the role of the *piantata* in producing forage for animals and faggots and, above all, the importance it previously had (and in other parts of the Po Valley preserved) for the cultivation of vines was much reduced.

In large zones of the Po valley, thus, in the second half of the eighteenth century and first half of the nineteenth century, the progress in the diffusion of rice fields and irrigated meadows was the agent of a true and proper agricultural revolution that brought deep changes even to the pattern of the agricultural landscape. In this pattern, opening a new irrigation canal or using water from a new spring often had a quite unsettling effect. The need for irrigation, and the new agricultural system that this implied, as well as the routing of irrigation canals, now required adjustments in the regular forms of fields and even the boundaries of farms and properties, according to new rigid schemata that the new techniques themselves imposed.

One sees, in Figure 14—taken from the *Note relative all'agricoltura milanese*, published in 1784 in an appendix to the *Elementi di agricoltura* of Mitterpacher—precisely how these schemata were configured in the works of agronomists of the second half of the eighteenth century. Even today, it is easy to find traces in the Lombard and Piedmontese landscapes of the tenacity and candor with which, in the second half of the eighteenth century and first half of the nineteenth century, these

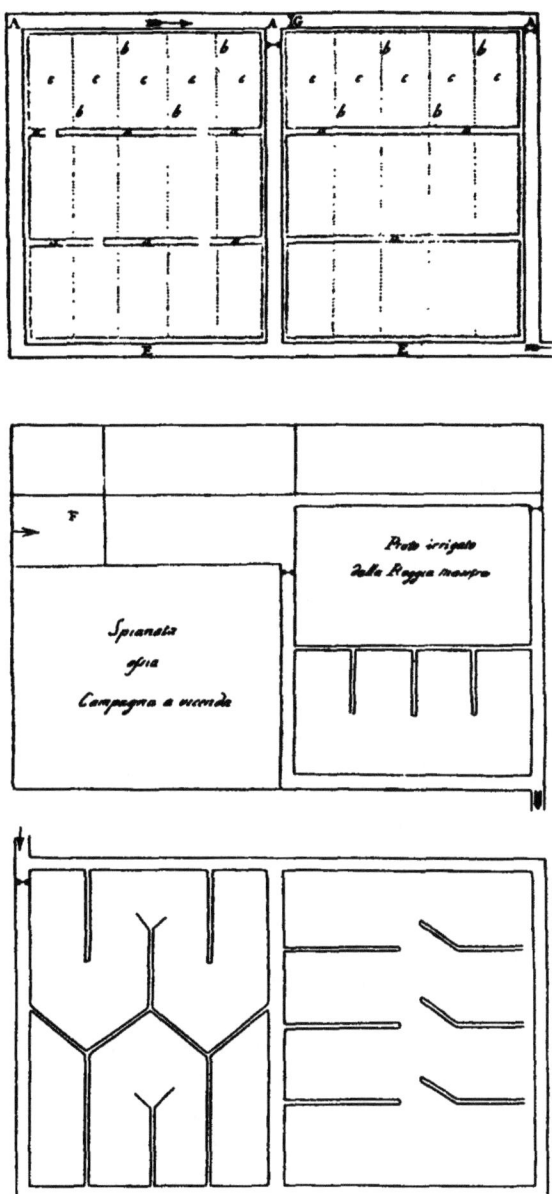

Figure 14. The geometric schemata of irrigation channels and fields in the landscape of the irrigated *cascina* (from the illustrations in *Note relative all'agricoltura milanese* of 1784).

schemata were translated into reality, overcoming every kind of technical, economic, juridical, and political obstacle, and overturning the traditional boundaries of feudal, clerical, and communal property. The agents of this process were middle-sized and larger renters, who commonly now no longer had only an intermediary role between great landed proprietors and sharecroppers in the Po valley, but instead acquired a decisive importance within agricultural enterprises, and in the rural economy of this region, through contributions of capital and initiative.

This phenomenon assumed, undoubtedly, an earlier and more general importance in the irrigated zones, but its importance spread, between the second half of the eighteenth century and first half of the nineteenth century, to a good part of the Po valley. In the nonirrigated zones, to be sure, the resistance of the old agricultural system of fallow—with its open field system, often, and the more precarious forms of landscape that this implied—was more tenacious. But even here the new systems of continuous rotation rapidly assumed a decisive predominance, through the introduction of forage, Indian corn, potatoes, and industrial plants in a regular crop cycle. And more and more widely these stamped the impression of their typical forms on the minute pattern of the agricultural landscape, making it more precise and less precarious than had been characteristic of the system of fallow.

Throughout the Po valley, it was not only the minute pattern of the agricultural landscape that was transformed and made more precise in this period, through the ordered network of irrigation channels, the alignment of plantations of trees, and the regular pattern of fields with rotating crops. On drained as well as irrigated land, the "agricultural revolution" of the triumph of agricultural systems of continuous rotation over the old system of fallow, brought, with the need for investment of large amounts of capital into agricultural undertakings, new functions and relationships, which were adapted to the new level of development that socially productive forces had reached in the agriculture of this zone. Estates that had embodied seigneurial property—with a traditional division into sharecropping farms—had to be reformulated and adapted to the new needs of capitalist enterprise, which imposed new forms both on the minute network of irrigation channels, plantations, and fields, and on the larger structures (one might say) of the agricultural landscape. There was a new distribution and configuration of the centers of enterprises, a new configuration of boundaries among enterprises (and also properties), and also a new dislocation of the whole rural habitat.

· 74 ·

The Landscape of the Po Valley: From the Sharecropping Farm to the Great Capitalistic Rented Holding

Between the sixteenth and eighteenth century, throughout Italy, the slow progress of science and agricultural techniques was tied basically to the progressive penetration of mercantile capitalistic relationships into the economy of the seigneurial estate. It is thus not surprising that the centers of radiation of this progress, and the centers of diffusion of the new literature and new agronomic techniques—Florence, Brescia, Padova, Venice—were in Tuscany and the dominions of Venice: in lands, that is, were the new capitalist techniques had begun to graft themselves most precociously and characteristically onto the old trunk of the seigneurial economy.

We have already indicated, however, the contradictions and limitations that the Italian mode of development of capitalism in the countryside involved, and these limits also strongly affected the rhythm of agricultural progress. The very tone of agronomic literature of the sixteenth through eighteenth centuries—directed as it was, essentially, to seigneurial groups—remained often aulic in form, and tied more to the humanistic tradition of Latin writers *de re rustica* (of rural things) than to the new conquests of experimental science and technology.

It was not by chance that a not unimportant part of this literature was written in verse up to the eighteenth century—and sometimes even in verse with artistic merit—according to a custom that began to decline, however, toward the end of the eighteenth century. In the first decades of the nineteenth century verse seemed bizarre for works on agronomy, which were addressed to a larger and more diverse public and were founded on the more modern results of the experimental sciences, and on recent conquests of technology.

The fact is that parts of the Po valley such as Lombardy and the Veneto, and then Piedmont and Emilia, clearly overtook Tuscany in the second half of the eighteenth century and first half of the nineteenth century in the Italian mode of the development of capitalism in the countryside. Thus, while Padova and Venice remained among the most important editorial centers of the new agronomic literature, the center of agricultural science and progress now clearly transferred itself from Florence to Milan and other towns of the Po valley.

This was not a transfer of an only geographical and spatial character. One sees, for example, at the end of the monumental twenty-six volume *Dizionario di agricolture* of Rozier, published by Crescini in Padova between 1817 and 1823, the list of associates to the publication of this work, whose translation permitted Italian agronomic culture to raise itself to the highest European scientific level. One should indicate immediately that need for greater access to European agronomic culture was felt above all in the provinces of the Po valley, from which, in fact, the overwhelming majority of subscribers came; many fewer were Tuscan, and almost none were Neapolitan. No less significant was the distribution of subscribers among different social strata. There were many names among the most famous houses of the landed aristocracy, which was becoming more bourgeois and had such a large part in the economic and political life of the Risorgimento in these regions, as it already had in the age of enlightened despotism and of Napoleon. This presence confirms that the capitalist development of agriculture was proceeding in an "Italian" direction even in the Po valley: that is, not through a revolutionary process, which might have liberated agricultural capitalism and its productive forces from the weight and encumbrance of feudal proprietary relationships, but by grafting new capitalist relationships onto the old seigneurial trunk.

In the Po valley, however, from the second half of the eighteenth century and even more in the first half of the nineteenth century, capitalistic investment in the landed economy, the crisis of the sharecropping economy, and the consequences of the "agricultural revolution" (which we discussed in chapters 67 and 73) imposed on the development of capitalism in the countryside—although still of an "Italian" type—a pace and energy that was much stronger than in Tuscany, for instance, and it reminds one more of what happened in the English countryside. The very limits of the "Italian" mode of development were exceeded. Agricultural capitalism subordinated to its own productive needs not only the size of enterprises and the layout of the boundaries that divided them, but often even the dimensions and boundaries of the more ancient properties; and following its own needs it remodeled the whole structure of rural society, promoting the rise of new antagonistic social classes that were different from the old ones.

Among these classes, the one that appeared most as the protagonist of the "agricultural revolution" in the Po valley—alongside the landed aristocracy that was becoming more bouregois—was the great and middle-sized capitalist renters, who assumed a decisive role in agricultural enterprises in these provinces. It was not by chance that exponents of this class were particularly numerous among the associates of Rozier's *Dizionario di agricoltura*. But their agricultural experience and needs soon found a more adequate expression in another monumental work, Gera's *Nuovo dizionario di agricoltura*, which was published by Antonelli in Venice between 1834 and 1845. This was no longer only a translation of foreign works; its twenty-six thick volumes of text and plates also contained valuable contributions by Italian practitioners and agronomists, which almost all referred to new agricultural systems and new productive relationships, as well as to the new forms of the agricultural landscape that were tending to affirm their predominance in the Po valley.

Among these forms, the one that sums up, one could say, in itself the results of the historical process we are examining was the *cascina*: the new center of enterprise of the great capitalist leasehold farm, which became central to the reorganization of the whole agricultural landscape of the Po valley. We indicated a characteristic landscape reflecting this new type of enterprise, one that had already begun to assume an increased economic and social importance in the second half of the eighteenth century, as shown in Figure 14; Plate 62 shows us how it is still today present in the reality of a rice-growing zone in Piedmont. The reader will not miss the importance of the changes that the *cascina* introduced into what we have called the larger structure, the macrostructure one might say, of the agricultural landscape. In increasingly large zones of the Po valley during the second half of the eighteenth and first half of the nineteenth century, it was no longer the seigneurial estate, with its traditional division into farms of sharecroppers, marked by the boundaries of these farms and punctuated by scattered peasant houses, that determined the layout of this larger structure. The minor unit of cultivation, the *podere*, which had been adapted to the labor force of one more-or-less-numerous family, was now subordinated to the major unity of the *cascina*, and was adapted to new technical and economic requirements of agriculture that were passing from the artisan to the manufacturing stage through important contributions of fixed and circulating capital. The *cascina* normally employed a wage labor force and, at the height of the season, a large number of day laborers.

As one can see, the whole structure of the agricultural landscape was dislocated by this development, which led, among other changes, to the concentration of masses of day laborers in rural settlements, from which they sought daily work in one establishment or another. The change in the macrostructure of the rural landscape and habitat, however, was only a reflection of not less important changes that were taking place in the social structure of the rural population itself. Among the mass of workers in the countryside, in the place of the sharecropper, the *colono parziario*, the *livellario*, the small renter, and the direct cultivator of the old seigneurial estates, the dominant figure now became the fixed-wage

Plate 62. The landscape of a rice field in the region of Vercelli in a painting by Enzo Gazzone.

worker, the *bracciante*, who was truly an agricultural proletarian dependent on the new capitalist enterprises. He was deprived not only of land and subsistence, but also of any means of exercising an independent role in production, and was thus obliged to sell his own labor power to the capitalist. In contrast to the situation of the old dependents of seigneurial estates—who were also without land but employed their own means of production and labor in little peasant family undertakings, the products of which (after paying the seigneurial rent) they could then take to market—the wage workers and *braccianti* were obliged to work for a capitalist agricultural enterprise whose products not they but only the capitalist could dispose; their labor became the only product for the market they still possessed.

We thus find ourselves confronted with the first affirmation of productive relationships characteristic of modern capitalism in the agriculture of the Po valley. The capitalist enterprise and the exploitation of a wage labor force were no longer (as they had been in previous periods) marginal or occasional facts, but became instead the fundamental norm of productive relationships. Requirements of space do not permit us to expand on the phases and forms of primitive accumulation, studied so deeply by Marx in the first volume of *Capital*, which—through the expropriation of the mass of direct producers and the concentration of the means of production in the hands of a class of capitalist entrepreneurs—is the historical presupposition of the process we are considering. It is enough to consider that already in the second half of the eighteenth century this presupposition created the crisis of the economy of sharecroppers that we discussed earlier in chapter 67. It is certainly true that the internal differentialization and impoverishment of sharecroppers, *coloni parziari*, *livellari*, and small renters provided the great mass of fixed-wage workers and *braccianti* who crowded into the large villages of the countryside to seek work. A significant contribution was also made to this labor market in certain zones of the Po by small independent owners from the hills and mountains as well, who were also expropriated, or forced in one way or another to supplement their meager incomes with wage work.

This worsening in the contractual situation of peasants with respect to previous centuries certainly had an important part in the proletarianization of large masses of direct producers in the eighteenth century. But no less important was the utilization of new productive means and circulating capital that—with the renewal of agricultural systems and consequent redimensioning of agricultural enterprises—were increasingly required in agricultural enterprises. Besides impoverishing the great mass of peasants, the rudimentary means of production and limited resources that had been sufficient in the past to cultivate a sharecropping farm were absolutely inadequate for the needs of the new entity of cultivation of the *cascina*, whose much larger dimensions were imposed by the new order, and where there was a much larger utilization of both fixed and circulating capital. It was thus not by chance that even the better-off groups of peasants contributed relatively little to the creation of the new class of large and middle-sized capitalist renters. More numerous, instead, were former managers or large renters who had previously had an intermediary function between the great seigneurial proprietors and the direct producers, and now assumed a much more decisive role in the agricultural enterprise itself. The new class of agricultural capitalists, merchants, tax collectors, and so on, who invested much capital in the new industry of agriculture, also made an important contribution.

With a more or less accelerated rhythm, but accompanied inevitably by an increased distress of the mass of the rural population, the processes briefly sketched out here developed in the different regions of the Po valley from the second half of the eighteenth through the nineteenth century: more quickly, generally, in the irrigated zones, where in the last decades of the eighteenth or first decades of the nineteenth century the large-scale rented estate became the dominant type of agricultural enterprise; but later instead, and often with a slower rhythm, in the dryer zones where the sharecropping farm was better able to persist. Throughout the Po valley, however, already in the period of the Risorgimento, the renewed pace of capitalist development of agriculture after the reforms of the revolutionary and Napoleonic period—while placing these provinces decisively at the forefront of agricultural progress—affected more and more deeply the larger structures of the agricultural landscape, and also dislocated the rural habitat.

. 75 .

Landfills in the Hills, and Arrangements a Prode *and* a Spina *in Tuscany in the Age of the Risorgimento*

In Tuscany, from the second half of the eighteenth to the second half of the nineteenth century, as in the provinces of the Po, first the Leopoldine reforms, and then those of the Napoleonic period, eliminated some of the obstacles that past feudal institutions and legislation had opposed to the development of capitalism in the countryside—through ecclesiastical mortmain, primogeniture, fideicommissa, and so on—which had favored a system of open fields over a large territory. Here also the overthrow of feudalism was not, to be sure, the result of a peasant revolution shaking the foundations of the old regime and destroying the great seigneurial estates. On the contrary, the "revolution from above" was carried out under the direction and close control of the old dominant classes, and particularly by groups in the landed aristocracy that were becoming more bourgeois and had a large part in the events of this period. Far from weakening the position of the great seigneurial estates, this further reinforced them, both qualitatively and quantitatively. On one hand, in fact, a large part of the land under ecclesiastical mortmain, which the reforms returned to circulation, became concentrated, or was again concentrated, in the hands of the old landed aristocracy, or of new large bourgeois owners; and the same can be said for a significant part of the land that the state recovered from fens and marshlands through improvements in this period. Even in a qualitative sense the position of the great seigneurial holdings was reinforced. And through the reforms of the Leopoldine and Napoleonic periods the greater part of these lands was liberated from *usi civici*, communal rights, that imposed the system of open fields, and many other bonds that limited the disposition of land or diminished the possibilities for cultivating it.

Thus more than ever, the way opened in Tuscany of the early nineteenth century for the Italian mode of development of capitalism in the countryside. Around the year 1830, one can estimate that out of 50,000 or so farms existing here, not less than 12,000 were grouped in 1,000 great *fattorie*; and even among scattered farms ownership was often concentrated in the hands of the landed aristocracy or of the new larger bourgeois owners. Contemporary sources are unanimous in revealing that precisely the development of the *fattoria*—the great seigneurial enterprise that grouped the administration of numerous farms—was one of the most important consequences of the reforms of the Leopoldine and Napoleonic periods, along with the increased importance assumed in these enterprises by investments of fixed capital (for rural construction and plantations) and circulating capital (*scorte vive e morte*—seed, animals, and equipment given to the peasants), which is amply documented by the bookkeeping of the period. Not less characteristic of the Italian mode of development of capitalism in the Tuscan countryside were the disturbing voices that arose, beginning in the first decades of the nineteenth century, to denounce the impoverishment of sharecroppers, among whom a more and more numerous and "socially dangerous" group of true agricultural proletarians increasingly split off: *pigionali*—agricultural wage workers—as they were called in Tuscany.

After a period of high prices of agricultural products in the Napoleonic period, a fall in prices—particularly for grain and wine, the chief products of Tuscan agriculture—followed in the second and third decades of the nineteenth century to aggravate the crisis of traditional sharecropping. And it was not by chance that in these years a debate developed in the Accademia dei Georgofili, and throughout the region, about ending or reforming the sharecropping system, which the most advanced exponents of the Tuscan capitalist ruling classes now denounced as an obstacle to the development of new socially productive forces in agriculture.

In some cases, in fact, here, as in the Po valley, there was a transition from the traditional system of managing seigneurial estates divided into sharecropping farms to a new system of management *in economia*, that is, rented out on a large scale. But in contrast to the provinces of the North, in Tuscany—and still more in Umbria, the Marche, and the Romagna—such cases were fairly isolated. There lacked here the renovating impetus that was fueled in the Po valley by the rapid development of education and decisive progress of new agricultural systems of continuous rotation: an impetus that broke the traditional pattern of seigneurial estates, overturning even the boundaries of old feudal properties and opening the way to a more rapid development of capitalist agriculture. In Tuscany instead, as in other regions of central Italy where the sharecropping system traditionally predominated, great holdings of feudal and seigneurial origin still preserved their predominance practically intact in the first half of the nineteenth century. In these regions the overthrow of feudalism and the evolution of agricultural systems reinforced, rather than shook, the "Italian" mode of capitalist development in the countryside.

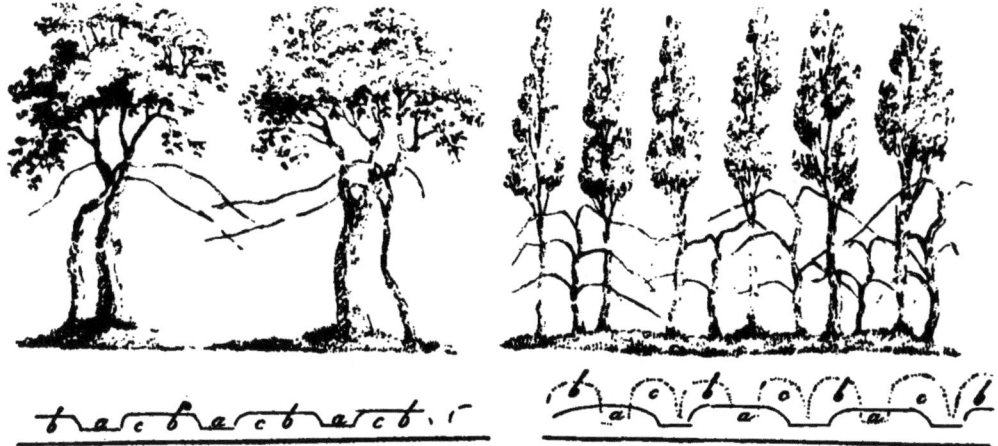

Figure 15. Rows of trees and arrangement in *porche* in the eighteenth-century *alberata* from an illustration in the *Tableau de l'agriculture toscane* by Sismondi.

In fact, even in Tuscany and other regions of central Italy, the spread of forage crops, Indian corn, and potatoes favored a certain progress of agricultural systems of continuous rotation. But blocked by the habits of mixed cultivation, which traditionally dominated the economy of the sharecropping farm, this progress could not take on the decisive rhythm that we have noticed in the provinces of the Po. In traditional seigneurial estates divided into farms, which continued, capitalist development and investment orientated themselves in the "Italian" way toward a further elaboration of traditional forms, like those of systematization in the hills and the *alberata*, rather than toward their revolutionary renewal.

In Tuscany and other regions of central Italy, in fact, the first half of the nineteenth century was when the forms of the typical *alberata* of Tuscany, Umbria, and the Marche were perfected and spread to new areas of the plain. Figure 15, taken from an illustration in Sismondi's *Tableau de l'agriculture toscane*, shows that at the end of the eighteenth century the rows of trees and arrangement of *porche* in the Tuscan *alberata* already assumed the characteristic forms that they preserve today in this region. For a good part of the central Italian plain, the normal arrangement now became that *a prode*—that is, with rather narrow rectangular fields divided into *porche* and trees festooned with vines in double rows along the sides. Already at the time of Sismondi the trees in these plots were becoming denser, often through the plantation of two rows of mulberry trees in the middle of the field; and from that time onward the width and length of the fields was generally reduced, with a consequent further development of drainage furrows between the hummocks, and with an increasing density of the cultivation of trees and shrubs.

In the plain, in short, the increasing importance of capital investment in seigneurial estates expressed itself in the first half of the nineteenth century in a further development of plantations of trees, in improved hydraulic works, in an extension of the landscape of the *alberata* over new areas, and in construction of new farms and peasant houses. One can calculate that in the great *fattorie* alone between 1830 and 1854, the number of *poderi* (farms) grew from 12,000 to 15,000.

Even in the zone of the hills increased capital investment was reflected, above all, in a further extension of traditional systematizations, like those with banks that grew in extent. In the first decades of the nineteenth century, agronomists such as Marchese Ridolfi complained that there was a true and proper "obsession with banks," erected often where geological and climatic conditions were not the most favorable for their full effectiveness. An estate agent of Ridolfi, named Testaferrata, and then Ridolfi himself later reacted usefully against this obsession, by elaborating and perfecting a new type of systematization in the hills that was more effective in the rational management of water and for the new possibilities that capitalistic agriculture now opened to the transformation of the agricultural landscape.

If the eighteenth century could be called the golden age of *colmate di piano* (landfills in the plain), the first half of the nineteenth century could well be considered the golden age of *colmate di monte* (landfills in the hills), of which the theory and practice went back to the time of Leonardo da Vinci. But only with improvements made by Testaferrata, and then Ridolfi, did improvements in the hills come to assume, in the new conditions created by the capitalist development of the seigneurial estate, an importance that spread well beyond the regional boundaries of Tuscany to have a deep effect on the forms of the Italian agricultural landscape.

"Don't let the useful teacher of rustic things," Averardo Genovesi, the predecessor of Carducci in the Liceo of San Miniato, wrote of Landeschi,

> . . . né fia che te di rustiche faccende
> util precettor lasci in oblio,
> se i colli, tua mercé, più non scoscende
> l'acqua con ruinoso mormorio;
> ma declinando placida discende
> da ciglio in ciglio con dolce pendio,

finché stretta in canali ai campi lassa
il tolto limo, li feconda e passa.

(... don't let the useful teacher
of rustic things be forgotten,
if the hills, your reward, no longer shed
water precipitously with a ruinous clamor;
but rather more peacefully descending
from bank to bank on a gradual slope,
until, restrained by canals, it gives its silt
to the fields, fertilizes them, and passes on.)

But as Marchese Cosimo Ridolfi revealed in his *Lezioni orali di agraria*, without denying the merits of Landeschi, Genovesi's tribute of praise should have gone instead to Testaferrata, the genial estate agent who thought up the branching arrangement *a spina*, which was then perfected by Ridolfi himself. In the systematization of the hills they had in mind, which was further developed and propagandized by Landeschi, the deficient management of water was the problem. One could not say, in arrangements already practiced at the time, that there was the effective utilization of the "silt," which Genovesi's ode mentions. Systems of *ciglionamento*, *tagliapoggio*, and *girapoggio* at most adapted cultivation to the nature of the hillside terrain; this had not become, as it did with landfills in the hills, a true and proper remodeling of the hillside surface.

"Make of these mountains and valleys," Ridolfi wrote, enunciating his program of systematizations in the hills that was adapted to the new spirit and new economic and technical possibilities of the capitalist seigneurial estate, "what an industrious man makes of a heap of clay. He models it into a vase, a head, a lion, and sometimes removes, or adds, sometimes cancels out, sometimes creates according to the dictates of his genius."

If one compares this statement of Ridolfi with the one of Landeschi, or to what Alamanni in the Renaissance had said about the "gentle spur," the "soft correction," the "careful composition," in which the whole art of agriculture consisted (see chapter 41 of this volume), one can understand the significance of the qualitative jump caused by the penetration of capitalistic relationships into the Tuscan seigneurial estate, even with regard to the elaboration of new forms of the agricultural landscape. With *colmate di monte*, in fact—and Plate 63, taken from Gera's *Nuovo dizionario di agricoltura*, illustrates the process and result—and also the branching arrangements *a spina*, there was no longer a concern only to impede the erosion of the hillside landscape through the disordered downhill runoff of water. As well, the very impetus of the

Plate 63. The operation of a hillside landfill (*colmata di monte*) in an illustration from the *Nuovo dizionario di agricoltura* by Gera.

water was now used fruitfully, not only to improve low and marshy lands, as in landfills in the plain, but also to remodel the precipitous hill profile, to depress peaks and outcroppings and fill up gullies with the turbid runoff from higher ground.

"Many think," Testaferrata used to say, "that it is useful to raise the level of the plains, but I alone have thought it equally useful to lower the hills and mountains." And it was truly to the merit of Marchese Ridolfi's estate agent to have not only enunciated the principle of *colmate di monte*, but also to have elaborated the best system of this improvement to the hills, which led to the branching arrangement *a spina*.

Constraints of space make it impossible to discuss further the complex mode of executing this type of hillside improvement. It is enough to say that important elements in *colmate di monte* were the dams placed in gullies parallel to the base dam, so that the lowest landfills became elongated and rose through reception of the silt that the water left above the dam. A skimmer (*sfioratore*) served to settle the silt from the water collected above the dam and kept it from going over the edge, while conducting ditches (*solchi portatori*) dug on the higher slopes carried rainwater laden with the maximum amount of soil from the peaks or outcroppings to be lowered or leveled. As for the branching arrangement *a spina*, this was the point of arrival of this hillside improvement. We can only say that it was an arrangement of the hillside with a one-directional slope, where the irrigation channels tended to be horizontal—aligned perpindicular to the greatest slope—and laid out in rectangular plots with acute (*spina chiusa*) or obtuse angles (*spina aperta*), depending on the concave or convex surface of the hillside.

The arrangement *a spina* was thus, in substance, a *girapoggio* with rectilinear fields. It spread in the early nineteenth century from the Fattoria of Meleto, which belonged to the Ridolfi in the province of Florence, through a zone delineated by Empoli, Montaione, and Castelfiorentino. It then jumped the regional boundaries more slowly to spread into the Marche and Emilia, and into other parts of Italy, where at present its preferred place is on clayey hills of moderate slope (15 to 25 degrees).

Among arrangements on hillsides, this can undoubtedly be considered the most elaborate, and also the one that corresponded best to the need to manage surface and subterranean water more effectively. Its relatively restricted diffusion, however, is an indication of the limitations the Italian mode of development imposed on the ability of capitalism to effect a deep and decisive reelaboration of forms in the natural landscape for the ends of agricultural development, of which only elimination of monopoly control of the land and a commitment to social planning might permit more decisive realization.

. 76 .

*The Overthrow of Feudalism in the South,
and the Agricultural Landscape of Open Fields
in the Age of the Risorgimento*

In the provinces of central and northern Italy, as we have indicated, beginning in the age of the communes, the evolution of agricultural systems had conditioned—although in the context of the larger entities of property and the seigneurial estate—the forms of stable peasant enterprises, which were most often a farm of sharecroppers. From Tuscany, Umbria, and the Marche to Lombardy, the Veneto, and Piedmont the legislation that overthrew feudal institutions, although diverse, was as we have seen finally effective: everywhere, however, it was confronted in the end with the massive reality of the stable peasant enterprise, from which not even capital could detach itself when it undertook the task of conquering the countryside and transforming agriculture. This involved, as in the provinces of the Po valley, a deep reorganization and even overthrow of the entire structure of the seigneurial estate divided into farms, to adapt it, in the new form of the great leasehold farm, to the productive and organizational needs of capital. Capitalism limited itself in other cases, as in Tuscany, Umbria, and the Marche, to penetrating the internal and external relationships of the multifarm unit, and subordinating its structure and the individual farms to its needs, although the farms preserved, from a formal point of view, their traditional structure intact. Whatever: the point of entry of the development of capitalism to overthrow feudalism in the countryside always remained the great estate and the seigneurial economy based on stable units of cultivation adapted to the work force of a peasant family.

Quite different, instead, from this point of view, was the situation in Naples, Sicily, Sardinia, and the coastal strip along the Tyrrhenian Sea in central Italy, from the Tuscan Maremma to the Roman and Pontine Agros. We have already indicated how here, in the Middle Ages, the weak impetus of communal society and the evolution of agricultural systems had generally conditioned not so much stability, but rather a prevalent precariousness of landed relationships and peasant enterprises, which were founded on widely diffused common rights of pasturage (*jus pascendi*) and sowing (*jus serendi*) on feudal and communal domains with a system of open fields.

It was not, to be sure, that even in these parts of Italy instances of less precarious landed relationships were entirely lacking. In the seigneurial form of property, such instances were most frequently found before the overthrow of feudalism on allodial lands (which the feudal lord held as private rather than as feudal property), or on preserves, which the baron abusively enlarged over the best lands of the feudal domain by closing them off and excluding them from the *usi civici*, common rights, of the local population. It was precisely on lands of this type that the capitalist evolution of the seigneurial estate was gradually affirmed already before the overthrow of feudalism, sometimes by planting trees and bushes on great preserves that were cultivated directly by day laborers, and sometimes by renting out the best preserves to large entrepreneurs who occupied themselves with the raising of large flocks combined with a precarious cultivation of grain.

In cases like these, however, the norms of landed relationships did not generally lead to the creation of stable peasant enterprises, but even excluded them through the prevalent utilization of day laborers, shepherds, or *terraticanti* (tenants for a rent in kind) on seigneurial landed preserves rented to large entrepreneurs. Less rare, by contrast, was the creation of stable peasant enterprises on allodial lands held with non-feudal titles by bourgeois or peasants, whose extent, nonetheless, except in a few regions, was limited in the zone we are considering; however, the preferred place for stable peasant enterprises remained the closed plots on which, thanks to the exercise of the *jus coloniae*, or long leases (*enfiteusi*), the local population was able to cultivate trees and shrubs on communal or feudal domains, or on ecclesiastical properties.

Thus on the eve of the overthrow of feudalism, most lands in the zone we are studying were feudal or communal domains, or ecclesiastical properties, on which—where forms of capitalist enterprise had not appeared through the employment of paid shepherds or day laborers—there was a net prevalence of precarious concessions of individual plots to a crowd of *terraticanti*, who worked them with primitive means of production and returned to the lord a certain part of the harvest. In contrast with what was now usual in the central and northern provinces of Italy, these precarious tenures were closely related to the persistent prevalence in the South and the Islands of agricultural systems of temporary clearings or fallow, and to a scarce diffusion of forage crops that—at this low level of development of socially productive forces—were also seriously impeded by climatic conditions. One can thus not generally speak of a stable peasant enterprise for this period and geographical region, except in the limited zones where there was allodial property, or where *Jus coloniae* and long leases permitted a spread of cultivation of trees and shrubs. The great mass of the rural population was generally made up of *terraticanti* who from year to year scratched a living from a myriad of tiny plots that were scattered variously according to the arrangement of fallow and pasturage on the immense extents of feudal, communal, and ecclesiastical latifundia.

The vitality of these precarious peasant economies—founded on a primitive cultivation of grain, which excluded forage from the crop cycle—depended necessarily on the prevalence of an open field system—a system that along with the right of pasturage exercised on fallow and stubble, and on communal and feudal domains, permitted the mass of precarious cultivators to integrate their meager agricultural output with small-scale family herding, and also, and particularly, to keep a minimum number of beasts of burden, which were indispensable for their agricultural pursuits.

Already before the overthrow of feudalism, however, this precarious equilibrium of peasant enterprises was seriously threatened by various historical agents. These included centuries-old exploitation by the barons, who tended to exclude their feudal lands from *usi civici* and reduce them abusively to preserves that were extended even over communal domains whose use and control they more and more frequently usurped. With the development of seigneurial enterprises of a commercial and capitalist type in the second half of the eighteenth century, the tendency to usurp and enclose communal and feudal domains notably increased, and the middle stratum of peasants, which on other occasions had appeared (and occasionally, although rarely, still appeared) at the head of peasant movements to win back rights usurped by the barons, was now more likely to push in the same capitalist direction.

We have already indicated in chapter 68 how during the course of the eighteenth century it was precisely the effort to enclose lands and liquidate promiscuous property rights that provided a fundamental unifying point between the reformist aspirations of the landed aristocracy, which was becoming more bourgeois, and the new middle stratum of the agricultural bourgeoisie. In the Kingdom of Naples (beginning with the first law of Joseph Bonaparte, in 1806), with a certain delay in Sicily (where the law of 1812 was not enforced, in reality, until after 1825, and particularly after 1838), and in Sardinia (in 1835), the legislation that overturned the feudal system intersected these conditions, and threatened to overturn the already precarious balance of the peasant economy. In the Kingom of Naples particularly, the legislation involved two distinct operations with regard to the system of landed property holding, which are worth exploring briefly to see their significance and results with regard to the ends that most concern us.

The first phase of the operation of suppression—which was realized with relative rapidity—consisted of the so-called mass division through which promiscuous rights, which local populations and the barons claimed on feudal domains, were abolished, and local communities were assigned a part of the domains themselves (which varied from one-fourth to three-fourths according to the importance of *usi civici*, common rights for wood gathering, pasturage, sowing, and so on, which had previously been exercised on these lands). The remainder was given to the ex-fief-holder in full possession, and he was also freed from any burden or *uso civico*. Even beyond this mass division of feudal domains, the laws decreed all lands in the kingdom free from any right of pasturage to which they may have been subjected, and to enjoy this freedom it was sufficient to enclose one's lands as a preserve.

The mass division of feudal domains, as one sees, far from favoring the creation of stable peasant enterprises, still further aggravated their traditional precariousness. Only for *colonie perpetue* (lands held with a permanent right of settlement) that were established on domains on the basis of the ancient *jus coloniae* was there recognition of a cultivator's right to remain on the land after payment of a quitrent to the ex-fief-holder. Beside these cases, no right of the cultivator was recognized on that part of demesne lands on which he had once exercised the *usi civici* of sowing, pasturage, wood gathering, and so on, which were assigned in full right to the ex-baron. The loss of these *usi*—whose cessation was, by itself, destined to overturn the already precarious equilibrium of the peasant economy—was certainly not compensated by assignment of the remaining part of the feudal domain to the local community through a division that all sources recognize was absolutely disproportionate to the real economic importance of the *usi civici* that the local population had previously exercised on feudal lands.

This was, really, rather than a division, a true and proper mass expropriation of the rights of those who farmed the fiefs. The mass division and assignment to the communities of at least a part of the old feudal domains was to have been followed, in the second phase of the acts of suppression, by an "allotment" of the domains themselves. But it was not by chance that this second phase of operations proceeded through the whole first half of the nineteenth century with extreme slowness. The old fief-holders, but above all the new landed bourgeoisie that had taken over administration of the fiefs, took advantage of it to usurp shamelessly the best lands assigned to the communities through the mass division. Thus were formed, as is known, many of the most famous landed patrimonies of the southern bourgeoisie. But even where the allotments were realized, the tiny plots assigned to direct cultivators became concentrated rapidly, in the majority of cases, in the hands of the ex-barons or, more often, in the hands of the new bourgeoisie, which also increased its territorial patrimony through the purchase of formerly ecclesiastical or feudal lands. Dumped onto the market, these lands could be purchased in extremely favorable conditions, and at a price that sometimes reached, even, only an index of 100 for every 30 of revenue!

In some cases these conditions benefited the better-off groups of cultivators, who were differentiating themselves from the mass of the peasant population by enlarging their patrimonies through trade in agricultural products and usury and who extended their farms by employing wage workers. But for the mass of cultivators, whose economy was overturned by the abolition or limitation of *usi civici*, the cultivation of the little plots left to them became, often, an unresolvable problem. When they lacked even use of the necessary implements for cultivation, or even the pair of oxen that was indispensable for any work of clearing, it became impossible, especially now that they did not enjoy common rights of pasturage on demesne lands.

It is no wonder that, in these conditions, the laws meant to overthrow feudalism had at first more negative than positive effects on constructing stable peasant enterprises, so that, in a certain sense, the social disaggregation of local populations in the South became even worse. One should not underestimate, to be sure, the significance of the perspectives that these laws opened for the capitalistic evolution of the seigneurial estate, the form of which was now often confused with that of the new great bourgeois landed enterprises. Not less significant, on the other hand, was the stability now achieved by peasant enterprises on *colonie perpetue*, or on the farms of those peasant-capitalists who now rapidly differentiated themselves from the mass of cultivators. In the course of our study, we will have an occasion to expand further on the consequences of these processes and their impact on the forms of the southern agricultural landscape in the period from the last decades of the Risorgimento to the first decades of Italian unity. Meanwhile, what still continued to prevail was less a reorganization than a further degradation of the southern landscape, which found its culminating manifestation in the massive extension of ill-advised clearings. These were already much advanced in the Napoleonic period and were then further favored by the laws overthrowing feudalism, which left the field open to the penetration of commercial and capitalist relationships into the forest economy. One cannot say, generally, that this deterioration—and what, more seriously, resulted from the extension of ill-considered clearings in the mountains—corresponded to a reorganization of the agricultural landscape in new and superior forms. Aside from the progress in cultivating trees and shrubs in the more progressive seigneurial and bourgeois estates, the agricultural system that continued to predominate absolutely remained that of fallow. And despite some progress of enclosure, still toward the middle of the nineteenth century (as Bianchini noted) the dominant system remained the system of open fields.

Plate 64. The *Roman campagna* by Enrico Coleman.

While in the age of the Risorgimento the agricultural landscape in the provinces of the Center and North had already begun to be reelaborated in more precise and complex forms, in the South and Islands instead, except in the zones where trees and bushes were cultivated, the landscape remained the formless, or ill-defined, one of the preceding centuries. And this landscape of bare and open fids spread in the mid-nineteenth century to new and vast areas that were subjected to the effects of unwise deforestation and clearings in the mountains.

A separate discussion is needed for the evolution of the agricultural landscape in this period of certain zones of the region we are examining, where for centuries agricultural systems of temporary clearings had predominated (as in the Maremma, the Roman Agro, the Tavoliere of Puglia, and the Corsi of Calabria), or where clearing by burning (as in the Pontine Agro) was combined with free-range breeding and often with transhumance as well. In the Roman Agro particularly—whose desolate landscape of pasturage and open fields is shown suggestively on the eve of Italian unity in the painting by Enrico Coleman reproduced as Plate 64—one cannot say that there were important signs of change in the system of landed property or agriculture in the age of the Risorgimento. The traditional system of temporary clearings continued to prevail along with the system of open fields, even if one notices within these

Plate 65. The landscape of the Maremma in an engraving by Fattori.

systems a certain progress of clearings and cultivation of grain at the expense of pasturage, and here and there some progress in planting trees and shrubs close to the inhabited centers.

The same considerations are substantially valid for the landscape of ponds and fens in the Pontine Agro, where the work undertaken in the last decades of the eighteenth century to excavate the great canal that now crosses it—the Linea Pio—was still uncompleted. Here free-range breeding of buffaloes continued to dominate, along with the precarious cultivation of grain in areas burned out from the thicket and marshy vegetation. These were also found in the similar environment of the coastal southern plains of the Tyrrhenian and Ionian coasts.

Largely unchanged in this period was also the landscape of a large part of the Tuscan Maremma, which we can view through the engraving by Fattori reproduced as Plate 65. Here, however, improvements had been undertaken by Leopoldo II in 1828, but on the eve of unification they were completed on a little less than 10,000 hectares out of the 50,000 that remained to be drained. But the progress of clearings already seemed more rapid. In the Province of Grosseto alone, between 1828 and 1843, thousands and thousands of hectares of cleared land were put under cultivation, and no less than 453 new farm buildings were built in the open countryside. On these new lands, to be sure, sowable fields without trees continued to prevail; but the landscape of woods, thickets, and pastures of the agricultural system of temporary clearings—which nonetheless continued to dominate in the greater part of the Maremma—now began to cede to a less formless one marked by a regular alteration of fallow.

The same story can be told, to a certain extent, in the Tavoliere of Puglia. One cannot speak here, as one can in the Maremma, of important works of hydraulic improvement. But the law of 1806, by which Joseph Bonaparte suppressed the *Dogana della mena delle pecore*, had established that the so-called stable cultivated lands would remain "rented forever to the peasants or current cultivators" on payment of a quitrent. As well, the law of 1808, through which the lands of the Tavoliere were liberated from "seasonal rights"—that is, from the "custom of summer pasturage"—must have promoted even here a certain progress of clearings. In 1816, a commission for the purpose could reveal that—out of 220,000 hectares of land surveyed in the Tavoliere—more than 60,000 hectares had been cleared; and with this extension of territory where there was now a regular alteration of fallow, there was clearly a corresponding reduction in lands where the traditional system of temporary clearings still predominated. As a result of this new orientation of the economy of the Tavoliere, otherwise, the currents of transhumant pasturage, although still important, became increasingly less. In 1816 the total number of head of transhumant animals in the Tavoliere had fallen to 1,233,600 (as compared with 2 million in the period immediately before the law of 1806), and the sheep and goats fell to 886,000 (as compared with 1.2 million sheep alone in 1789).

With further development through the Risorgimento, until the new Italian law of 1865 came to accelerate its rhythm, the process begun in the early nineteenth century continued to develop in the Tavoliere. An important step was marked in 1838 by the foundation of the colony of San Cassano (now San Ferdinando). More generally, one can indicate already in this period the importance assumed in the Tavoliere as in other regions of the South and Islands by great capitalistic enterprises for production of grain founded on an alteration between grain and fallow, and these now also employed a growing mass of *braccianti*.

. IX .

ITALIAN UNIFICATION

· 77 ·

The Railroads in the Italian Agricultural Landscape in the Age of the Risorgimento and Italian Unification

In the age of the Risorgimento and Italian unification, and then up to our own days, a growing importance was assumed by a new agent that operated with great power to effect a basic reelaboration and broad geographical redistribution of the forms of the agricultural landscape. The beginnings of railway construction in Italy seemed at first to be of quite modest significance. The first railroad, between Naples and Portici, was inaugurated, as is well known, in 1839; the following year the Milan-Monza line was opened; and Plate 66, taken from a contemporary print, shows us, in 1846, how the construction of the first tracts of the Imperial Regia Strada Ferrata Privilegiata Ferdinandea Lombardo-Veneta, the line that linked Milan with Venice, began to invade the landscape in a form that is familiar today in every part of Italy. But still in 1861, one should not forget that only 2,136 kilometers of track were in operation in the peninsula, while another 1,933 kilometers were under construction, and another 1,355 were contracted or projected.

Up to the time of unification, in fact, the fragmentation of Italy into small states, for the most part under foreign domination, contained the development of the railroad net within relatively restricted limits. And precisely for this reason, debates about the problem of the railroads assumed an increasing importance in the pressure for unification of the Italian bourgeoisie. Thus already in 1870, from a little more than 2,100 kilometers in 1861, the length of track in operation rose to more than 6,000, to pass 8,700 in 1880, 13,000 in 1890, and reach nearly 16,000 in 1900.

But it was not only for its direct impact on the forms of the landscape that the development of the Italian railroad net assumed that decisive importance for the purposes of our investigation, to which we referred at the beginning of this chapter—even though, toward the end of the nineteenth century, the railroad and the railway station had became integral elements of the landscape in a good part of the peninsula. What is more important was the indirect effect of the railroads as an agent in the reelaboration and geographical redistribution of forms of the landscape. One could say that—after the invention and spread of new means for

Plate 66. The Imperial Regia Strada Ferrata Prinvilegiata Ferdinandea Lombardo-Veneta (Imperial Royal Lombard-Venetian Railroad), from a contemporary print.

utilizing draft animals in the last centuries of the Middle Ages, which multiplied their uses in the service of man and contributed much to the impetus of the communal period—no invention, like the railroad (or the steam engine generally), contributed as much to the development of productive forces and the transformation of the Italian agricultural landscape; and now the effects of this deep technological revolution were not absorbed gradually through the course of several centuries but concentrated within the space of a few decades.

The indirect result of the building of the railways made itself felt on the forms of the Italian agricultural landscape, and their geographical redistribution, above all through its decisive influence on the penetration of commercial and capitalistic relationships into Italian agriculture. This was all the more decisive in that the period of railroad construction coincided with that of political unification and the removal of internal customs barriers. Thus the two conditions essential for the formation of a national market in agricultural products occurred at the same time, and the consequent regional specialization of crops was now also regulated by the laws of competition and capitalist profit.

Without expanding on the other important consequences of the development of the railway network, we will limit ourselves to documenting its influence on the regional redistribution of crops (and thus on the forms of the agricultural landscape) with some statistical data. One can see, for instance, what happened for some forms of cultivation, like that of vines, olives, and grain, of which the geographical distribution was deeply affected. Using mean production in the years 1870–74 as an index of 100, in 1890–94 the production of wine in northern Italy fell to 93, that is a reduction of 7 percent. Wine production seems to have only slightly increased in central Italy (where the index barely reached 104), while it grew to 160 in the continental South, and to 142 in the Islands.

Still more evident was the increasing specialization of cultivation of trees and shrubs, and of olives in the South. In the past, two regions of central and northern Italy had produced olive oil in particular abundance: Liguria and Tuscany. But when the production of 1870–74 is set to 100, in 1890–94 the production of olive oil in these regions fell respectively to 25 and 66, while it rose to 111, for instance, in the southern region of Puglia.

The elimination of internal customs barriers, the decreasing cost of transport, and the influence exercised by development of the railway net to further the penetration of commercial and capitalist relationships into agriculture thus rapidly subordinated agriculture to the laws of competition and capitalist profit, obliging farmers to specialize production—which was now destined to a national market—with reference to the environmental conditions that were most fitted for each crop. While we see that the South specialized increasingly in the cultivation of trees and shrubs, grain production in this region decreased. When the production of grain in 1870–74 is set at 100, in 1890–94 it fell to 95 in the Islands, and to 57 in the continental South, while it rose to 123 in northern Italy, which thanks to an extension of cultivated land and also a rapid increase in production per unit, now became the granary of Italy.

But the development of the railway network (and also of steam navigation) now subordinated agricultural production and the redistribution of forms of the agricultural landscape not only to the laws of capitalist competition in the national market, but also in the European and world market. The Mont Cenis tunnel was opened in 1871, and the Saint Gotthard in 1882, which linked the Italian market directly to western and central Europe; and in only six years, between 1881 and 1887, the American tonnage charges for European ports fell by 35 percent! This entry of Italy into capitalist competition in agricultural products on the world market, while involving it in the agricultural crisis in the last decades of the century—powerfully accelerating the evolution of capitalistic concentration in the Italian countryside—also provoked a further national and regional specialization of crops, which translated itself into a reelaboration and geographical redistribution of the forms of the landscape. This happened, for instance, in the spread of the cultivation of citrus fruits in the South, and in the forms of the Mediterranean garden as well as those of the plantation. Italy, which still in 1871–74 exported only 24 percent of its citrus fruits, already exported 43 percent in 1884–87, and 53 percent in 1894–97!

Limitation of space does not permit us to expand further on the indirect influence that the construction of the railways exercised in this and other ways to effect a deeper penetration of commercial and capitalist relationships into the Italian countryside, and a consequent transformation of the agricultural landscape. It is enough to indicate, again, that this development also conditioned the rapid detachment of more and more branches of artisan and industrial activity (spinning and weaving, manufacture of agricultural tools, and so on) from the agricultural economy, and their emergence as autonomous activities of large-scale mechanized industry, which now began to appear in Italy as well. And with the rise of the first great and middle-sized industrial centers, the habitat of the rural population, along with the more general forms of the agricultural landscape, was also deeply affected and dislocated.

As our investigation continues, we will have occasion to return to some aspects of the decisive influence that the development of the railroads had on the reelaboration and geographical distribution of the forms of the agricultural landscape. But we cannot end this section without recalling the negative effect that the lowering of transport costs had—in the forest economy, which was now also dominated by the laws of profit—in the rapid dilapidation of the forest and woodland patrimony of united Italy, and the consequent catastrophic deterioration of the mountain landscape.

. 78 .

The Piantata *in the Dryer Zones of the Po Valley in the Age of the Risorgimento and Italian Unification*

We have already indicated, in chapters 64 and 73, how the development of works of irrigation since the Renaissance had begun to condition the evolution of the most ancient forms of plantations along the Po into two profoundly different types of agricultural landscape. We will return, in the following chapter, to the ulterior evolution of the irrigated landscape of the Po, which was now dominated by the new forms imposed on it by works of irrigation, and characterized—in addition to the progressive decline of the cultivation of vines—by the alignment of rows of trees along the small and local roads, the banks of canals, and the boundaries of the great irrigated fields. We can thus no longer speak here, in the traditional sense of the term, of a *piantata* involving rows of trees festooned with vines; and we will return to this type of landscape in due course, particularly with regard to the works of irrigation that characterized it and conditioned its forms.

Of no less interest during this period, however, was the evolution of the other type of landscape that was typical of the nonirrigated zones of the Po valley, and that—if one remains faithful to the traditional definition of the *piantata*—now also revealed a tendency to expand more rapidly and to assume a greater dynamism in its forms. It is true that here, in contrast to what happened in the irrigated parts of the Po, there was for the moment a less impetuous development of capitalism, which had not broken the traditional pattern of the sharecropping system. Sharecropping even continued to expand to some extent onto new land cultivated through clearings and the resumption of works of systematization and improvement. But the new employment of capital in these works, and in the functioning of sharecropping enterprises, not only conditioned an ulterior extensive expansion of the system, and of the planted landscape typical of it. It also promoted a noticeable intensive development that, despite the obstacles presented to a full development of new productive forces by this system and its traditional type of landscape, profoundly modified its forms, and its very content, which became more and more characteristic of the capitalist mode of production.

One might consider, for instance, in the expansion of the landscape of the *piantata*, a province like Ferrara, for which we have available reasonably accurate statistical information. In the last years of the eighteenth century, a commission appointed to revise the cadastre of tax assessments revealed that in a productive area of 150,000 hectares (from which were excluded as agriculturally unproductive the vast territories of the valleys of Comacchio, Codigoro, Mesola, and the Prepostura of Pomposa) no less than 58,000 comprised productive land that was not cultivated. Land regularly utilized in agriculture totaled an area of not more than 93,000 hectares, of which 35,000 comprised sowable fields without trees (*seminativi nudi*) and hay meadows (*prati falciabili*), and 58,000 comprised sowable fields with trees and vines (*seminativi alberati e vitati*). In short, a large part of the territory of the province—productive but uncultivated land, and valleys with marshes with reeds for fishing (*da canna e da pesca*)—was not yet reduced to regular cultivation at the end of the eighteenth century. But on the cultivated land, the landscape of the *piantata* clearly predominated over that of sowable fields or pastures without trees.

Twenty-five years later, the information provided by the cadastre of 1835 (which reflects in reality the situation as it was about 1825) shows the first results of the lively employment of capital in the work of clearing, which had been typical here as in much of Italy, in the Napoleonic period. The territory covered by the cadastre (in which were also counted 89,000 hectares of marshes with reeds for fishing) was 243,000 hectares, and the productive part seems to have increased only a little (from 150,000 to 154,000 hectares) with respect to the late eighteenth century. But much greater was the increase in the extent of regularly cultivated land: the spread of clearings and improvements in the hydraulic system had permitted, in twenty-five years, an extension of this land from 93,000 to 119,000 hectares. But the landscape of the *piantata* had not yet had time to increase on this newly cultivated land: while the area of sowable fields without trees, and pastures, had increased by 17,000 hectares, from 35,000 to 52,000, that of sowable fields with trees and vines had increased only by 9,000, from 58,000 to 67,000 hectares. On regularly cultivated land, in short, the landscape of the *piantata* was still prevalent, but—in this first rush of clearings—the rhythm of plantations did not equal that of clearings, so that this prevalence was proportionally attenuated in comparison with the late eighteenth century.

Fifty years later, when Scelsi published his *Statistica della provincia di Ferrara* in 1875, the total surface under regular cultivation in the region of Ferrara was already much increased, partly through improvements since 1825 that had reclaimed about 5,000 hectares of new land from the marshes with reeds for fishing. In total, the area under regular agricultural cultivation had increased from 93,000 hectares at the end of the eighteenth century, and 119,000 in 1825, to some 165,000. But this extension of cultivated land had been followed by a similar extension of plantations, which from 58,000 hectares at the end of the eighteenth century, and 67,000 in 1825, had increased to some 103,000 hectares. These were more or less densely planted with trees and vines. However, from 1825 onward, the area of sowable fields without trees, hay meadows, and rice fields had increased only a little, passing from 52,000 to 64,000 hectares.

About 1870, in short, in comparison with the late eighteenth century, the extent of the *piantata* had about doubled in the region of Ferrara, passing from 58,000 to some 103,000 hectares. And almost everywhere in the dryer provinces of the Po, if the first two decades of the nineteenth century had been a period of large-scale clearings that had rapidly extended the area of cultivated land, the following decades, almost up to the end of the century, marked besides a further progress of clearings and the beginning of great works of improvement, an age of increased employment of capital in the plantations of trees and shrubs, and in the work of systematizing farms—particularly in Emilia, the Veneto, and the region of Mantova—with a consequent rapid extension of the landscape of the *piantata*. At the beginning of the twentieth century, when further spread of plantations of trees and vines practically stopped, the information from a less certain statistical agricultural inquiry permits us to clarify the nature and orientation of this final development. Throughout northern Italy, land destined for mixed cultivation of woody and herbaceous plants, with a total area of 3,166,000 hectares, was a little less extensive, now, than the sowable fields without trees, which had expanded to 3,568,000 hectares.

But beside the fact that the balance between land with trees and without varied quite considerably from region to region (it was highest in the Veneto and Emilia, less in Piedmont, and the least in Lombardy), it no longer adequately expresses the true extent of the typical landscape planted with trees and vines, in that the picture now also contained mixed forms of cultivation with mulberries, and other trees that were now excluding vines from the irrigated zone, and from others as well. A better idea of the extent of the traditional landscape of plantations with trees and vines is provided, meanwhile, by information about the extent of mixed cultivation with vines, which particularly in zones of the plain was often practiced in its traditional form in these regions. In general, Emilia was in the lead with 831,000 hectares of mixed cultivation with vines, followed by the Veneto with 667,000. Piedmont with 230,000, Lombardy with 201,000, and Liguria with 46,000 hectares lagged behind. If we limit ourselves to the landscape of the plain, which was more typically planted with trees and vines along the Po, the importance of Emilia and the Veneto becomes still more evident, with an area of mixed cultivation of vines of 581,000 and 507,000 hectares respectively, in comparison with only 156,000 in Lombardy and 37,000 in Piedmont.

During the Risorgimento and Italian unification, in short, and in the nineteenth century generally, the characteristic landscape of the *piantata* of the Po with trees with vines enlarged rapidly in Emilia, the Veneto, and in smaller zones of Piedmont and Lombardy, such as in the zone south of Mantova, so as to become typical of this dryer part of the Po valley, whereas the cultivation of trees with vines, and the traditional landscape of the *piantata*, declined and diminished in the irrigated part of the Po valley, where it lost a large part of its importance.

Not less important, as well, between the Risorgimento and Italian unification was the internal dynamics of forms of the *piantata* in the dry part of the Po valley. We have already had the opportunity to describe how these forms developed between the sixteenth and eighteenth century, in chapters 64 and 73. But beginning just at the end of the eighteenth century, they entered a new phase of intense dynamism, and it is worth pausing for a moment on this development.

Beside the evolution we have just described of the *piantata* in the irrigated zone of the Po, in certain sectors of the dry zone there began to appear and spread, in the late eighteenth and early nineteenth centuries, a type of landscape that replicated certain forms of the traditional *piantata*, even if it was quite different in internal structure. We are referring here, particularly, to the spread of the cultivation of mulberry trees in the hills and the upper Lombard plain, which were already important centers of Italian silk in the first half of the nineteenth century. But even if, in this zone, the regular alignment of mulberry trees (and sometimes also of vines grown with them in fields destined for cultivation of grain) repeated certain typical forms of the traditional *piantata*, the different nature of the arable soil here made superfluous the drainage channels and systematic hydraulic works that were an essential constituent element of the older type of agricultural landscape. For the vines as well, for the same reason, it was not generally necessary here to implement a rigorous arrangement in festoons, which in the classic *piantata* hung from the trees that provided their living support.

From the point of view of extensive development, the affirmation of the landscape of this atypical *piantata*—its forms are shown in the drawing reproduced in Plate 67—was undoubtedly of notable importance in this period. But still more significant, if less visible at first sight, was the evolution of a differentiation in the form of the *piantata* in large zones of the dryer part of the Po, where this type of landscape had traditionally expanded.

Up to about the end of the eighteenth century in most of the dryer part of the Po, as we have seen, drainage ditches for water, and even

Plate 67. The landscape of the atypical *piantata* in *Countryside at Monte Orobio* by Cesare Breveglieri.

embankments (*cavedagne*) had been essential constitutive elements of the *piantata*, along with the rows of trees with vines. These had a subsidiary function in the runoff of water, but they also marked the turn of the plow at the heads of fields, and gave a space for carts. As for the arrangement of the land within the fields, this was temporary: for working it, the field was divided systematically time and again into strips, or hummocks, separated from one another by small irrigation and drainage furrows. The strips, depending on their greater or lesser width, took the name in different provinces of *pracioni, presoni, prace, quaderni, vanegge, porche*, and so on.

We have already had occasion to indicate (in chapter 36) some of the chief agronomic deficiencies of arrangement in *porche*, which were only attenuated by arrangements *a prace* and *a pracioni*. To the deficiencies already discussed one must add those derived from the progressive *scolmatura* of the field—that is, the concave profile it tended to acquire through the accumulation of earth and debris around the edges, along the surrounding ditches. With this, the water ended up flowing toward the center of the field, where it stagnated and damaged the crops, and it was necessary to drain it off through the excavation of a drainage ditch in the center.

With the progress of agricultural systems of continuous rotation, however, and with their prevalence over the traditional system of fallow—but above all with the spread of crops such as Indian corn and hemp—experience taught farmers of the Po, already before the end of the eighteenth century, that temporary extensive arrangements of the *porche* type were inadequate to the needs of the new agricultural system. Thus, already toward the end of the eighteenth century, Arthur Young, in his famous *Travels in France and Italy*, noted in the region of Modena, for instance, the spread of the practice of *baulatura* (convex grading of fields), which then in the first decades of the nineteenth century directed arrangements in the *piantata* of the Po toward permanent intensive forms that were much more elaborate and differentiated than the traditional temporary extensive ones had been.

Only with the French Revolution and the Napoleonic occupation, however, and with the movement of ideas and initiatives that this produced—and with the fiscal burdens of this period on rural holdings, which stimulated improvements that might lighten the tax burden on revenues—did the reelaboration of the traditional form of the *piantata*, even and precisely regarding its internal arrangement, generalize itself through an employment of capital and labor. This assumed truly imposing proportions through the nineteenth century. Experience, already mature in the late eighteenth century, was then elaborated by the modern school of agronomy, which in the early nineteenth century—from Filippo Re to Carlo Berti-Pichat—already founded its doctrines on the most recent discoveries in the physical, chemical, and biological sciences, and contributed in a decisive measure to the diffusion of rational agricultural techniques. Among these, a particular importance was assumed along the Po by a deeper working of the soil, and by turnover plowing. But in the heavy low-lying soil of the dry part of the Po Valley, crops like Indian corn and hemp, and then other industrial plants, gave high yields only when not only careful cultivation, abundant fertilizer, and rational rotation, but also a true mattress of earth, to a depth of thirty-five to forty centimeters, could absorb up to half the volume of rainwater, and with the proper preparation this also allowed for runoff of excess water. This end called for permanent and intensive arrangements—with various types of *baulatura* of fields longwise, crosswise, or in different directions—which now tended to generalize themselves in this region.

With plow and wheelbarrow, thus, through the nineteenth century and beyond, and through a movement of earth that involved hundreds of thousands of hectares in surface area and millions of cubic meters in

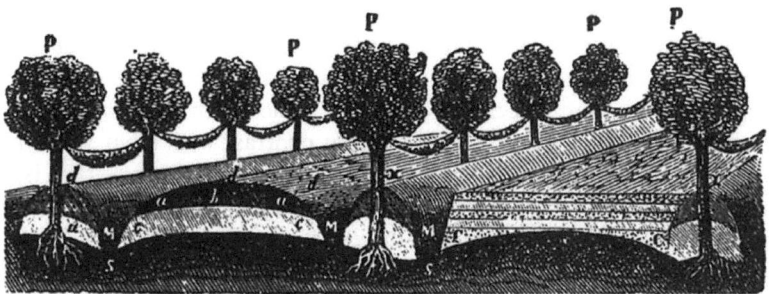

Figure 16. The work of systematization of the terrain for the Bolognese *piantata* in the early nineteenth century from the *Istituzioni di agricoltura* by Berti-Pichat.

volume, these new forms of the *piantata* were elaborated and differentiated in the dry part of the Po. They still today structure the three fundamental underlying types of agricultural landscape that dominate this region, which had the following characteristics.

The first type—arrangement *a cavalletto*, or *alla bolognese* (with its variant *a strena*, or *alla ferrarese*)—still today dominates the two provinces of Bologna and Ferrara, in a zone bounded by the line Somoggia-Crevalcore-Ferrara-Bologna-Somoggia. This is illustrated in its nineteenth-century form in Figure 16, taken from volume 3 of the *Istituzioni di agricoltura* by Carlo Berti-Pichat. As our illustration shows, the form of the landscape continued in the *piantata alla bolognese* to be characterized by wide and elongated rectangular fields, which were reserved for herbaceous crops, and at the heads of which an embankment served, besides the passage of carts and the turn of the plow, as a place for the runoff of water. But on land with imperfect natural drainage, which was likely in this area, a greater need to provide for drainage was met through excavation of two ditches that ran parallel to the rows of trees with vines and marked the limits of the *cavalletto* or *strena*: the hummock on which these rows were planted. The ditches were placed three to six meters from the fields used for herbaceous crops.

As appears clearly from the diagram of Berti-Pichat, in both the field and the *cavalletto* of the *piantata alla bolognese*, significant movement of earth assured the *baulatura* of the arable soil, which is presented to us in the diagram in the longitudinal type, but which more and more frequently was perfected during the nineteenth century in this region as a multidirectional *baulatura*, which became typical of plantations in the region of Bologna.

The longitudinal *baulatura* by contrast, generally remained typical, up to our own days, as the underlying type of the *piantata* in the region of Emilia and the Romagna (for its current form, see Plate 77). Today it is dominant, with its variations, between Parma, Cesena, and the line of the Po, and diffused also from the Romagna to the regions of Modena, Reggio, Parma, and Piacenza as far as the zone south of Mantova. In the *piantata* of Emilia and the Romagna, in contrast with what was typical of the *cavalletto* of the plain of Bologna, the planted area—that is, the strip of land four to six meters wide reserved for the cultivation of trees with vines—was not bordered by double drainage ditches that divided it clearly from the fields destined for herbaceous crops. Instead, the runoff of water was entrusted to the embankment at the heads of fields, and to secondary collecting ditches, generally in the perimeter, around the boundaries of the plot.

If, as we have indicated, longitudinal *baulatura* generally became typical of the *piantata* of Emilia and the Romagna. Those, instead, that were arranged crosswise were prevalent in arrangements *a cavini*, that became the underlying type for plantations characteristic of the plain of Venice. Here, the hummock of earth occupied by trees with vines was generally four to five meters wide, as in the *piantata* of Emilia and the Romagna, except that unlike the arrangement *a cavalletto* it was not bordered by double ditches separating it from the fields used for herbaceous crops, which were 35–50 meters wide and 60–100 long. Typical of arrangements *a cavini* was a high *baulatura* arranged crosswise that assured the runoff of water toward narrow embankments (of not more than 2.0–2.5 meters), called *cavini*, that ran along the heads of the fields.

It should be indicated as well that, in the plain of Venice, besides the arrangement *a cavini*, there was also a general diffusion of an arrangement *a prode*, which, with its rows of trees with vines placed side by side along the ditches that divided the fields, often repeated certain typical forms of the *alberata* of Tuscany, Umbria, and the Marche. Without expanding, however, to a more detailed illustration of this form, what we have already said in these pages will be enough to give an idea of the importance that the internal dynamics of the landscape of plantations—besides its further extension—assumed in the nineteenth century. From these extensive and intensive developments, the landscape of the dryer part of the Po valley was transformed and differentiated, with types and subtypes not much different from those that still dominate the region today. Until about 1870, to be sure—up to the beginning of the period of great works of hydraulic improvement—the increasing employment of capital and labor in the systematization of plantations was not enough to create a true and proper agricultural revolution in this region, like the one that from the late eighteenth century conditioned the affirmation of the typically capitalist productive relationships of the irrigated zones of the Po. In the dryer zones, more often, while the traditional landscape of the *piantata* continued to be extended and perfected, a less impetuous capitalist development of agriculture remained for the moment generally contained in the also traditional forms of sharecropping. With the growing investment of capital in works to transform the land and to exercise agricultural industry, to be sure, sharecropping itself began to acquire a more clearly capitalistic character. But only the launching of the great works of improvement in the decades after Italian unification would effect, even in the dryer zones of the Po, a decisive subordination of productive relationships in the countryside, and of landed property itself, to the new requirements of capital. Meanwhile, the irrigated zones of the Po valley remained decisively in the lead of agricultural progress in Italy, as we will see in the chapter that follows. There, already for decades, the agricultural revolution had found a form suited to its full development in the capitalist agricultural enterprise.

· 79 ·

The Agricultural Landscape of the Irrigated Zones of the Po Valley, and Rice Fields

The prevalence of agricultural systems with continuous rotation, which had begun to appear in a good part of the Po valley toward the end of the eighteenth century, became general and exclusive, one could say, in the age of the Risorgimento and Italian unification. We have already indicated, however, that the renewal of forms of the landscape, types of enterprises, and productive relationships within enterprises, which were inherent to these transformations of the agricultural system, assumed a particular importance in the zones of the Po where they were hastened by the development of works of irrigation. It was, in fact, in the irrigated part of the Po that the crisis of the old system of sharecropping soonest and most rapidly advanced toward its conclusion; and here, as well, the *cascina* first affirmed and generalized itself as the new dominant type of enterprise controlled by great renters, and employing capital and a productive organization that repeated, in the agricultural sector, the dimensions and forms that characterized the age of manufacturing in the industrial sector.

In the irrigated part of the Lombard plain, this process had already reached a point, in the last decades of the eighteenth century, where England itself, through Young's famous *Travels through France and Italy*, could find teachings and examples here for its own "agricultural revolution," and for the perfecting of its own "high farming." And for the whole first half of the nineteenth century, up to the point of the new impetus to capitalist development conditioned by the development of great works of hydraulic improvement in the first decades of Italian unification, the lead in Italian agricultural progress and capitalist development remained those regions of the Po valley where the effect of the new works of irrigation had spread most rapidly.

It is the worth the effort, meanwhile, to follow more closely the spread of the new forms of agricultural landscape in this period that were typical of the irrigated zones of the Po—which were essentially, as we have already indicated, those of irrigated meadows and the rice fields. Aside from the progressive and rapid transformation of stable into rotated rice fields, one cannot say that these forms underwent substantial changes as to their type in the age of the Risorgimento and Italian unification. These have already been illustrated in chapters 64, 73, 74 and

78. What is more interesting to demonstrate here was the progressive and rapid spread of these forms, which now more and more widely took over entire sectors of the Po valley.

One can refer, for instance, in this spread of the landscape of rice fields, to precise information available for the region around Vercelli. Still at the beginning of the eighteenth century, the cultivation of rice here only involved 7,254 hectares, that is only 7 percent of the total area, of which at least 32.5 percent was uncultivated, marshes, pastures, and woods. About the middle of the eighteenth century, the area planted with rice seems to have only slightly increased, to 7,365 hectares; but in the second half of the century it expanded rapidly so that in 1809 it exceeded 30,000 hectares—that is, 25 percent of the total area—while the total in woods and uncultivated land was reduced to less than 20 percent. About 1860, Dionisotti provides figures for the area devoted to rice fields in the zone of Vercelli that is not much more than the Napoleonic one, but changes in territorial subdivisions and different criteria used to designate land devoted to rice do not permit precise comparison. It is certain, however, that after unification the development of works of irrigation and the growing regional specialization of crops helped to produce a further rapid expansion of rice fields around Vercelli. In 1884, out of 33,917 giornate of land served by the Società d' Irrigazione Ovest-Sesia, 19,895 (that is, 58.66 percent) comprised rice fields, and in 1901 this had grown to 26,156 (that is, 71 percent). Throughout the region of Vercelli, in the first years of the twentieth century, Pugliese calculated that not less than 50 percent of the total area was used for rice fields.

No less significant is the information regarding the spread of the landscape of rice fields through the Po valley. About 1860, the total extent of rice cultivation was calculated to be 144,907 hectares in Italy generally, of which by far the largest part was distributed through the provinces of the Po, and more precisely:

42,429	the Ancient provinces
62,392	Lombardy
13,127	Emilia and the Romagna
25,298	the Veneto
143,246	all the provinces of the Po

In the Ancient provinces (near Milan) and Lombardy, these were now almost exclusively temporary, or rotated, rice fields; while in Emilia and the Veneto stable rice fields established in fens were still significant, and rice fields in landfills had spread in areas undergoing improvement. At the beginning of the twentieth century, and then according to the

Cadastre of 1929, the total surface of rice fields in Italy appears to have only slightly increased in comparison with 1860 (from 144,907 hectares to 145,500 about 1910, and 148,964 in 1929), but there was a large increase in the production of *risone* (unhusked rice) from less than two million quintals about 1860 to an average of four and a half million at the end of the first decade of the century, and more than seven million in 1929. This important increase in yields was due, aside from the now universal diffusion of rotated rice fields instead of stable ones, to technical progress in cultivation: fertilizers, rotation, and spread of the practice of transplantation. At the beginning of the twentieth century, otherwise, and still more in the decades that followed, regional and provincial specialization in rice cultivation made further progress. In 1929, more than a third of the area planted in rice was in the province of Vercelli (53,981 hectares), and our Plate 62 provides a picture of the form the landscape of rice fields now assumed there. Thus, with the province of Pavia (42,027 hectares), Milan (18,638 hectares), and Novara (18,360 hectares), a limited region of the Po valley now accounted for almost all of Italian rice production, which had completely disappeared from central and southern Italy, and was much reduced even in the Veneto and in Emilia, whose more modern landscape of rice production is illustrated in Plate 68.

With regard to the landscape of rice fields, however, still more than its spread in the age of the Risorgimento and Italian unification one must indicate its progressive concentration in determined sectors of the Po valley, where (because of a more concentrated territorial specialization of cultivation) its forms now assumed a decisive significance. At the most general level this meant the changes it brought to the system itself, in the landscape, in agricultural relationships in the Po valley, and also in the progressive and rapid spread of irrigated meadows. And, as well, still at a general level, the various types of arrangements for irrigation affected what particular crop (rice, meadow, or other), or what crop course, was most favored for increased productivity.

Since the first centuries of the Middle Ages, as we have seen, and then in the age of the communes and the Renaissance, by digging canals to divert the river network, and the typical springs in zones with underlying waters, the provinces of the Lombard plain—followed at a distance by those of the Piedmontese and Venetian plain—had been in the forefront of Italy in extending and perfecting systems of irrigation. After a period of stasis in the seventeenth century, and even because of this, the second half of the eighteenth century and first decades of the nineteenth century had marked a new advancement of initiatives and works of irrigation in these provinces. Thus, already in 1847, Cattaneo, in his famous

Plate 68. An Emilian rice field in *Mondine 1940* by Aldo Borgonzoni.

letters on the *Istituzioni agrarie dell'Alta Italia*, calculated that in the plain between Milan, Lodi, and Pavia alone, almost 250,000 hectares (that is, eight-tenths of the cultivated surface) enjoyed the advantages of irrigation, while for the whole Lombard plain this had increased to more than 400,000 hectares, that is, about half of the cultivated surface. The canals of first order alone, which derived their waters directly from the Ticino and the Adda Rivers, extended in that period to more than 200 kilometers, and these were also matched by a much larger development of canals of a second, third, or fourth order, whose subordinate network was already so complex that between Genivolta and Casalmorano, for instance, the road from Cremona to Bergamo had to cross thirteen bridges in the brief space of 160 meters in a place that acquired the name "Tredici Ponti." In the 250,000 hectares irrigated in the plain between Milan, Lodi, and Pavia alone, Cattaneo calculated that in the previous decades and centuries not less than a billion cubic meters of earth had been moved to the end of leveling and arranging fields to permit irrigation, and that this had employed labor worth not less than 400 million francs, as compared with the 200 million francs in work for the excavation of canals, and the billion for the total cost of the principal and accessory works of irrigation.

For the Lombard plain as a whole, Cattaneo calculated that, of the 30 million or more cubic meters of water that flowed through the countryside daily during the summer, about three-quarters came from the river system, while a lesser part came from underground springs. But whatever the source of the water, its use in irrigation, and the arrangements this use required, were already so conspicuous that Cattaneo could rightly speak of the land of the irrigated Po as an "artificial fatherland," a fatherland, one could say, where nine-tenths of the soil was the work and product of the men who had moved it.

A few years later, on the eve of unification, in 1865, a report of Torelli to the minister of agriculture provides a more precise statistical documentation of the new extent of irrigation of the Po valley in the age of the Risorgimento and Italian unification. The province of Milan (with 248,000 hectares of irrigated surface—83 percent of the total surface) was always by far in the lead of all the provinces of Italy both absolutely and relatively; followed at a distance by Pavia (with 122,000 hectares irrigated—37 percent of the surface), Brescia, and Cremona. In the Lombard provinces generally, where less than twenty years previously Cattaneo had estimated an irrigated surface of 400,000 hectares, this now extended over some 646,000, with an increase that—if it cannot be known precisely, given the approximate estimate of Cattaneo—was certainly rapid and conspicuous.

Important also was the extent, already at the time of Torelli's figures, of the irrigated zone in Piedmont, which extended in 1865 over 405,000 hectares: with 156,000 hectares of intensive irrigation—24 percent of the cultivated surface of the province—the region of Novara, particularly, was close to the level of the Lombard provinces. The contribution of first importance that Cavour, through both his personal and governmental initiatives, had in the rapid development of irrigation in Piedmont is well known. It was not by chance, that in his early years, when he was already dedicated to agricultural matters on his ancestral lands but was still distant from the cares of politics, Cavour was in close contact with the illustrious Giovannetti, to whom he owed the section on the management of waterways in the Albertine code of 1837. But decisive in this area was above all Cavour's initiative in the formation of the Associazione d'Irrigazione dell'Agro Ovest-Sesia that—founded in 1853 with 3,500 members—represented (as Cavour himself revealed in a speech to the Chamber) "a new fact, not only for this state, but I dare say for all of Europe . . . the greatest application of the spirit of association there has ever been to agriculture." In 1884, the association Cavour founded was still governed for the most part by the statutes he inspired, and it had provided for the irrigation of more than 23,000 hectares. On the eve of World War I it was composed of 60 syndicates, with a total irrigated zone of some 50,000 hectares. It was in Cavour's merited honor that, in 1862, in recognition of an undertaking that had transformed the agriculture of an entire region, the Chamber gave the name "Canali Cavour" to the whole system of irrigating tunnels and ditches in this area.

Torelli's figures for 1865 do not cover the Venetian provinces, which had not yet at this point become part of the Kingdom of Italy. But in the preceding decade some reliable information for these provinces is provided in a volume by De Bosio, published at Verona in 1855. In these years, continuing a tradition that had roots in the old dominion of Venice, not less than 148 syndicates, with a zone of improvement that extended over at least 616,518 hectares, were involved with the management of waterways. But in the majority of cases these were commissions for conservation of water (21 in number), for drainage (77 in number), or a combination of conservation and drainage (21 in number). Only 15 syndicates, with a total of 64,602 hectares, were concerned exclusively or chiefly with diverting and distributing water for irrigation. Only a small part of the land included in these zones, however, was in fact irrigated.

Thus in the decade just before unification, the Venetian provinces were still much behind the Lombard and Piedmontese ones in the extent of works of irrigation. Here waterworks for conservation and drainage still clearly predominated, as they did also in the provinces of Emilia. But even in Emilia the figures of Torelli indicate an area of improvement that reached almost 26,000 hectares in the province of Bologna, and 11,000 hectares in Piacenza, but that did not exceed 67,000 hectares in the region as a whole.

The last decades before Italian unification, in short, marked a last period of vigorous initiation of works of irrigation, which spread to new zones in Lombardy and Piedmont, but much more slowly in the provinces of Emilia and Venice. It should be indicated, however, that the figures cited by Torelli and by De Bosio seem to refer, more than to the area in fact irrigated, to the total extent of the land in the water system; and one must keep this consideration in mind, when comparing them with the figures of the Direzione Generale di Agricoltura that we shall now cite for the first years of the twentieth century.

For the years immediately before 1905, in fact, the more accurate figures of the Direzione d'Agricoltura give, for Lombardy, which remained always in the forefront of progress in this matter, an irrigated area of 644,000 hectares, and an area that could be irrigated of 154,000, with a combined extent in this zone of 798,000 hectares, in comparison with 644,000 in 1865 and 400,000 in 1847. Even if the data are not exactly comparable, they provide an index of the pace with which works of irrigation continued to spread in this region. For Piedmont, the area under irrigation jumped to 341,000 hectares, and what might be irrigated to 154,000, showing an extension of the irrigated surface from 405,000 hectares in 1865 to 495,000 in 1905. Piedmont remained in second place for Italy in the extent of works of irrigation. In third place was Venice, with 98,000 hectares irrigated and 117,000 more that could be, 215,000 hectares in all in 1905 in comparison with 64,000 in 1855. Here the impetus to extend the work of irrigation was relatively more noticeable after unification, and the impetus to expand was relatively more important than in Piedmont and even in Lombardy; even if, in absolute figures and the proportion of total cultivated surface, the irrigated area remained rather modest. In the same period (or more exactly between 1865 and 1905), the irrigated zone passed from 67,000 to 129,000 hectares (of which 68,000 were irrigated and 61,000 could be) in the region of Emilia, which was chiefly involved in great works of hydraulic improvement during the first decades after unification, rather than in efforts at irrigation.

In general, in 1905, the irrigated zones extended over an area of 1,651,000 hectares in Piedmont, Lombardy, the Veneto, and Emilia (of which 1,151,000 hectares were irrigated and 500,000 irrigatable), in comparison with 1,181,000 in 1855–65. A further indication—and we continue to emphasize the lack of precision of statistics of this kind—is given by information published by the Ministero dei Lavori Pubblici in 1931, according to which the area effectively under irrigation in these four regions reached 1,164,000 hectares. According to these data, the effective irrigated area in Lombardy (with 482,000 hectares) now only slightly exceeded Piedmont (480,000 hectares), and this was followed at a distance by Venice (125,000 hectares) and by Emilia (77,000 hectares).

As one sees, the data provided by these statistics can be interpreted only as indicative, and they seem sometimes even contradictory. But in general, one has the impression of a notable progress in the expansion of works of irrigation in both old and new zones of Lombardy, and in Piedmont, already in the first half of the nineteenth century, and particularly in the last decades before unification. After unification, the progress continued in Lombardy and Piedmont and became more rapid in Venice, which had remained backward. In Emilia, the expansion of irrigation was contained within more modest limits, while, in the first three decades of the twentieth century, there was a particularly notable increase in the area irrigated in Piedmont, which now equaled Lombardy in this regard.

There is no doubt, however, that in the age of the Risorgimento and unification the extension of improved zones and of irrigated surface in the provinces of the north Italian plain brought with it decisive transformations of systems, relationships, and the agricultural landscape, in the sense already indicated, particularly with regard to an expansion of the landscape of irrigated pastures and a concentration of the landscape of rotated rice fields. In the irrigatable zones, on the other hand, the expansion of new agricultural systems contributed generally, from the mid nineteenth century onward, to precipitating a crisis in the old system of sharecropping, and to a more rapid evolution of all agricultural productive relationships in a more typically capitalist sense. In the dryer part of the Po valley, as we have seen, this process was delayed and slower; but it took on a new impetus in the last thirty years of the nineteenth century with the expansion of works of hydraulic improvement. Thus, in general in these provinces, the technical and capitalist development of agriculture acquired a more decisive importance and more rapid rhythm than in other parts of Italy.

Not by chance, in 1880 and then again in 1882, the agricultural proletariat of the region of Cremona, a typical province of the irrigated zone of the Po, was the first to take recourse to an organized strike in the Italian countryside. In this second year, 1882, agricultural strikes expanded from Cremona to other zones of the irrigated Po, into Lodi, the lower region of Brescia, Vercelli, and then into the neighboring provinces of the dryer Po: for instance around Parma. In 1884, the agitation grew in importance in the region around Mantova and in Polesine. Following the just mentioned progress of works of improvement in the last decade of the century, the epicenter of agitation by the agricultural proletariat, and the socialist movement in the countryside, then tended to move from the provinces of lower Lombardy toward Emilia. Agricultural strikes and the socialist movement, for that matter, had now become social phenomena that—in both the dryer and the irrigated Po—intersected and involved not only *braccianti* and wage workers, but all categories of laborers in the countryside, including, often, sharecroppers, small renters, and even small owners.

In fact, statistics on the types of enterprises do not always adequately represent the degree that the effective subordination of agriculture to capitalism had reached. More significant seem other statistics that show the decisive and growing part that the provinces of the Po—where this

Agricultural Products of the Po Region in 1860 and in 1909–13
(percentages of the total for Italy)

	1860	1909–13		1860	1909–13
Grain	27	48	Oxen and cows	43	63
Indian corn	56	68	Sheep	12	8
Rice	88	98	Pigs	21	47
Wine	42	36	Silk cocoons	57	85

subordination was most advanced in the last decades of the nineteenth century—assumed in the progress of the techniques of Italian agricultural production.

To begin with, let us look, in the accompanying table, at the distribution of some agricultural products. As one sees, in all fundamental agricultural products, the transformation of agricultural systems, development of technology, and growing subordination of agriculture to capital assured to the four Po provinces an increase in the productivity of agricultural labor, which was not replicated in other regions of Italy that were more backward in capitalist development. The capitalist agriculture of the Po provided to the national market a rapidly growing proportion of the principal agricultural products. An exception, among the products cited, were only vines (whose cultivation became restricted in the Po valley for reasons already indicated earlier), and the breeding of sheep, which also decreased in importance in the Po with the shift from extensive agricultural systems to intensive ones.

Not less significant, at the end of this period, are data relative to the productivity of agricultural labor in the Po provinces, as compared with the means for Italy as a whole.

Gross agricultural production in 1921 (in lire)

	Per Hecatre of Agricultural and Forest Land	Per Capita Employed in Agriculture
Piedmont	414	1,030
Lombardy	494	982
Veneto	367	702
Emilia	493	1,011
Means		
Po region	439	931
Italy	302	777

In all the regions of the Po, as one sees in the second table, the value of gross agricultural production per hectare of cultivated land and forest remained, about 1921, clearly superior to the mean for Italy as a whole; and the same was true of gross production per capita of the population employed in agriculture. Only the Veneto was an exception, where capitalist agriculture was less advanced and there were also unique problems of depression in the postwar years. In general, however, the four Po regions (which irrigation, works of improvement, transformation of agricultural systems, and the impetuous capitalist development had brought to the forefront of agricultural progress), produced about 1921—although with only 32 percent of the agricultural and forested territory of Italy—some 47 percent of Italian gross agricultural production. And this proportion would be even greater if we referred not to agricultural production in general, but only the part of it destined for commercial exchange.

. 80 .

The Alberata *of Tuscany, Umbria, and the Marche in the Risorgimento and Italian Unification*

One cannot say, for the age of reforms and later for the first half of the nineteenth century, that the agents of reelaboration of the agricultural landscape operated in Umbria and the Marche with the same effectiveness that we have been able to demonstrate for the neighboring region of Tuscany. In chapters 70 and 71, we saw that in the Grand Duchy of Tuscany, although it still lacked the impetus of the "agricultural revolution" (and the revolution in landed relationships) of certain sectors of the Po valley, the policy of absolutism had created, if nothing else, certain presuppositions in favor of the Italian mode of capitalist development in the countryside, and in favor of a similar evolution of systems and forms of the agricultural landscape. In the provinces of Umbria and the Marche, by contrast, which had environmental conditions not much different from those of Tuscany, the inept, corrupt, and backward administration of the papal legates frustrated any even moderate initiative in agricultural, social, cultural, or political progress. In the first half of the nineteenth century the countryside of the Marche and Umbria, not to mention the Roman and Pontine Agros, remained in a state of backwardness and stunted growth, which contrasted with the relatively flourishing state of Tuscan agriculture.

This did not mean that the basic impulse to agricultural and capitalist development was entirely without effect on the evolution of forms of the agricultural landscape even in the Marche and Umbria in the first half of the nineteenth century. Even here, to be sure, as we have seen, a certain progress was shown by the extension of clearings and plantations of trees, and in the spread of less primitive forms of cultivation in the hills, that began to be adopted on the example of nearby Tuscany. But for the whole first half of the century this progress was generally slow and sporadic. Still on the eve of Italian unity, the information provided by Maestri for Umbria and the Marche gives us a picture of the distribution of uncultivated and cultivated land not much different from the one revealed by the papal cadastre in the first decades of the century.

For these reasons we can thus justly say that the evolution and diffusion of new forms of the agricultural landscape, which are shown in the more precise statistics of the first years of the twentieth century and up to our own days, appeared to the largest extent in the decades following

Distribution of Cultivated Land in the Marche
(in thousands of hectares)

	1830	1860	1910	1929
Sowable land	506	476	632	616
Without trees	(311)	—	(243)	(228)
With trees	(195)	—	(398)	(388)
Olive groves	16	19	—	—
Specialized trees	—	—	7	—
Meadows	18	18	—	24
Pastures	267	272	172	123
Woods	156	168	100	113
Uncultivated and waste	49	—	61	77
TOTAL	1,012	953	972	969

the political unification of the country. In the second half of the nineteenth century, freed from pontifical administration and included now in a larger economic and political framework, these provinces saw the way open to a kind of capitalist evolution of agricultural relationships, and a more rapid pace of agricultural progress, which carried them in a few decades to a level not much inferior to that of nearby Tuscany.

One can compare in the accompanying table, for example, the information for the Marche from the pontifical cadastre (for the period before 1830), those provided by Maestri (for the period 1860–64), and those of the statistical inquiry of 1910 and the agricultural cadastre of 1929.

As we have already warned, in the thirty years between 1830 and 1860, aside from minor changes resulting at least in part from changes in the boundaries and criteria of the surveys, the proportional relationship between the surface occupied by cultivated land, woodlands, pastures, and wasteland did not undergo substantial changes in the Marche. The agricultural landscape of the region remained for the whole first half of the nineteenth century marked by forms that had been delineated already centuries before; and in 1860, as in 1830, cultivated land made up only half of the territory. The other half was abandoned to woodlands, pasturage, or waste.

Only between 1860 and the first years of the twentieth century, after Italian unification, do our statistics indicate a new and noticeable extension of cultivated land in the Marche, which passed from 495,000 hectares in 1860, to 639,000 in 1910, and 628,000 in 1929. As one sees, the maximum extent of cultivated land in this region was reached precisely in the first decades after unification and then it stabilized. But this

Distribution of Cultivated Land in Umbria
(in thousands of hectares)

	1830	1860	1910	1929
Sowable land	280	288	430	413
Without trees	(139)	—	(156)	(164)
With trees	(141)	—	(274)	(249)
Olive groves	42	43	—	—
Specialized trees	—	—	12	13
Meadows	11	11	—	13
Pastures	420	276	248	105
Woods	83	263	230	225
Uncultivated and waste	33	17	58	80
TOTAL	869	898	978	849

was not a question only, one should note, of an extension of the agricultural landscape to new parts of the Marche that had been abandoned earlier to pasturage or woodland. Deforestation was accompanied, in fact, by an even more rapid progress of plantations of trees and shrubs, whose most typical forms—as earlier in Tuscany—were imprinted on the agricultural landscape in the first decades after unification through an expansion of the typical *alberata* of Tuscany, the Marche, and Umbria. Thus, while the extent of cultivated land increased in the Marche from 522,000 to 639,000 hectares between 1830 to 1910 (that is, by more than 100,000 hectares), the extent of sowable land without trees decreased from 311,000 to 243,000 hectares, while that of sowable land with trees increased from 211,000 to 396,000 hectares. As in Tuscany, now here as well, the forms of the *alberata* of Tuscany, Umbria, and the Marche clearly dominated the regional agricultural landscape.

A development quite similar to the one in the Marche, occurred in Umbria, as appears from the second table. Here as well, as one sees, the expansion of cultivated land, which did not increase noticeably between 1830 and 1860, was realized with an accelerated rhythm in the first decades after unification, and then it stabilized in the first years of the twentieth century. And here as well, more significant than an expansion of clearings, was the further diffusion of the characteristic forms of the *alberata* of Tuscany, Umbria, and the Marche. Between 1830 and 1910, while the extension of sowable land without trees only increased from 139,000 to 156,000 hectares, that of sowable land with trees increased from 141,000 to 274,000 hectares. Thus now also in the provinces of Umbria, as in the Marche and Tuscany, the forms of the *alberata* impressed themselves on the regional landscape with a qualitative

Distribution of cultivated land in Tuscany
(in thousands of hectares)

	1830	1860	1910	1929
Sowable land	—	567	1,214	1,009
Without trees	—	—	(553)	(454)
With trees	—	—	(661)	(554)
Olive groves	—	155	—	—
Specialized trees	—	—	71	74
Sowable land	649	—	—	—
Meadows	—	26	—	39
Pastures	—	454	455	104
Woods	541	697	909	813
Uncultivated and waste	448	243	135	255
TOTAL	1,638	2,142	2,410	2,290

and quantitative significance that much surpassed that of sowable fields without trees, which tended to predominate in other sectors of peninsular Italy and the Islands.

As for Tuscany, according to what we have been able to discover, already in the first half of the nineteenth century the extension of clearings and plantations of trees assumed a quicker pace. However, still in 1830, in this region as a whole, vast expanses of sowable land without trees and uncultivated land in zones of the Maremma and the mountains of Tuscany kept the percentage of cultivated land in the total area (39 percent) only a little greater than that of Umbria (with 36 percent) and quite a lot less than the Marche (51 percent). But already between 1830 and 1860, in contrast to the other two regions, the total cultivated area increased in Tuscany from 649,000 to 722,000 hectares, an increment that can only partly be explained by changes in boundaries. (See the third table.)

Even here, to be sure, the rhythm of clearing further accelerated after unification. In 1910, as appears from the table, the area of cultivated land in Tuscany had increased further, from 722,000 to 1,285,000 hectares, and on these, sowable land with trees (661,000 hectares) clearly predominated over sowable land without trees (533,000 hectares), while the cultivation of specialized trees and shrubs accounted for 71,000. One must indicate that in Tuscany, rather more than the Marche and Umbria, changes in territorial boundaries and criteria of the survey give these figures a value that has to be considered only indicative.

In total, however, excepting zones like the Maremma and the Tuscan mountains, already in the last decades of the nineteenth century the agricultural landscape of the three regions under consideration seems to have been elaborated in forms, and with a relief, basically similar to the present one. More than ever, Tuscany, Umbria, and the Marche had become the classic region of sharecropping, whose diffusion was closely tied to the characteristic *alberata* of these provinces. According to the agricultural survey of 1910 for the three provinces considered together, the area covered by cultivated land reached 2,366,000 hectares, that is, more than 54 percent of the total surface, only 40 percent was made up of sowable land without trees, and 56 percent was made up of sowable land with trees, while the remainder was accounted for mostly by the cultivation of specialized trees and shrubs.

In these regions of central Italy, to be sure, the pace of investment and of the capitalist development of agriculture was far from reaching what we have found in the Po valley, which especially in the irrigated Po reached the point of breaking through the traditional structure of the sharecropping system. But here, instead, precisely the expansion of sharecropping, and the forms of the *alberata* inherent to it, marked the penetration of capitalist relationships into the countryside, and also the progressive subordination of agriculture and landed property to the interests of capital. Thus, one can see from Plate 69—which reproduces a cadastral map of the early nineteenth century that is preserved in the State Archive of Perugia—the form taken by the landscape in the region of low hills near Lake Trasimeno, near Magione. Aside from a few pieces of land, where the lines of trees of the *alberata* were already appearing, the greater part of the land was accounted for by ecclesiastical properties, which are indicated on the map by name, where there was an exclusive predominance of sowable land without trees. In the decades after unification, through surveying and alienating land once under ecclesiastical control, the spread of the *alberata* over large areas earlier dominated by sowable land without trees marked a notable advancement in the subordination of agriculture and landholding to the interests of capital. But this subordination realized itself often in a traditional compromise between old relationships of an ecclesiastical and feudal type and the new relationships characterized by the capitalist economy. And this compromise undoubtedly weakened the pace of agricultural progress. In the Marche and Umbria, as in Tuscany, this relative stasis of techniques and agricultural relationships expressed itself in the landscape, in the crystallization of the traditional type of *alberata* of which Plate 69 provides a picture from the early nineteenth century that, although it has expanded to a larger territory, has remained substantially unchanged up to the present. In contrast to what happened in the age of the Risorgimento and Italian unity in the landscape of the *piantata* of the Po valley, one cannot say that the *alberata* of Tuscany, Umbria, and the Marche underwent a similar process of internal elaboration of form, which re-

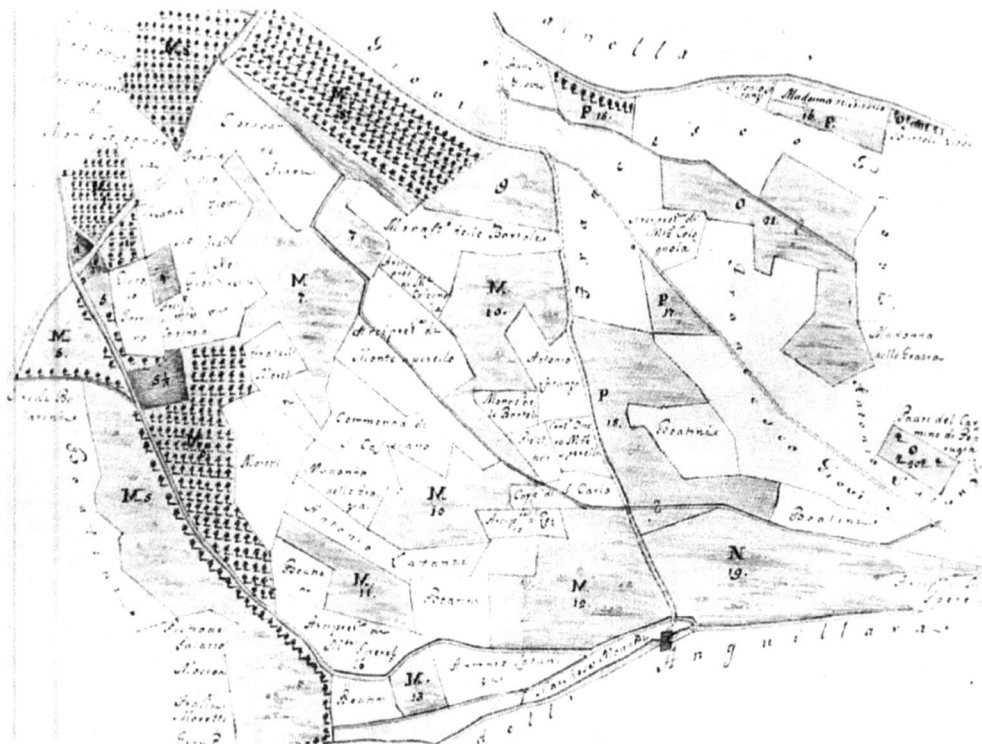

Plate 69. The extension of the *alberata* of Tuscany, Umbria, and the Marche at Magione, in the zone of plain and hills near Lake Trasimeno, from a cadastral map of the early nineteenth century.

mained that of arrangement in *porche*: rows of trees festooned with vines (eventually disposed *a rivale*—two parallel rows) along the sides of the fields that were rather narrower, generally, than those of the Po valley. These were, in short, all forms that we have seen develop in these regions in the Middle Ages. In the first decades after unification, to be sure—in Tuscany, but even more in the Marche and Umbria, where they gained momentum only then—the extension of the traditional *alberata*, with a corresponding prevalence of agricultural systems with continuous rotation over those of fallow or temporary clearings, and of a system of closed fields over open fields, masked the effective stasis of agricultural progress, which had assumed an extensive rather than intensive character. But already at the beginning of the twentieth century, when this extensive phase was essentially complete, the *aurea mediocritas* (golden mediocrity) of the sharecropping system and the *alberata* of Tuscany, Umbria, and the Marche—although it remained more than ever the ideal of the new landed bourgeoisie, which had become preoccupied with the first signs of serious social tension in the countryside—revealed the restrictive and impassable limits, within which it restrained the possibility of agricultural progress. This occurred with a quite different impetus where, as in the Po valley, more modern and typically capitalist productive relationships had cleared the way.

. 81 .

The Landscapes of the South in the Risorgimento and Italian Unification

In contrast with what happened in the irrigated provinces of the Po—and, with slower effect, in other provinces of central and northern Italy—the evolution of forms of the agricultural landscape during the Risorgimento and unification in the provinces of the South and the Islands seemed at first to be dominated, not by the agents of an agricultural revolution, but instead by those operating in the direction of a large-scale (if not deep) revolution in landownership.

Around 1860, in fact, the mass division of ex-feudal domains seemed almost concluded in the South, with assignment to communities of 600,000 hectares of land (out of the 750,000 eventually assigned to them at the end of this operation). If to these 600,000 hectares one adds another 520,000 of patrimonial lands, the landed patrimony that the laws ending feudalism had assigned to the communities appears to have been rather large. But between 1806 and 1860, out of this large patrimony, only 205,000 hectares had been divided among 116,000 receivers of allotments, while the remainder generally remained subject to traditional *usi civici*, or had been usurped by private citizens. Between 1861 and 1899, the pace of allotments became more rapid, and in the course of less than thirty years 193,000 hectares were divided among 202,000 holders. But, still more rapidly than in the preceding decades, the small holdings assigned to agricultural laborers in the southern countryside became concentrated in the hands of a few *galantuomini* (nouveaux riches). It was precisely in this way, besides the usurpation of communal domains, that the landed patrimony of the new southern landed bourgeoisie grew. At Eboli, for instance, in a few years, three families, who still today are the largest landholders in the region of Salerno, got into their own hands almost all the allotments of the original holders. At Teramo, out of 7,260 allotments resulting from partitioning domains, no more than 2,777 remained a few decades later in the hands of those who had first received them; at Barletta, in the space of thirty years, two-thirds of the 800 allotments assigned to individuals without land had passed into the hands of the larger owners; and one could thus continue. Even today, otherwise, in the agricultural landscape of the South, one can find traces of allotments in ruins of dry walls and rubble, that permit one to reconstruct their ancient boundaries, although they are now concentrated again in the hands of a few large landowners.

In general, at the end of the operation that ended feudalism in the South, 741,362 hectares, obtained through the dissolution of promiscuous rights, and 521,832 hectares of property from demesne lands were assigned to the local populations, for a total of 1,263,194 hectares. In this enormous mass of property, out of the 407,941 hectares that had been reserved for the *usi civici* of the populations themselves, 461,296 had been allotted, and had largely returned to the great landowners, while not less than 393,957 hectares were granted directly to usurpers, through the so-called procedure of conciliation.

But not only ex-feudal lands were involved in this huge redistribution of landed property that developed in the first decades after unification in the South, and to a lesser degree in other parts of Italy. No less important in this process was the liquidation of ecclesiastical lands, which also went to a large extent in this period to increase the landed patrimony of the new landowning bourgeoisie. Overall, the results of this process can be summarized in the accompanying table.

More than two and a half million hecatres of land were involved, situated for the most part in the Italian South, in Lazio, and in the Islands, which during the first decades of the new kingdom went largely to increase the landed patrimony of the bourgeoisie. It was precisely this process that gave the salient bourgeois character to landed property in the South during this period. Even in these provinces, to be sure, although to a lesser extent than in other parts of the peninsula, there was a continuing erosion of noble lands by the new landed bourgeoisie. But the larger part of the landed patrimony of the southern bourgeoisie was constructed not too much at the expense of noble property as at the expense of ecclesiastical lands previously subject to mortmain: that is, in the last analysis, at the expense of the great mass of peasants. Typical of this process was the case of the Roman Agro, where, still in the period of formation of the new kingdom, out of a total of 188,000 hectares, at least 104,000 (that is, 55 percent) were in the hands of nobles, 56,000 (that is, 30 percent) were lands subject to mortmain, and only 28,000 (that is, 15 percent) were in the hands of bourgeois owners. At that time, eight great proprietors—patrician families or institutions—owned more than half of the Agro. A few decades later, by contrast, at the beginning of the twentieth century, bourgeois property had jumped to 75,000 hectares, that is to 40 percent of the total; but its extension was not, for the most part, so much at the expense of noble land—of which the total had decreased by only 4,000 hectares—as at the expense of lands previously subject to mortmain, which had fallen to only 13,000 hectares, that is, from 30 to 7 percent of the total.

Redistribution of Landed Property

	Hectares	Allotments
Ecclesiastical lands	750,000	176,953
Ecclesiastical lands (Sicily)	190,000	20,300
Ancient demesne lands	300,000	97,990
Lands in customary tenure (Sardinia)	470,000	?
"Conciliated" demesne lands	393,957	?
"Allotted" demesne lands	461,296	600,000
TOTAL	2,565,253	895,243

Elsewhere, certainly, even in the South, the formation of new bourgeois properties developed in a less linear manner, and even the erosion of noble lands contributed to a larger extent. On the whole, however, the figures we have cited for the Roman Agro can be taken as being reasonably representative of a situation that was only paradoxical in appearance: it was typical of the South. The considerable expansion of bourgeois property did not correspond to a similar reduction of feudal property, which even often continued to dominate, with its imprint, all the social relations of a given place, and the new bourgeois property thus ended up by assuming a semifeudal character.

Even, and precisely, because of the reasons we have just indicated, the huge redistribution of property rights that was carried out in the South after unification did not produce a true "agricultural revolution," similar in importance to the one that had developed in the Po valley from the second half of the eighteenth century. Still, this does not mean that the reorganization of property rights had no effect on the forms and types of the agricultural landscape in the South and Islands. We have already indicated, for instance, that traces are still visible today of the mark that the allotment of domains made on the landscape of the South, with the remains of dry walls and rubble intended at the time to mark and enclose the little plots assigned to individual holders. In general, we know that these little plots were rapidly absorbed and reconstituted by the new middle-sized and larger properties of the bourgeoisie of *galantuomini*. But even in this case, which was the most frequent, and wherever a new bourgeois or peasant property was constituted through the dissolution of promiscuous rights, the usurpation of domains, or even the erosion of a noble latifundia, a system of closed fields tended to replace the traditional system of open fields, which, as we have seen, had continued to predominate up to unification in the South and Islands even after the end of feudalism. The noble latifundium, itself, was often adapted in this direction to the new needs and juridic norms typical of a society that was now largely dominated by capitalist property relationships. Thus even here one could now generally speak, even in the South and Islands, of a system of properties that were closed—although not always in a system of closed fields—with their boundaries more severely marked and supervised than in the past.

This progressive extension of the system of closed fields—or at least of properties—generally corresponded in the South, if not to a rapid spread of agricultural systems with continuous rotation, at least to a decisive affirmation of the agricultural system of fallow, even in zones where up to unification a system of temporary clearings was still common. This now remained only in the Maremmas, in the Roman and Pontine Agros, and in small zones of Calabrian, Sicilian, and Sardinian latifundia. Typical of the new system and the agricultural landscape of the South and Islands was the system of fallow, that remained well into the twentieth century, although it was now almost totally abandoned in central and northern Italy. Of the 2,206,000 hectares of fallow with or without pasturage that was surveyed in the total 13,237,000 hectares of sowable land in the agricultural survey of 1913, almost all was located in the provinces of the South or Islands, besides those in the Maremmas and in the Roman Agro, which were more similar agriculturally to the South than to the other provinces of central Italy.

The progress of the system of fallow and of a regime of closed fields (or at least properties) is not sufficient, however, to characterize the evolution of agricultural landscapes in the South in this period, if one does not keep in mind the other agents that also acted to change their forms. In the system of closed fields, there was an increasing importance, besides the system of fallow, of continuous rotation that had been limited in the past to particular zones, like that of the Campagna of Naples or to a few more progressive properties in other zones, but that now spread much more widely, with the more frequent substitution of cultivation of beans, for instance, for bare fallow. But still more important was the influence exercised by two other agents of transformation in the southern landscape in this period, which worked, one toward its degradation, and the other toward the elaboration of new forms.

For these, consider the information in the table on land use in the South and the Islands. Despite the lack of precision in the data, which is particularly serious for the earliest figures, and of certain variations in the criteria of classification, two facts seem evident from this table. On one hand, especially in the first decades after unification, the frightening extent of deforestation, which had already advanced in the South after the end of feudalism, now had a fearful impact of erosion on the southern landscape. In a little more than fifty years, the area of woods was reduced by almost half, as a result of deforestation and unplanned clearings,

Distribution of Cultivated Land in the South and Islands
(in thousands of hectares)

	1860	1911	1929
Sowable land with or without trees	5,570	6,470	5,340
Specialized trees and shrubs	276	1,078	1,609
Permanent meadows and pastures	2,999	3,088	2,669
Woods and chestnut groves	2,094	1,371	1,277

which now threatened the integrity of arable soil and the inhabitants of entire provinces. On the other hand, this degradation of the southern landscape evoked also a process of reelaboration and extension of some of its traditional forms, which in the past had been confined to quite small zones and now spread to larger ones. This second process expressed itself particularly through an increase in the area destined for the specialized cultivation of trees and shrubs, which exceeded a million hectares in 1911 and a million and a half in 1929.

One cannot say, however, that this rapid expansion of plantations in the South involved substantial innovations in forms and types of the agricultural landscape. In the Roman Agro, as in the zones of southern Italy and the Islands, where transhumant pasturage still predominated in an agricultural system of temporary clearings—of which a painting by Sartorio provides a documentation in Plate 70—the landscape still replicated substantially the forms (or rather absence of well-defined forms) that we have illustrated for previous decades and centuries with the paintings by Poussin and by Coleman reproduced in Plates 48 and 64. Elsewhere, the blackening or reddening of fallow plots, and the stockades, hedges, and ditches that bounded properties and fields, more and more frequently interrupted the monotony of the traditional landscape, which now appeared less desolate and deranged and began to be articulated with more precise forms. But even here, these forms, although they now spread to a larger territory, replicated a type that was not entirely absent in the South even in much earlier periods. More noteworthy was the innovation of rudimentary systematization with banks and grading, that—with the extension of allotments and clearings—now spread on even the steepest slopes of the hills and mountains. Throughout the southern Appennines, one can still today find traces of such arrangements, sometimes still used, but more often abandoned because of soil erosion or unfavorable trends of prices, crops, and costs of cultivation. Elsewhere there were less rudimentary arrangements in the hills in banks and terraces, which were destined not so much for herbaceous crops and poorer shrubs as for the cultivation of more valued trees and

Plate 70. *Flocks in the Roman Countryside* by G. A. Sartorio.

Plate 71. A landscape of the Mediterranean garden with works of construction in *Caprile* by Rosina Viva.

Plate 72. Telemaco Signorini, *The Olive Grove.*

shrubs, for citrus fruits particularly. These arrangements in the hills sometimes even assumed the character of true and proper works of construction of the type that one can see spread out around Amalfi, along the Sicilian coast, and elsewhere, whose typical landscape is shown in the painting by Rosina Viva, reproduced as Plate 71.

All these forms of the South, however, even when they assumed an importance and extent unknown in the past, did not represent substantial innovations as to type or, even when they were perfected, as to the structure or techniques they represented. The same can be said for the other two types of landscape that now began to spread through new zones of the South and Islands with the spread of cultivation of trees and shrubs. The great plantations of vines, olives, almonds, and so on—whose landscape is presented in this period in a painting by Signorini, reproduced as Plate 72—replicated substantially the traditional forms of the sixteenth-century preserve, even if their productive organization was now that of a large capitalist enterprise. Thus even in these regions, such as around Bari and Velletri, and in many other places, where the cultivation of trees and shrubs spread rapidly in this period, through

Plate 73. Filippo Palizzi, *Country Road*.

enfiteutic leases to the direct cultivators, the layout of the landscape underwent, to be sure, a deep transformation, but it developed, generally, into the traditional type of the Mediterranean garden, of which a suggestive image is provided for this period by a painting by F. Palizzi, reproduced in Plate 73, and for a more recent period by R. Guttuso, in Plate 80.

This landscape of the Mediterranean garden—from its first appearance in Sicily under the Greeks up to the present—was imprinted with a suburban physiognomy by the dividing walls and the contiguous placement of houses and rustic storage sheds, and this now gained significance even in places more distant from the chief urban agglomerations of the South. It was precisely in the dislocation of the urban habitat that one can perhaps find the deepest transformation of the southern landscape in the period we are examining. From towns perched on hills, that the local population had been obliged to inhabit for economic and productive reasons, and also because of malaria and brigandage, farmers now more often descended toward the plain, closer to the seacoast, the railway stations, and the major roads. The habitat of the centralized town remained characteristic of a large part of the South, but here and there it began to swarm into the surrounding countryside. And this greater dispersion was furthered not only by the large-scale reorganization of property rights and construction of a few nuclei of more stable enterprises, but also by the purchase of land and construction of houses in the countryside by the Americans, that is, by emigrants who returned to their native land to invest their savings.

In general, however, even in this period, the precariousness of peasant enterprises, traditional to the agricultural economy of a large part of the South, was worsened rather than alleviated; and this worsening reflected the social disaggregation that was already progressing in these provinces. After integration into the unitary state, the South suffered more and more both from the development of capitalism and from its own insufficient development. The "Southern question," already latent for centuries, now exploded into more evident contrasts. Rooted in the last period of involution of southern feudalism, and in the failure to liquidate this system effectively, it was now further complicated by the new overbearing centralized state, which served the interests of a new feudalism of banks and monopolies. The very extension of the landscape of plantations, and the Mediterranean garden, which seemed to promise a renewed agricultural prominence to the South, found insuperable limits—and worse than those set by the geological and agronomic conditions of large areas not adapted to the cultivation of valuable trees and shrubs—in the foreign, commercial, and fiscal policy of this new feudalism, which often frustrated the market outlet for southern agricultural products. This very policy—founded on the *pactum sceleris* (bad deal) of the tariffs on grain, through which the industrial monopolists of the North won over the compliance of the southern owners of latifundia— was also inauspicious for the southern economy in another way. Not only did it effectively suppress the possibility of industrial development, but—by favoring an exaggerated increase in grain production—it also raised decisive obstacles to the progress of the "agricultural revolution." Agricultural progress in the South could not be founded on the spread of plantations of trees and shrubs alone; it also needed to orientate itself around the integration of forage crops and livestock breeding into the productive cycle of agriculture.

· 82 ·

The Landscape of Campi a Pìgola: *Irregular Fields in United Italy*

Throughout Italy, once political unification was achieved, and the customs barriers were taken down, which until then had obstructed the formation of a single national market, the mode and relationships of capitalist production assumed a further importance and put their stamp even onto the semifeudal institutions that an incomplete democratic-bourgeois revolution had often left almost intact. With regard particularly to agriculture, the mode of development that Lenin called, precisely, the "Italian" (or "Prussian") mode continued to dominate the countryside, where new relationships and forms of capitalist exploitation were often grafted onto the old trunk of feudal exploitation and oppression, complicating and increasing its rigor. Thus not only in the South, but throughout the peninsula—as Engels had justly revealed in his famous letter to Turati—Italian society still suffered at the end of the nineteenth century, both from the development of capitalism and from a retardation of this development, impacted, as it was by the feudal residues that an incomplete democratic-bourgeois revolution had allowed to survive in the countryside. In these conditions, particularly serious for the economy and for Italian society, as remains the case even today, was the weight of land rents, whose effects had certainly not lessened with a greater participation of new large-scale bourgeois landholders in the monopoly on landholding, which was no longer an exclusive privilege of nobles and the clergy. To counteract this new and old de facto monopoly was the possibility that the mass of peasants might have access to property, and to stable and viable agricultural enterprises, which might permit them to take an autonomous initiative in agricultural capitalist production. With the "Italian" mode of capitalist development in the countryside, the national market for the chief agricultural products remained more than ever dominated by the market production of the great seigneurial estates that had been more or less transformed in a capitalist sense, while relatively modest was the market contribution that went beyond the limits imposed by the family needs of peasant enterprises.

There were, to be sure, exceptions to this norm: for instance, in the contribution of small peasant enterprises to the production of wine, and of fruit and vegetables. It was typical, in fact, of the "Italian" mode of capitalist development in the countryside for there to be this extreme variety and complexity of situations and relationships, as was understandable where, as in Italy, a peasant revolution had not come to make a clean slate of the feudal residues that had succeeded in accommodating themselves to new capitalist relationships. But it was not only a question of this historical variety of local situations. The persistent monopoly of great landholders, and the hated burden of land rent—in short, the "Italian" mode of capitalist development in the countryside—produced by themselves not only a slackening of technical progress and growth in productivity of agricultural labor, but also, through the misery and backwardness of the mass of peasants, a serious retardation in the formation of an internal market for large-scale capitalist industry. From this came the vicious cycle in which the Italian economy and Italian society floundered, from unification through the end of the nineteenth century, and up to the present, when the new intrigues of monopoly capital came to further complicate this cycle. The "Italian" mode of capitalist development in the countryside, and agricultural backwardness, kept the absorptive abilities of the internal market for industrial products within narrow limits and seriously delayed the development of large-scale modern industry. But, as well, the delay in developing industry—which was further aggravated at the turn of the century by a new monopolistic development of finance—impeded a normal absorption by industrial establishments of the work force liberated by the progress of technology and social division of labor in the countryside. Thus was formed in the Italian countryside and throughout the nation what can rightly be called an artificial excess of population. Even in Italy, in short, through a normal process in capitalist society, millions of units of labor were made redundant to the needs of agricultural production. But for the reasons just mentioned, instead of finding employment in industry or in other productive sectors, this went to increase the army of the permanently unemployed, or it was pulled from its native soil by the tide of emigration.

If, in this regard, artificial overpopulation had clear and typical social effects, no less important for the evolution of agriculture and Italian society in this period became the artificial overpopulation that could be said to have been latent in the countryside. In contrast to other countries, where normal capitalist development permitted agricultural productivity to increase with a diminishing agricultural population, successive censuses register practically stationary agricultural production, and an agricultural population that did not diminish, or even increased.

This growing mass of laborers, who were excluded, in effect, from the process of productive agriculture through the progress of technology and capitalist development, remained desperately attached to their little plots of land, where they utilized rudimentary tools and barely productive unskilled labor so as to escape complete destitution in a situation where capitalist development did not provide them with employment in industry. On the land of seigneurial estates, or in the margins of the great modern capitalistic enterprises, this mass of underlying and artificial overpopulation grasped at the larger or middle-sized properties to obtain at whatever price a tiny piece of land where they could in some way employ their exuberant labor. Whenever savings accumulated through years of foreign emigration permitted these laborers to buy a small allotment, they paid an exorbitant price for it. Whenever—as more often happened—even this means was closed to them, they competed desperately among themselves, offering a rent or division of the crop for the precarious enjoyment of some piece of land that was entirely disproportionate to any capitalist revenue, and that can only be understood as a measure of the rapaciousness of usury.

In these conditions, the very development of normal modern relationships in the great capitalist enterprises was often retarded and distorted. Rather than one large capitalist rent, the landowners preferred the precarious concession of small plots to a crowd of small renters or peasants from whom they could extort a usurious rent that, altogether, was higher than a large-scale capitalist one. Even more, in zones where this phenomenon developed the most, the existence of these usurious rents became an attraction even for lesser middle and lower bourgeois, who now in preference invested their revenues and patrimonies in the acquisition of land that could be profitably exploited through concession to small cultivators who were driven by desperate competition to offer exorbitant usurious rents in exchange for any possibility of employing their labor. This gave a new impetus to the accumulation of middle-sized and small bourgeois landed properties, which were often neither less hated than the great ones nor less unproductive and parasitic. And the social groups based on these had a not small part in the bureaucratic structure of the new unified state, and in the translation into practical reality of its stingy and backward policies. And it was not by chance—when the economic and political circumstances of the First World War seriously shook their underpinnings—that these groups provided, as well, numerous adherents, and important means of operation, to the Fascist reaction.

Meanwhile, already at the turn of the century, the phenomena we are considering was not unimportant in influencing the evolution of agricultural relationships, and also the forms of the Italian landscape. In the southern latifundium, the fragmentation of great seigneurial estates into ever smaller plots conceded to direct cultivators in exchange for more and more usurious rents, did not generally induce—given the precariousness of the concessions themselves—lasting changes in the form of the agricultural landscape. Wherever instead, through concessions of enfiteutic leases or alienation of marginal land of the estate, the direct cultivators proceeded to the cultivation of trees and shrubs, these often took, as we have already seen, the traditional forms of the Mediterranean garden. Elsewhere, however, and here and there also in the South, the creation of new properties or less precarious peasant enterprises introduced new and different forms into the landscape that replicated those of the *campi a pigola,* irregular fields, that we have already seen in Renaissance suburban landscapes, and that now expanded to larger zones.

A typical case is a zone in the Piedmontese hills, which has remained to the present a zone where the landscape of irregular fields has clearly predominated. As was revealed at the end of the past century in a study by Einaudi,[1] small and relatively stable and viable holdings and small peasant enterprises were born here from the division of ecclesiastical and noble estates, in which an intermediary group of Jews had an important part. These plots, which were proportioned in accordance with the peasants' ability to work, proved to be lasting because of particular environmental conditions, and particularly because of the opportunities the cultivation of high-quality vines gave to peasant families who had more labor power than capital. This modern Piedmontese landscape differed, however, from the Renaissance landscape of irregular fields because of its historical origins. Rather than being a casual coincidence of individual efforts to clear the land, it was born from a systematic division. No plan for colonization that was inspired by the collective interests of a given community gave this regular form to the agricultural landscape, as in the case of ancient Roman colonization, or modern projects for improvement. If there was a plan, it was a quite casual and individualistic one conditioned only by the profits of an intermediary, and the need this intermediary felt to be adaptive to the variety of work capacities and financial resources of the prospective owners. Whatever: this more modern landscape of irregular fields differed from the Renaissance one through a greater mean size and less accentuated irregularity in the form of individual plots. In contrast to the landscape of the Mediterranean garden, on the other hand—which these modern Piedmontese *campi a pigola* almost unconsciously reflected—this hill landscape was distinguished not only by a greater mean size of plots and greater regularity of form, but also by the nonobligatory, but universal, presence of cultivation of trees and shrubs, and their lesser density, variety, and exclusivity,

[1][Luigi Einaudi, "Le società cooperative di lavoro fra braccianti, muratori ed affini in Italia," *Credito e cooperazione* (1 March 1896); "La cooperazione nell'agricoltura Italiana," *Credito e couperazione* (15 September–1 October 1896).—Trans]

so that less significance was assumed by the dividing lines and defense of individual plots, and the peasant community remained concentrated, rather than scattered near the plots themselves.

But the modern landscape of irregular fields did not express everywhere in this period, as it did in the Piedmontese hills, the creation of peasant properties and enterprises with a relative autonomy and stability. This landscape generally remained the expression of a forced reflex toward the countryside caused by the artificial relative overpopulation, which did not find its most profitable employment in a society where normal industrial development was both retarded and distorted. In the South and the Islands, often, this reflex expressed itself in the predominance of other forms of landscape; but throughout the hilly zones of central and northern Italy—where the tradition of less precarious peasant enterprises is still a living one—the expansion of irregular fields often expressed in this period, and still today, a desperate effort to preserve a possibility of exercising autonomous or supplementary labor on the margins of the great seigneurial estates of the plain, with all the gradations and variety that local situations provided. One can add the part assumed in the diffusion of the forms of this landscape, especially in suburban areas, by the interest we have already indicated that groups of the urban middle and lower bourgeois showed for purchasing single plots, from which they received excessively high rents or other payments. Thus the form that the landscape of *campi a pigola* took from this period up to the present is not surprising—from the hills around Asti, the Acquese, the Monferrato, and the Laghe, to those of Novi, Broni, and the zone beyond the Po at Pavia, around Bergamo, and those that stretched from around Ancona on the Adriatic to Monte Amiata in southern Tuscany.

. 83 .

Improvements in the Po Valley, and the Agricultural Landscape of the Larga *in United Italy*

"The work of improvements," declared Minister Finali in 1876, inaugurating the Concorso Regionale Agrario at Ferrara, "is one of the greatest resources of modern Italy. The improvements in the Ferrarese will be the example and herald of similar undertakings." In this province of Emilia, we have already seen how the extent of cultivated land, that hardly reached 93,000 hectares at the end of the eighteenth century, had surpassed 119,000 about 1835, and 165,000 in 1875; while between 1825 and 1870, the area of low-lying marshland decreased from 89,000 to 84,000 hectares. Still around 1875, however, of the 165,000 hectares cultivated in the zone of Ferrara, about 45,000 were natural pastures and fixed rice fields, so that one can speak of regular cultivation in a narrow sense for about 120,000 hectares. But after unification, the rhythm of hydraulic improvement intensified. The area of marshlands decreased from 89,000 hectares in 1870 to 78,000 in 1906, 60,000 in 1925, and 56,000 in 1950, while the area of land under regular cultivation passed from 120,000 hectares about 1870, to 170,000 about 1906, 180,000 in 1925, and 190,000 in 1950. Thus the area under cultivation practically doubled in this province from the beginning of the nineteenth century to the present.

A similar development of the cultivated area occurred in zones like the one around Ravenna, where there was also rapid development at the turn of the century of great works of hydraulic improvement. Here still often environmental conditions (elevation above sea level and other factors) advised application of the traditional methods of landfills, although these were elaborated and perfected according to the dictates of a more up-to-date technology; while in the zone around Ferrara, and elsewhere, beginning just in the years around 1870, a new utilization of mechanized pumps, which spread rapidly, opened the way to improvement of marshlands where their particular condition and elevation would have made them difficult to improve for cultivation using traditional techniques.

For Emilia as a whole, to be sure, the dynamics of improvement of new land for cultivation did not reach, proportionately, the level of the zones of Ferrara or Ravenna that had particular hydraulic conditions. But even when this larger region is considered, the area reduced to regular cultivation, which Maestri estimated to be 1,452,000 hectares about 1860, passed to 1,452,000 hectares in 1929, and to 1,521,000 in 1950: an increase of 357,000 hectares, which was due largely and more or less directly to works of improvement and to hydraulic systematization that was undertaken after unification.

Thus while in the southern provinces the extension of land under cultivation after unification was realized more often at the cost of ruinous deforestation and unplanned clearings and of a serious deterioration of the arable soil, and while in Lombardy and Piedmont it was more often achieved through the development of works of irrigation that permitted an extension and intensification of cultivation, in Emilia instead—and in a more limited way in some zones of the Veneto, Lombardy, and the dryer zone of the Po—it was great works of improvement through hydraulic arrangements that chiefly permitted the extension and intensification of cultivation. This was an "agricultural revolution" that deeply changed the productive and social physiognomy of the region, and carried it to the forefront of Italian agricultural progress.

As we have already indicated, this was not only a question of extension of the cultivated area. Consideration of a few statistics gives a more precise idea of the form and extent of the "agricultural revolution" of Emilia in the decades after unification. In the period between 1840 and 1860, for instance, a combination of information provided by Galli and Maestri makes it possible to calculate approximately the production of wheat in the provinces of Emilia at 2,500,000 hectoliters, about 7 percent of Italian production as a whole. In 1909–11, by contrast, the mean grain production of this region had jumped to 6,782,000 quintals (about 8,900,000 hectoliters), or about 14 percent of national production. In the space of a little more than fifty years grain production in Emilia had jumped from an index of 100 to 356, and this more than threefold increase was for the most part due to a huge increase in yields per unit of cultivation rather than to an increase in the area cultivated in grain. In 1936–39, the production of wheat in Emilia had further increased, to 10,692,000 quintals, more than 14 percent of national production. Thus in the period between 1909–11 and 1936–39, although the rhythm of absolute progress in yields of grain per unit was still remarkable in Emilia, progress had slowed relative to other parts of Italy. In the years 1947–50, finally, Emilian grain production rebounded and it now exceeded 16 percent of national production.

But still more significant for the agricultural progress of Emilia are data relative to the production of industrial crops, whose development was one of the most typical signs of the spread of modern techniques, as it was of the evolution of commercial and capitalistic relationships in agriculture. In 1840–60, according to the data provided by Galli and

Maestri, the area sown with hemp—which was the only industrial crop widely grown in Emilia—occupied 58,842 hectares, on which however cultivation of hemp alternated with grain, so that it was grown annually over a lesser area. Production was calculated at 193,000 quintals annually. In 1909–16, hemp was cultivated over an area of 45,000 hectares in Emilia, but yielded a production of 495,000 quintals, two-and-a-half times that of 1840–60; in 1936–39, however, the area planted in hemp reached 49,000 hectares, and production of 631,000 quintals, more than triple that of 1840–60, and it was 56 percent of national production.

But after 1860, other industrial crops besides hemp spread rapidly in Emilia, as is well known. To speak only of the principal crops, one recollects that already in 1909–16 sugar beets were cultivated in Emilia over an area of almost 21,000 hectares, and this passed 54,000 hectares in 1936–39, with a production first of 7,216,000 and then of 13,566,000 quintals: this last figure represented 42 percent of national production. And a notable importance was assumed here also by the industrial production of tomatoes, which extended over an area of 4,610 hectares in 1916, and 11,000 in 1936–39, with production of 795,000 and then 2,636,000 quintals, or 26 percent of national production.

The area destined to these valuable and exacting industrial crops in Emilia thus rose from about 30,000 hectares in 1840–60 (although, given the alternation with grain production, more might be added to the effective total of industrial crops to the 58,842 hectares of hemp). The area destined to valuable industrial crops thus jumped here from at least 30,000 hectares in 1840–60 to 70,610 in 1909–16, and 114,000 in 1936–39. With addition of the industrial growth of fruit trees—which was unknown in Emilia in 1840–60, but occupied more than 10,000 hectares in 1929—one can say that, between 1840–60 and 1936–39, the area destined to the most exacting industrial crops had more than quadrupled in Emilia to the point of occupying a percentage of cultivated land (almost 9 percent) that was unmatched in any other region of Italy. The region had a high percentage of the national production of industrial crops, varying from 26 percent for tomatoes, to 42 percent for sugar beets, and to 56 percent for hemp.

This progress, it should be understood, was tied, in addition to the extension of works of improvement and systematization, to deep plowing and use of chemical fertilizers, in which Emilia passed, during this period, to the first rank, putting it side by side and often in the lead of the most advanced parts of Italy. But this could not have happened without an analogous progress of agricultural systems, the spread of forage crops and their insertion into the crop cycle, and the integration of breeding into the productive agriculture. Here we reach a subject that assumes a particular importance in the evolution of the Emilian agricultural landscape, which is our specific concern.

On the eve of unification, as we have already indicated, the cultivation of forage crops was not widespread in this region, except for mostly permanent meadows that occupied an area of 148,000 hectares, and 211,000 hectares of pastures. In 1929, by contrast, according to data from the agricultural cadastre, the area in permanent meadows had fallen to less than 82,000 hectares, and pastures to 106,000. But the production of forage crops appears to have increased enormously, and to have become integrated into the growing cycle through the spread of rotated meadows, which now occupied some 439,000 hectares, to which one can add another 25,000 hectares destined for the cultivation of intermediate plants.

In contrast to what had happened earlier in Lombardy between the late eighteenth and early nineteenth century, it was not from the development of irrigation that the "agricultural revolution" in Emilia departed between the 1860s and World War I, but rather from works of hydraulic improvement. But the agricultural renewal here, in the end, inevitably expressed itself through the insertion of forage crops and industrial plants into the growing cycle, and the integration of breeding into productive agriculture. Works of irrigation themselves took on an increased importance at a later phase of evolution of the agricultural landscape of the region. About 1860, the irrigated surface hardly reached 56,000 hectares, and it had grown by only 68,000 by 1905, but in the years between 1922 and 1930 alone it increased by more than 66,000 hectares.

We have already indicated that it was not only through its most direct, but also, and still more, through its indirect consequences, that the development of works of hydraulic improvement gave a decisive impetus to the Emilian "agricultural revolution." In Lombardy between the late eighteenth and early nineteenth century, the development of works of irrigation and the spread of irrigated meadows and rice fields conditioned the impetuous capitalistic development of agriculture to the point of breaking through the limits of the old system of landholding. Thus also in Emilia, between 1870 and World War I, works of hydraulic improvement became the decisive agent of a rapid capitalist development of agriculture, with an enormous increase in capital invested in works of improvement or in agricultural industry, and the formation of a mass agricultural proletariat, into whose ranks flowed numerous laborers from other regions, who were attracted to Emilia precisely by the need to carry these works forward.

The technical, economic, social, and historical conditions, in short, in which the impetuous capitalist development of agriculture in Emilia was realized were thus quite different from those that had characterized the analogous development of Lombardy from the mid-eighteenth century. As in the irrigated part of the Lombard plain, it is true that, even in the open plain of Emilia, and more generally the land recently improved, capitalist development overturned the old system of landholding and subordinated to its needs the very forms of agricultural enterprises—which assumed, precisely through cultivation by wage laborers, typically capitalist economic forms. But where the effect of new works of improvement was only indirect, and the impetus of capitalist development was less impetuous, the first candid subordination of the forms of agricultural enterprises gave way, in contrast to what had happened in the different historical circumstances of Lombardy, to a growing diffidence on the part of the landowning classes, who became preoccupied with the revolutionary pressure of the mass of agricultural workers, and turned against the wage system, and any form of cultivation that seemed to let the masses escape the direct economic, social, and political control of the dominant classes. Thus, the capitalist development of agriculture continued often to be contained here within the traditional form—although it was now clearly inadequate—of the sharecropping system.

The whole period between 1870 and the outbreak of World War I was filled with this explicit polemic within the Italian ruling classes (and not just in Emilia) against the development of new agricultural relationships in the countryside of this region. If these provided results more adequate to the impetus of social production, they also questioned, precisely, the forms of class domination. Around the year 1900, the center of gravity of the movement to exert the rights of the mass of agricultural workers, which had first developed in the provinces of the Lombard plain, moved toward Emilia—which had become a dangerous "red region." In the postwar years after World War I, and then under Fascism, the polemic against "red Emilia" remained central to the preoccupations of the Italian privileged classes, which now confronted it with the violence of Fascist squads and vigilante courts.

In these conditions, and much more than in eighteenth-century Lombardy, one could say that the productive impetus of the development of socially productive forces in Emilian agriculture was entrusted, basically, to the strivings of the mass of agricultural laborers, and first of all of the *braccianti*—day laborers. The form of this study does not permit us to deepen the documentation of this thesis, whose validity is otherwise largely recognized even by scholars distant from the workers movement. In a few cases, as in this one, in fact, Marx's well-known affirmation seems evident and significant: the revolutionary class is the first and most decisive productive force in a given society. And in exercising this decisive force, which produced technical, economic, and social progress, the Emilian proletariat was conquering its mature national class consciousness.

We will not enter here into a deeper examination of this aspect of the problem, which we have treated elsewhere. Let us limit ourselves, rather, to studying the effect of the processes we are examining on the form of the Emilian agricultural landscape, and look briefly at some statistical data, that we can illuminate in more detail in this regard.

The first and most obvious incidence of the development of the great works of hydraulic improvement in Emilia after Italian unity is given, naturally, by the new and more obvious significance assumed, in this region, by agricultural landscapes such as the *larga* (open plain) of Ravenna, or the lowlands of Ferrara, which became the theater of grandiose strivings of day laborers. By *larga* one means, in Emilia, the vast extent of land in the plain, that generally included the zone of recent improvement, and was provided with a large network of hydraulic systematization, but was not arranged in farms, and was as yet without trees. Plate 74—which reproduces a plan of the Gallare holding of the Istituto di Fondi Rustici from a report of this great joint-stock company—gives a more concrete image of the landscape of the *larga*, as it appeared in a zone of the lowlands of Ferrara. In 1878, the marshlands later included in the Gallare holding were sold by the Commune of Comacchio to two private individuals, who initiated the hydraulic improvement of the land. This continued and was brought a good ways forward by the bankers Klein from Vienna, who in turn, however, were obliged by a judicial ruling in 1888 to give over their administration to a receiver. Finally, in 1892, after expropriation and an auction, the Banca d'Italia took ownership to satisfy its own credits in the matter. The Istituto di Fondi Rustici acquired the holding from the Banca d'Italia.

The history of this holding is typical for the problems it had in common with many other improvement projects carried out often by Italian or foreign stock companies that took over from those who had first initiated improvements with technical experience but often insufficient capital. The zone of the *larga*, as we have already indicated, was in fact a preferred zone for large and very large capitalist enterprises, where the management of landed property was most often subordinated to the needs of capital, which presented itself here often—given the size of the financial involvement required—in the most up-to-date form of joint-stock capital, which was typical of finance capitalism. Extending over an area of 3,700 hectares, which were included within the drainage area of the Consorzio Scoli Polesine di San Giorgio, the Gallare holding was divided in 1906—as also appears from our illustration—into six sectors, each of which was assigned to an agent, with an assistant, who

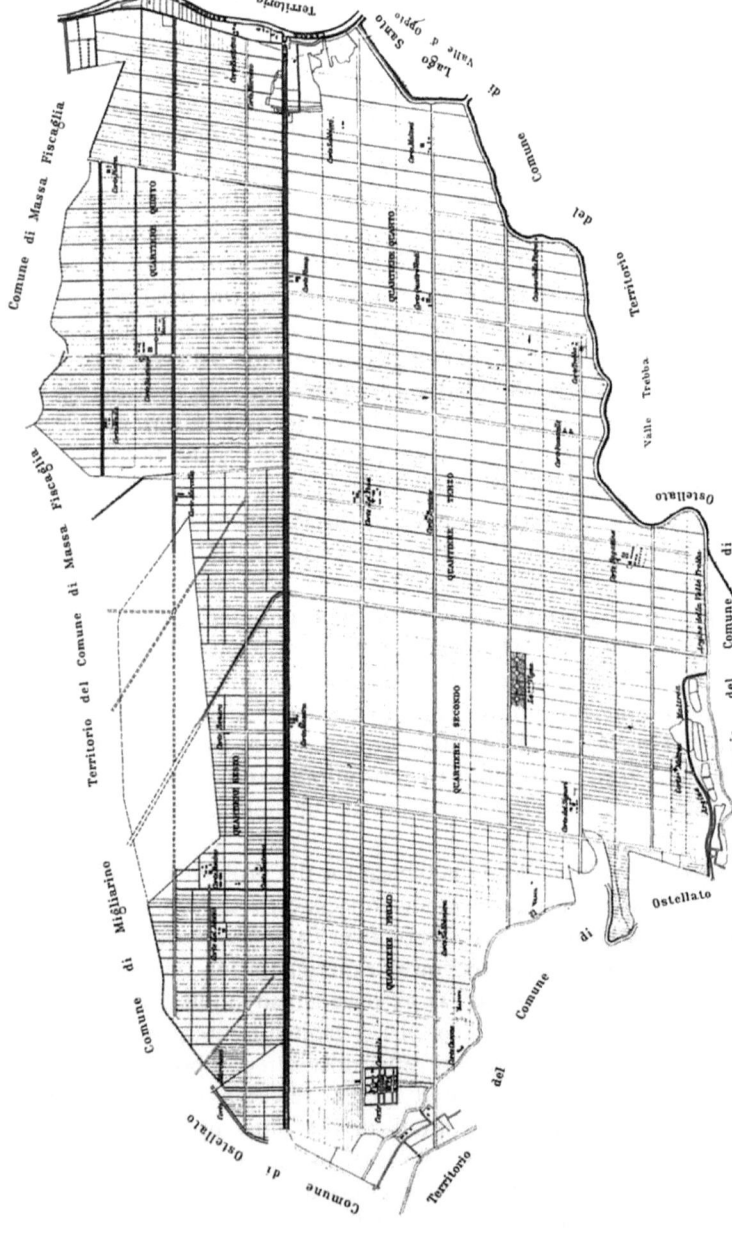

Plate 74. The landscape of the *larga* in the early twentieth century, in a map of the Gallare holding of the Istituto di fondi rustici.

supervised the economic management of the holding in the interest of the Istituto di Fondi Rustici. The drainage consortium, for its own part, arranged the drainage of water through the three great centrifuges of the Stabilimento Idrovoro di Marozzo, which were situated along the northeast boundary of the holding. These were capable of pumping 10 cubic meters of water per second, more than enough to handle the usual 3.60. At the onset, the holding was further divided into large subdivisions of 10 hectares that were bounded by longitudinal and transversal water channels; but gradually, with the progress of improvement and hydraulic arrangements, the size of subdivisions was reduced, and they were covered with a network of smaller drainage ditches. In the year 1906–7, these already measured 1,000 kilometers for the holding as a whole, while internal longitudinal, transversal, and minor roads had reached the extent of 102 kilometers.

We will not linger further on the description of this holding, which can serve as an example characteristic of the landscape and social relations of a zone of the *larga*. We will only add that each of the great sectors into which the holding was subdivided had an administrative center, called a "Corte," consisting of buildings for habitation, stalls, and storehouses where the harvests were collected. For grain, for instance, these increased from 20,889 quintals in 1887–88 (with a mean of only 9.8 quintals per hectare) to 59,795 in 1906–7 (with a mean of 26.1 quintals per hectare).

Even this rapid technical, productive, and social evolution can be thought typical of the environment of the *larga*, where the impetus for improvement most completely subordinated landed property to the needs of capital, and gave its revolutionary efficiency a freer rein than was permitted elsewhere by the residue of feudal obstacles. But with regard to the more specific theme of our study, we should indicate that now, again, a landscape like that of the *larga* was imprinted with regular and well-defined forms through the implementation of a true and proper plan. The landscape was no longer abandoned to casual individual initiatives, but was inspired by technical needs for improvement, and promoted by a powerful association of capital.

This was certainly not yet a public initiative, or subordinated to the public interest. The Beccarini law on improvements, which dated from 1882 and was not otherwise lacking in value, could not overcome the presupposition of private interests. Even later legislation, for that matter, came up against the impassable obstacles in contemporary society against the elaboration of a true and proper plan for the landscape, when it was confronted with the objective (and much more effective) laws of capitalist profit and the monopoly of landed property. But within these limits, there is no doubt that precisely within the context of the *larga*—

and, in general, in great works of improvement—the ability of a human association to imprint itself on the forms of the agricultural landscape showed itself with particular significance. This was an indication, in the actual society of that time, of what such association could realize, if it were no longer subordinated to the interests of capitalist profit.

For the meantime, however, this subordination posed limits not only to a more perfect elaboration of forms and social relationships in the landscape of the *larga*, but also to its further extension. This was continually challenged, not so much by the progress of hydraulic and agricultural systematization in zones of less recent improvement, as by the efforts made by the landowning classes—for the social and political regions already indicated—to expand the sharecropping system and the landscape of the *piantata* even on the land that was newly reclaimed for cultivation: the forms of the *piantata*, in their origin and functional meaning, seemed so closely tied to this system.

For the provinces of Ferrara, Bologna, and the Romagna, in general, where the dynamics of these processes were the most lively, their global result, despite the poverty of statistical data, can be seen concretely through the following figures. About 1840, the data elaborated by Galli provide an extent of sowable land without trees, and hay pastures, for these provinces of about 200,000 hectares, while the landscape of the *piantata* in the same provinces extended over 360,000 hectares. About 1909, the landscape of the *piantata*, in the data from the agricultural survey, appears to have enlarged to 466,000 hectares in these provinces; while, in 1929, the agricultural cadastre shows it reduced again to an area of 372,000 hectares, in comparison with 338,000 hectares of sowable land and hay pastures without trees. In the region of Emilia as a whole, between 1909 and 1929, the landscape of the *piantata* shrank from 968,000 to 831,000 hectares, while the area of sowable land and hay pastures without trees increased from 396,000 to 591,000 hectares.

As one sees, the first period of impetus for clearings and systematization, which lasted until about 1870, was marked by an unchallenged absolute and relative new extension of the landscape of the *piantata*. And with the initiation of great works of hydraulic improvement, about the end of the nineteenth century and beyond, in some provinces, there continued to be an absolute extension of plantations, but this began to be challenged, relatively, with the more rapid extension of the landscape typical of the open plain on land of recent improvement. Still more: in the region of Ferrara, for instance, already before the end of the nineteenth century a true and proper regression of the *piantata* began, which affected not only its extent, but also the density of the plantations of trees and shrubs. Between 1909 and 1929, this process of absolute and relative decrease in the landscape of the *piantata* spread and generalized itself to the whole region of Emilia. There did not lack, naturally, chance reasons for this regression. In the region of Ferrara, for instance, already after the freeze of 1879 and the consequent mortality of vines, viticulture was abandoned in many places; and a more general cause for the reduced import of viticulture on Emilian farms was the spread of vine diseases, particularly phyloxera. But it would be a mistake to limit oneself to these, shall we say, chance causes: through them were expressed deeper and more lasting reasons for the regression of viticulture and the *piantata* in Emilia, reasons to which we will return in the concluding chapter of this volume.

Up to this point, however, we have wanted to show how the contrast between the *larga* and the *piantata* was not only a matter of landscape, but also expressed a much deeper contrast between technical, productive, organized progress adapted to the needs of a more candidly capitalist economy, and the limits that the dominant capitalist classes themselves imposed on this development in the interest of protecting their class dominance. It was not by chance in the years of the Fascist dictatorship when this attempted political and social reaction of the privileged groups became more open and vocal and subordinated the whole apparatus of the state to its ends—against the landscape, against productive organization, and against the people of the *larga*, the mass of day laborers—that the state, the organizations of directors of enterprises, and the official schools of agronomy and agricultural economics were all engaged in a polemic and relentless struggle to defend and further spread the restoration of individual farms. One cannot say that this attempt was without negative results, in the sense of a certain slackening and, sometimes, involution of productive relationships that were more clearly adapted to technical progress. Even through the two decades of Fascism, however, and still more in the period after World War II—as had already occurred in the years between 1900 and 1922—the typical landscape and productive relationships of the *larga* continued to spread, not only in Emilia, but throughout the improved zone of the Po, from Cremona and Mantova to Parma and Moglia, from the Burana to the Venetian lagoon, the Basso Piave, and the Bassa Friulana, leaving its mark on vast territories in these regions of the North.

. X .

AN AGRICULTURAL PANORAMA OF CONTEMPORARY ITALY

· 84 ·

The Agricultural Landscapes of Contemporary Italy

What we have revealed of the process of elaboration of the principle types of landscapes of Italy in the preceding chapters, will permit us to treat the larger contours of an agricultural panorama of contemporary Italy more rapidly in this concluding chapter, as observed from the particular viewpoint of the forms of the landscape. It is necessary to indicate also how much the evolution of these forms in the period in question was conditioned not only by techniques, but also by the dominant productive relationships in the countryside, and particularly by a certain outcome of the struggle that the mass of labors and small agricultural producers have fought for their social liberation, and for the progress of Italian agriculture. An even rapid account of the development of these processes, which have been contradictory in many aspects, is indispensable for clarifying the nature of the agents that have had a decisive role in recent developments of the process of transformation and reelaboration of the Italian agricultural landscape.

Among these agents, particular attention is due, as we have already indicated, to the struggle by laborers and small agricultural producers. For the first time, in the decade and a half that has passed since the liberation of Italy at the end of World War II, the organized actions of peasants have shown an ability to orientate themselves systematically not only toward immediate objective goals, but also toward the realization of structural reforms that might assure stable social progress in the context of an effective policy of agricultural development. Among the most visible results of these struggles to influence deep changes in the agricultural landscape, have been those that led to the practical liquidation, in the greater part of the countryside, of the traditional agricultural systems of temporary clearings and bare fallow, which up to the eve of World War II had remained typical of large tracts of the economy of latifundia in the Tuscan and Roman Maremmas, in the South, and in the Islands, and that had become even more extensive during the war itself. Thus, still in 1950, sowable land "in repose" extended to at least 1,205,000 hectares (of which 190,000 were in Lazio, 365,000 in the South, and 516,000 in the Islands), as compared with 1,200,000 hectares in 1929. This landscape of the traditional latifundium, at Melissa, which was rescued from desolation through a bloody peasant struggle, is shown in

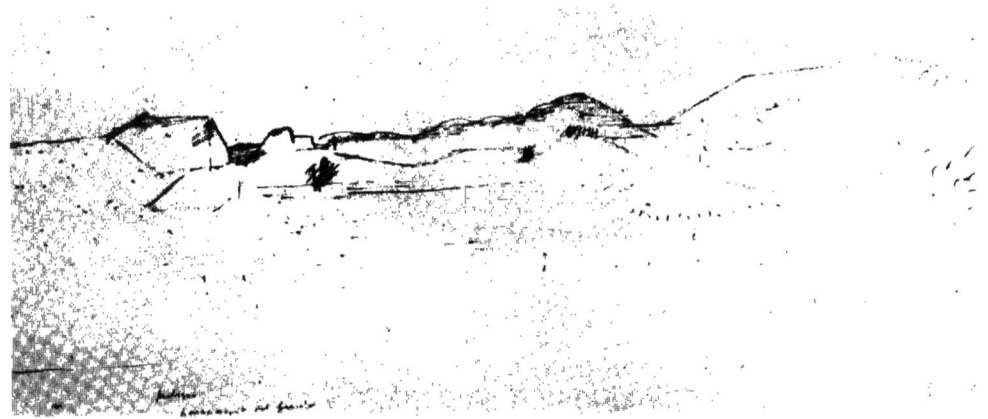

Plate 75. Ernesto Treccani, *Melissa*.

Plate 75 through the vivid image of a suggestive drawing by Ernesto Treccani. Already by the year 1950, however, the great movement to occupy uncultivated or poorly cultivated land, which was legalized by the Gullo decree, had reduced to an uncertain, but certainly significant, degree the extent of sowable land "in repose" with respect to the maximum (undoubtedly greater than 1929) reached at the end of the war. In contrast to what (to a much lesser degree) had happened after World War I, however, in the post–World War II period —under the leadership of united peasant organizations—the movement to return wastelands to cultivation did not exhaust itself, but developed into a great movement to reconquer the land (in which one of the culminating moments is reevoked in the painting by Guttuso reproduced as Plate 76). The victories of this movement are inseparably linked to the name of Ruggieri Grieco. Directly and indirectly, these victories undoubtedly introduced changes of a permanent and structural nature into the economic and social reality of the Italian countryside, and into its agricultural landscape. It is enough to consider, in this regard, that under the pressure of this impetuous mass movement the Italian governing classes, precisely in the years of the advancing capitalist restoration, could not avoid introducing legislation for agricultural reform, through which, although in a limited and distorted way, owners of latifundia saw the extent of their domains reduced by 762,000 hectares, which were mostly assigned to 109,000 families of *braccianti* and to poor peasants.

One can easily understand what deep changes these first victories in the struggle for the land introduced into the Italian agricultural landscape. It was not a matter only of the virtual liquidation of agricultural systems of temporary clearings and bare fallow, which had their greatest

Plate 76. *Occupation of uncultivated lands in Sicily* by Renato Guttuso.

concentration precisely in the lands of latifundia subject to expropriation. Lacking data for the extent of lands "in repose" for the nation as a whole comparable with those already cited for 1929 and for 1950, it is enough to remember that on lands expropriated by the reform commissions—where lands "in repose" had accounted for a considerable part in the past, or even (as in the case of the estate of Flumendosa) the estate's whole extent—the proportion designated as "in repose" had already fallen in 1953, as a result of the movement to occupy uncultivated lands, to 8.2 percent. It fell to 4.1 percent in 1956, and it fell further more recently to an almost negligible figure. Lands that had yielded, in the hands of the owners of latifundia, a mean gross marketable product certainly much below the 71,000 lire per hectare revealed in 1953, in the hands of the new cultivators rose to 95,000 lire in 1956, and 115,000 in 1958. And deeper changes that the peasant conquest of lands induced into the agricultural landscape of the ex-latifundia are significant: besides the disappearance of "repose," new forage and industrial crops in crop rotation and new plantations of trees and modern breeding played an important part in the enterprises of those to whom the lands were assigned.

But in reality, the positive results of the peasant struggle, in the sense of a progressive liquidation of the systems of temporary clearings and bare fallow, along with the forms of the agricultural landscape inherent to these, spread well beyond the area controlled by the reform commissions. Under the pressure of these struggles—and it was only through fear of occupation or expropriation—even in latifundia that escaped expropriation through the limited and one-sided legislation, owners were sufficiently pressured to conceal, if not other signs, at least the most obvious signs of their productive and social backwardness. Thus, all together, the landscape of the region of traditional latifundia visibly changed. But, as we have already warned, it was not only, or so much, the progressive liquidation of agricultural systems of temporary clearings and bare fallow that conditioned the more visible and decisive changes in the landscape of the areas under reform. These resulted rather from the introduction into the shapeless form of the latifundium of elements of organization, a plan for the landscape that (although it was often in fact discussed and criticized) nonetheless still documents what human association was able to accomplish when it concentrated its energies in a conscious and planned way on a particular terrain, however naturally and historically obtuse this might be, to impress on it forms adapted to the need for economic and social progress. But still more, however limited its imposition and distorted its application, the landscape plan of reform always remained a plan, conditioned and reoriented in a progressive direction, which the mass struggle induced and induces, and subject to changes in techniques and agricultural systems, or in relationships of property and production. Much more than was true of plans for improvement of a traditional type, however, a plan for reform, like the one the mass struggle succeeded in imposing on the Italian governing classes, had a direct effect on the structures of the Italian rural economy and society, on the orientation of collective life, and on the relocation of human settlements in the countryside. And in this way, the changes induced by the struggle for reform of the Italian agricultural landscape appear, under many aspects, not only as the most conspicuous, but even as those which most concretely proposed the prospect of a modern, democratic, and socialist form of development in Italian agriculture.

This last consideration is particularly valid with regard to what can be considered the central problem of this development, which—in the concrete environmental and historical conditions of Italy—was to provide new and original forms by integrating hundreds of thousands of peasant enterprises, both individual and associated (but founded on the realization of the principle of "the land to those who work it"), into large enterprises able to respond to modern technological and economic conditions. In the zones under improvement, reform centers had the task of coordinating the individual initiatives of those who were assigned lands according to a general plan, and of resolving fundamental technical and economic functions in associative or cooperative ways (plans for cultivation, mechanical operations of improvement, work, harvest, the industrial transformation of agricultural products, purchases and sales, financing, and so on). The mass struggle seemed to have succeeded in putting into operation, at least in a tentative way, precisely the type of relationship between individual peasant initiative and great cooperative enterprises, which could correspond to the concrete environmental and historical conditions of Italian agriculture. And despite all the limitations and problems that resulted from the instrumental character of current legislation and the antidemocratic, conservative, and reactionary administration of the reform commissions, this type of relationship has confirmed its potential effectiveness.

It was thus not by chance—considering the failures in carrying out the timid plan for agricultural reform, which was conceived as a diversion to the increasing radicalization of the peasants—that the Italian governing classes, the party with a relative majority, and the governing bodies directed by these soon did not limit themselves to calling a halt to the work of general reform, to which they were solemnly committed by the republican constitution and their own reiterated declarations. They even decisively reversed the course of their own policy, and progressively demobilized the commissions, concentrating their blows, particularly, on the reform centers that were supposed to guarantee essential conditions of survival and development to the small individual enterprises of those to whom land had been assigned. We are confronted here, in short, with one of the most clear manifestations (as we have already indicated) of the contradictory way in which changes in productive and property relationships operated in the postwar period, and still operate, in transforming the agricultural landscape of Italy: gradually introducing into it elements of modern organization under pressure from a progressive struggle of the masses, and then, conversely, disaggregating it through contortions and plots when the influence of the agents of capitalist restoration and monopolistic expansion succeeded in prevailing.

Already on the eve of the passage of the first reform laws, in fact, the negative effect of these agents appeared with the attempt to substitute institution of a Cassa per il Mezzogiorno (State Fund for the South) for the commitment to a general agricultural reform. This was designed to check the impetus of the struggle for land, breaking unity of action by offering to groups of peasants who were relatively better off the possibility of individual land ownership. Already of itself—despite the land hunger of Italian peasants—limitation of the efforts at reform to an area of less than 5 percent of the cultivated and forested territory of Italy served inevitably to heighten the effort of agricultural workers and small producers to find some way to assure, if nothing else, at least the possibility of individual access to property, through whatever means and at whatever price. This more and more—with the development of mechanized processes under the direction and in the exclusive interest of the dominant groups, and with the consequent worsening of overpopulation in the countryside—seemed the only true guarantee of minimum employment for the direct agricultural producer and his family. Even before initiation of the new legislation for the Cassa per il Mezzogiorno and peasant property [in 1950—Trans.], however, the absentee landowners (who were frightened by the spread of the peasant movement and the threat of expropriation) had begun to find it to their advantage to alienate the more or less marginal parts of their holdings to direct cultivators. By this means—which was to the financial advantage of the absentee ex-proprietors, whose generally quite high prices were guaranteed by the land hunger of the peasants—the subsequent legislation for the Cassa per il Mezzogiorno and peasant property hastened the extension of land under direct cultivation in these years, which reached, probably, in its quantitative extent, the 762,000 hectares included in the reform legislation.

Thus, through the direct or indirect effects of the peasant movement, property and productive relationships deeply changed in the direction of a break in the traditional backward structure of the Italian agricultural economy over a total area of about a million and a half hectares. But, aside from any other economic and social consideration, that we will not expand on here, it is easy to see how diverse and contradictory the effect of this large extension of the cultivated area was in its impact on the forms of the agricultural landscape, given the variety of ways in which it occurred. Where the extension was in fact effected through a reform plan, however limited and distorted, elements (which we have already indicated) of modern organization were introduced into the agricultural landscape, which are visible to even the casual observer. Traveling along an autostrada, or looking from the window of an airplane, one can see the now defined and precise contours of a landscape once dominated by the formless desolation of the latifundia. Where, in contrast, in the interest of the absentee ex-proprietors the extension of cultivated land was abandoned to the spontaneous "land hunger" of the peasants, or to an individual and casual access to land, the result, rather than producing a new and more organic elaboration of the forms of the agricultural landscape, has been an extreme fragmentation, to the point of a true and proper deterioration. This has often expressed itself in the further extension of a landscape of irregular fields on an ever reduced scale, where the parcellation and dispersion of plots has not rarely preceded their abandonment by the peasant cultivator, whether he was a proprietor, renter, or *colono*.

The agents and processes that we have indicated did not limit their contradictory effect to the zones where the peasant struggle had succeeded, in one way or another, in changing the system of landholding. The tyranny of space does not permit us to expand on the influence that other great aims of the peasant struggle of these years—like those to transform payments in kind, agricultural contracts, improvement, mechanization, and so on—had in changing the agricultural landscape of Italy. It is enough to mention, here, that the generally positive effects of these struggles to attain a productive democratic development of the agricultural economy was felt not only in the zones of expanding peasant ownership, but also—and to a greater extent—in agricultural enterprises of individuals or associates that have spread over more than half of the cultivated surface of Italy. But this is not to say that the dominant groups of Italy in confrontation with the positive effects of the mass struggle, have given up pursuing their antisocial objectives, conditioned only by the law of greatest profit, or that—even in the sector of larger agricultural enterprises—their policy has not induced into the forms of the agricultural landscape elements of disaggregation that are basically analogous to those we have seen with regard to peasant holdings. The further spread over large areas—alongside the limited islands where the networks of farms have grown—of the landscape of ever more fragmented and dispersed irregular fields (*campi a pigola*) that lack fully adequate size to meet the needs of modern techniques and economic procedures has been frequently only the prelude to further disaggregation of the agricultural landscape. This is abundantly documented by the preoccupying spectacle of the regression, in large areas, from cultivation to pasturage and woodland, by the depopulation of entire villages and valleys, and finally by the abandonment of hundreds and thousands of farms in every province of Italy—particularly in the mountains and high hills, but now more and more often in the plains as well.

Space does not permit us to show, in detail, the extent to which these contradictory processes have been influenced by mechanization, the orientation of policies of public investment, the adhesion of Italy to the European common market, and the sudden reversal of the grain policy of the Italian governing classes. These are themes that we have had the opportunity to discuss in other and more appropriate places, to which the reader who would like to study them more deeply may refer. It is enough to indicate here that the comprehensive result of these policies has been the expulsion in the past few years of more than a million laborers and small producers from the productive process of agriculture; an increasing retardation of agricultural development and revenue in comparison with industry; and further aggravation, even within the agricultural sector itself, of the disbalance between North and South. The general situation of Italian agriculture, in short, has reached a dramatic state, which the president of the Council of Ministers [Ammintore Fanfani—Trans.] recently called *preagonica* (preagricultural).

How can one explain a situation of this kind, in the context of a general economic trend usually considered to be particularly favorable to Italy? We must necessarily limit ourselves, even in this regard, to a simple statement of the situation that may have some value in putting our inquiry into context relative to the recent evolution of the forms of the Italian agricultural landscape. The reality of the matter is that the peasant struggle in the countryside did not fail to have a significant effect on relationships of power within the governing classes. In some ways this followed, and in others it contradicted, relationships derived from the laws of development of capitalist society in its current phase. A serious blow was undoubtedly struck at the class of absentee great landowners, who clearly lost economic and political weight in national life. To a certain extent, the redimensioning of traditional influence exercised by the large landowners complicated the role of some agricultural capitalists who were not entirely associated with the economically and politically dominant elite. But in general, the decisive process for the reconfiguration of power relationships in the countryside has been the increasing subordination of agriculture, considered as a whole, to the new superpowers of industrial, commercial, and financial monopolists, who—with the breakup of the government of national unity after World War II, which had barely initiated its work of democratic reconstruction, and with the initiation of a true and proper capitalist restoration—have found its characteristic expression in a clerical-political monopoly [the Christian Democratic Party—trans.]. The more and more complex interaction of interests of the economic and political powers, and this underlying clerical influence, has given new importance to the forms and methods of statist monopoly capitalism in the countryside. In this situation, the power of the state itself, in its fiscal policy and foreign trade, its policy of public investment, and its policy of price controls, has served more and more closely interests that are hostile to, or at least estranged from, the agriculturally productive groups and agriculture.

Still more: despite the positive results that the mass struggle had produced in these years, through the liquidation of certain of the most jarring feudal residues in the countryside and the progressive extension of peasant property, the dominant group of monopolists—who in their very origins were alien to strictly agricultural interests—has sought its greatest profits not only, or so much, in the productive development of agriculture, which might permit them larger margins of profit, than from a true, proper, and systematic pillaging of agriculture. This has been effected through control of the circulation and distribution of agricultural resources and products needed by farmers, through fiscal policy, investment, credit, and so on. Control of such commercial monopolies as the Federconsorzi [the state agency that distributed agricultural machines, fertilizer, and other resources—Trans.] have become essentially instruments for assuring a solidarity of interests among the major agricultural capitalists, and with the dominant capitalistic and clerical party.

One is thus not surprised, in these conditions, that the progressive tendencies induced into Italian agriculture (and into the very forms of the Italian agricultural landscape) by the struggle and successes of the rural masses in the immediate postwar period became intertwined and largely counteracted by the regressive ones we have just mentioned. This has already, in extreme cases, culminated in instances of true and proper deterioration of agriculture and of the agricultural landscape. And these cases have assumed such an extent and significance as to put into crisis— on the more strictly economic level, as their mass political base—the whole agricultural policy of the Italian governing classes. There was an explicit recognition of this crisis on the eve of the administrative elections of 1960, in the already cited declaration made by the Honorable Ammintore Fanfani to the convention of sections of the Confederazione dei Coltivatori Diretti (Confederation of Farmers), and his convocation of a Conferenza Agraria Nazionale (National Agricultural Conference), where a new line of agricultural policy would be elaborated, to replace the one that it was necessary to recognize was inadequate and had failed.

This is not the place, it is well understood, to expand on the more realistic alternative that this realization (and the very convocation of the Conferenza Agraria Nazionale) promoted, which was not only an alternative agricultural policy, but also an alternative of power, an alternative appealing to social groups not representing monopoly capital, large-scale agricultural capitalism, and the great landed properties barricaded around the political monopoly of the clerical party, and to whom alone could be entrusted the salvation and effective progress of agriculture. But what we have indicated will be sufficient, we hope, to put into context and explain what we will say in a more analytic way in the following part of this concluding chapter about the contradictory character of the recent evolution of different types in the Italian agricultural landscape.

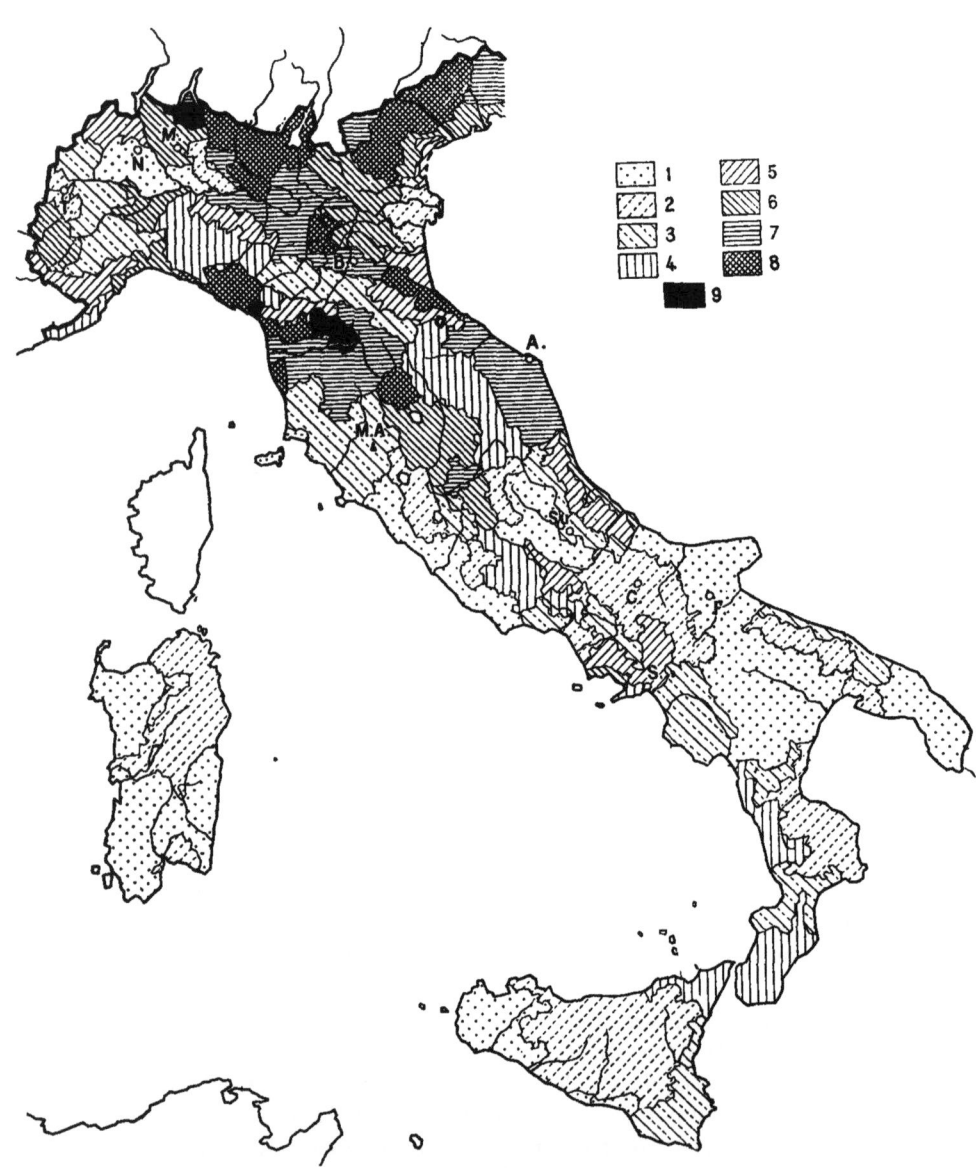

Figure 17. The relationship between zones of sowable land with trees and sowable land without trees in Italy [circa 1960] (darker areas are zones with trees, lighter areas are zones without trees). N = Novara; M = Milan; B = Bologna; A = Ancona; M.A. = Monte Amiata; Su = Sulmona; C = Campobasso; S = Salerno.

A glance at Figure 17 provides a summary picture of the geographical distribution of these different types in various zones of peninsular Italy and the Islands. From a clear predominance of sowable fields without trees—indicated by simple dots or dashes—in most of southern Italy and the Islands, one passes, in central and northern Italy, to a landscape dominated by sowable land with trees or shrubs, which are indicated in the map by darker hatching. This summary picture confirms the importance that is still preserved in a large part of Italy by the forms of the *piantata* of the Po valley, and the *alberata* of Tuscany, Umbria, and the Marche, of which we have explained the historical evolution to readers through the course of our study.

When we speak of sowable land with trees or shrubs, however, we should not identify this prevalence only with the typical *piantata* or *alberata*. A closer examination of statistical data available is needed to provide further specificity as to the true extent and geographical distribution of this type of landscape, of which we have revealed the origin and importance.

Consider, first of all, the *piantata* of the Po (of which one typical form is shown in the painting by Borgonzoni, reproduced in Plate 77), characterized—with its different variants—by longer and broader fields, important hydraulic arrangements, and the association of herbaceous crops with vines, which are raised high up on live supports. As a first approximation, we have seen that an index of the extent of this landscape might be taken from data in the agricultural survey relative to the mixed cultivation of vines in the Po valley, from Piedmont to Lombardy, Emilia, and the Veneto. For these four regions as a whole, the surface with mixed cultivation of vines passed from 1,932,000 hectares in 1911 to 1,480,000 in 1929, 1,344,000 in 1949, and 1,196,000 in 1957. One cannot say, naturally, that this extent of mixed cultivation of vines coincides exactly with the landscape of the *piantata*. Cases are known where mixed cultivation is practiced even in forms that do not conform to the more typical ones we have described. The truth of the matter, however, is that in the great majority of cases the typical form of the *piantata* of the Po with mixed cultivation of vines was practiced in these provinces. That is true, particularly, for areas of the plain for which, however, the area planted in vines—which fell from 1,280,000 hectares in 1911 to 1,043,000 in 1929, and 928,000 in 1949—reflects even more precisely a progressive diminution in the context where the landscape of the *piantata* continued to dominate. One can thus demonstrate that after having reached its maximum extent in the decade immediately after the turn of the century, this landscape, with its characteristic festoons of vines between live supports, is now limited to a progressively more restricted zone; and if this has been particularly true of Piedmont

Plate 77. The *piantata* of the Po in a painting by Aldo Borgonzoni.

and Lombardy, a drastic reduction of the *piantata*, which has accelerated in recent decades, can be demonstrated also for Emilia and the Veneto, where this type of landscape is nonetheless still the predominant one.

What were the reasons for this progressive decline of mixed cultivation of vines in the Po valley? There is no doubt—as we have already indicated in the preceding pages, with regard to the province of Ferrara for instance—that causes that might seem tangential, like the blight of vines as a result of an exceptionally cold winter, sometimes contributed to the decline of the landscape of the *piantata*. But these, in substance, were only epiphenomena, through which deeper technical, social, and historical factors expressed themselves, and on which we should pause briefly.

We have seen, in fact, how initially the very origin, and then fortune, of the *piantata* of the Po valley was due, at that time and then almost up to our own day, to the need to adapt viticulture and wine production to environmental conditions, and that in this region the conditions were not particularly favorable. In a period when the currents of trade and the commercial development of agriculture were rather limited, it was natural that local populations—who had become accustomed to the use of wine in other geographical contexts—should resort to the ingenious system of the *piantata* to make viticulture possible even in the low-lying heavy soil of the Po valley. But it was no less natural—once a national market for agricultural products was formed, and transport, trade, and exchanges from region to region became easier—that viticulture would tend to concentrate itself again in environments whose geography, climate, and soil provide more objectively favorable conditions.

But this was not all. We have seen the part that a traditional lack of essential forage crops for agricultural enterprises played in the diffusion of the *piantata*. Until the introduction of forage into crop cycles, and the spread of artificial meadows, it was natural that boughs of trees from the *piantata* had an important function in integrating forage for feeding beasts of burden and cattle. It is understandable that with the spread of modern crop systems, where forage was integrated into the productive cycle of agriculture, and with the wider use of concentrated artificial feed, this reason to extend the *piantata* also lessened. The same can be said for other essential functions that the trees of the *piantata* filled, and in part still do, to provide the farmer with firewood and wood for building that was essential to his enterprise and domestic economy. Even this function, clearly, has been progressively reduced, partly through the prevalent use of iron, cement, and so on, for construction and rural tools, and partly through the increasing spread in these provinces of electric energy and, still more recently, of liquid and gas fuels, whose use has revolutionized the peasant's domestic economy, as well as the economy of the individual agricultural enterprise.

It is true that the monopolistic prices farmers have had to pay for such new fuels, and for electric energy, iron, and cement, often reimposed on them the use of wood from the *piantata*, which has nonetheless gradually diminished. One must add, as well, that the traditional forms of the *piantata* with trees and vines has become inadequate to new productive needs through the increasing use of machines for cultivation and harvesting, not to mention the more and more urgent need to develop works of irrigation. Thus one can truly say that the traditional landscape of the *piantata* of the Po valley has entered into a crisis whose development should be studied carefully, foreseen, and managed, so as to create new environmental conditions more adequate to the more rapid development of the agriculturally productive forces in these provinces. The productive and technical groups, and the agricultural workers, who—through the centuries, and particularly in the nineteenth century—knew wisely how to elaborate the arrangement and forms of the *piantata* and bring these to the highest perfection, would undoubtedly have at their disposal what is needed to undertake, in historical perspective, this work of adjusting the traditional form of the landscape, and the organization and dimensions of farms, to the new technical, productive, and social needs. It is enough to consider, however, the close connection between the just mentioned technical and economic processes, and (for instance) the now universally recognized crisis of sharecropping, to understand—more than ever—how this adjustment might require deep changes in property and productive relationships, to which the dominant groups might put up new resistance. Thus, once again, even for this reason, it is precisely to the capacity and will of agricultural technicians and workers, and to the success of their efforts, that the first line of adjustment should be entrusted.

Meanwhile, it is not infrequently the case of the typical landscape of the *piantata* of the Po—and one is illustrated in Plate 78—that vines, destroyed by an infection of phyloxera, are not replanted. Thus, already by this means, an objective process of adjustment of the traditional *piantata* to new market and productive conditions has begun. In other cases, the *piantata* had already assumed atypical forms in the past, with simple lines of trees (particularly mulberries) that were not necessarily festooned with vines but were associated with herbaceous crops. During the past century, with the rapid spread of the cultivation of mulberries, this type of landscape became typical of large zones of the Piedmontese and Lombard hills and higher plain, for instance. And as an index of the spread and dynamics of the landscape of the atypical *piantata* (which more rarely, and particularly in the Veneto, also extended into the plain), let us consider the surface of the hill zone covered with sowable fields with trees but not associated with vines, that passed from 289,000 hectares in 1907–10, to 240,000 in 1929, and then continued to diminish in the following decades as the raising of silk worms declined, and consequently, also the cultivation of mulberries. But—to remain in the context of landscapes related or associated in some way to the traditional *piantata* —it is worth pausing on the fate of the landscape of rice fields along the Po, and also of rotated meadows, that were often irrigated.

We have already seen how, particularly from the second half of the eighteenth century, the cultivation of irrigated meadows and rice fields had spread on new lands brought into cultivation, especially in Lombardy, but also in Piedmont and the Veneto, and often at the expense of the typical *piantata*, which adapted poorly to the development of irrigated crops. But we have also seen that certain aspects of the traditional *piantata* persisted here, even in the more typical landscape of rice fields and irrigated meadows, whose boundaries were also marked, generally, by lines of trees, embankments, irrigation channels, and so on, that remind one, in a certain way, of some forms of the French *bocage*. But

Plate 78. *Branch Cutter* by Aldo Borgonzoni.

aside from the dimensions of fields, which were generally larger in the irrigated landscape of the plain than the *piantata*—and the absence of a relationship between trees and vines—the irrigated zones of the Po were characterized, logically, by a network of canals and embankments, along which were placed lines of essential trees, and arrangements typical of the irrigated cultivation of rice or forage crops. An image of this type of landscape—as it presented itself most typically in the rice fields of Vercelli or Novara—is provided by a painting by Gazzoni in Plate 62. In Emilia it was not uncommon for rice fields to be contiguous with the traditional *piantata*, and they are still today inserted into its landscape, as is shown in the painting by Borgonzoni reproduced as Plate 68.

There is not doubt, however, that the landscape of the irrigated *piantata*, with rice or forage crops, rapidly expanded during the nineteenth century and thereafter, even though lack of comprehensive statistical data for the past century makes it difficult to present this progressive extension in dynamic figures. One must say that, even beyond the already irrigated zone, landscapes of this type are often found in the Po valley where the cultivation of rotated meadows extended beyond the association with vines. In lack of other more precise information, an approximate but slightly exaggerated index of the extent of this type of landscape can be summarized from the data of the agricultural cadastre of 1929, for zones utilized in the plain of the Po valley for cultivating rice and rotated meadows, always keeping in mind that the expression "irrigated *piantata*" also included nonirrigated lands, and thus has a purely conventional value. With these limitations, we can show that in 1929 the landscape of the irrigated *piantata* occupied four zones of the Po, over a surface of about 1,026,000 hectares, of which 137,000 were used for the cultivation of rice, and 889,000 for rotated meadows. These data, we repeat, have only an indicative value, and are quite approximate, especially with regard to Emilia (296,000 hectares) and the Veneto (224,000 hectares), where in reality rotated meadows and grasslands—included here in the conventional calculation of the extent of the irrigated *piantata*—were often inserted in a quite bare landscape, and were not even marked by trees along the boundaries of the large fields. For Lombardy, instead (362,000 hectares), and for Piedmont (144,000 hectares), the conventional figures reflect more closely the true extent of the landscape of irrigated *piantata*, which precisely in lower Lombardy, and in such Piedmontese provinces as Vercelli and Novara, found its preferred place.

Judging from the forms of the irrigated *piantata*, however, the extent of artificial meadows in the Po valley has become not only a basic constituent element of the landscape, but also a decisive element in the agricultural progress of these provinces. Between 1948 and 1957, in the four regions of the Po, the area covered by rotated meadows and grasslands passed from 2,008,000 to 2,276,000 hectares, with an increase clearly superior to that of permanent pastures, or meadows and pastures, which increased in the same period from 1,300,000 to 1,406,000 hectares, although—since this increase was due to abandonment of cultivation and depopulation of whole valleys—it would be difficult not to attribute this in 1957 and the years that followed to significant regression, rather than progress. We are confronted here with a typical case of contradictory developments, which we sought, in the first pages of this chapter, to clarify as to causes and agents. These are easily identifiable in this case, in a deliberate and declared policy of the government and governing class oriented toward the return of vast territories to pasturage, especially in the mountains and high hills, which is considered to be more profitable from the antisocial point of view of capitalist "productivity."

In general, however, in northern Italy—to return to our review of the principal types of landscapes—despite the progressive extension of artificial meadows, the landscape of mountain meadows and pastures (*malga* as they are called in various parts of the Alps) also preserves a traditional importance. One can calculate approximately the extent of this landscape, using mountain meadows and pastures as an index. From the data of the agricultural cadastre these appear to have been 1,431,000 hectares, distributed with 477,000 in Piedmont, 274,000 in Lombardy, 402,000 in the Trentino, and 278,000 in the Veneto. In Liguria and Emilia the surface in meadows and mountain pastures was 72,000 and 100,000 hectares respectively. But—rather than the Alpine type of *malga*—these were the poor and thin pastures of the Appennines, of the Mediterranean type that we will have an occasion to discuss further.

In the Alpine territories, as well, the more pastoral than agricultural landscapes are significant for a productive nexus that ties them to the more properly agricultural landscape of the valleys and plains through the practice of *alpeggio* (taking herds up to pasture in the mountains). It should be noted, however, that there were also truly agricultural landscapes in the mountainous zones of northern Italy, like those characterized by terraced arrangements with banks, grading, and lunettes commonly found in the underlying hills. Even in Alpine pastures and meadows—the landscape of the *malga*—there are arrangements utilizing rudimentary banks. But no statistical source makes it possible to ascertain the precise extent of these various types and subtypes of northern landscapes, of which we can only indicate the most typical centers of diffusion. Similar difficulties appear in ascertaining the extent and development of the landscape of irregular fields, which had a particular importance in the Piedmontese hills, while (as we have already indicated) a bit everywhere in these regions, from the plains to the hills to the mountains, the further extension and fragmentation of this landscape preceded, and has continued to precede in recent years, the tendency of cultivators to abandon farms.

This review, summary though it is, of the principal changes taking place in the agricultural landscape of northern Italy would be deficient, however, if we did not say something about two processes that assumed a particular significance at the end of the recent period. One was the new prominence assumed in some northern provinces by the landscape of modern plantations of fruit trees (particularly apples, pears, and peaches), and the important concentration of production of fruit in regular and systematic alignments of trees, which has replaced their traditional and disordered mixed cultivation. Throughout the northern and central provinces (where this phenomenon has even begun to assume a certain, although secondary, importance in some places), the surface occupied by specialized cultivation of fruit trees, in this new type of landscape, has passed from only 40,359 hectares in 1936–39, to 61,842 in 1949–50, and 120,440 in 1957–58, thus approximately tripling in twenty years. The increase has been all the more significant in that, through concentration in a few provinces (that of Ferrara, for instance), its impact on the form of the landscape is quite evident. But again, the contradictory character of agents operating within Italian agriculture sheds light on the contrast between these data and those relative to the spread of this type of landscape in the South and the Islands, which, if anywhere, was its preferred place in the past. Even if when we add the figures for citrus fruits to those of specialized fruit trees in the South and Islands, we find that the landscape of plantations of fruit trees expanded in the South from 71,307 hectares in 1936–39, to 84,258 in 1949–50, and 96,286 in 1957–58, an increase, as one sees, much less than in the North, an indication that the South and Islands have been loosing their centuries-old predominance.

It is not difficult to understand what meaning these figures have as symptoms of a worsening of the balance between North and South, with regard both to the development of the modern agricultural landscape and to the level of commercial development through the investment of public and private capital in agriculture, and also to the revenue of the rural population. Similar considerations are valid, in part, elsewhere for another process, whose importance for recent changes in the landscape of the North we have already indicated for the irrigated part of the Po valley. This is the development of works of irrigation whose incidence is not limited to zones of the traditional *piantata*, but extends to the whole irrigable surface, over which—even for fields and plots without trees—a network of canals extends, with their characteristic pattern and visible relief. Throughout the countryside, the irrigated surface has grown quite rapidly in recent years, passing from 2,185,000 hectares in 1950 to 2,769,468 in 1958, and 3,272,453 in 1960–61. But the irrigatable surface of the South and Islands—which increased from 273,200 hectares in 1950 to 413,618 in 1958, under pressure of the popular movement for agricultural reform—has remained at 545,688 hectares, with an increase of less than 131,070 hectares in the last four years [1956–60—Trans.], while in northern Italy the parallel increase has been 319,870 hectares, thus further emphasizing the already established importance of systems of irrigation in this part of Italy.

For central Italy, meanwhile, the more modest increase realized in these four years in the extent of irrigatable surface, which has increased by 51,045 hectares, has nonetheless had a notable effect on the form of the agricultural landscape, partly because a rather smaller area, in absolute terms, has been involved than in northern Italy, and partly because much of this has involved hillside lakes, whose typical forms assume a significance, even to the least attentive observer, that is self-evident. On the whole, however, the regions of central Italy—excepting regions like the Tuscan and Roman Maremmas, which have had a more southern evolution—are those which have recently presented the greatest immobility of forms of the agricultural landscape and, if anything, a tendency to involution and deterioration related to the deep crisis of sharecropping. One might consider, to begin with, the traditional distinction between open surfaces with, and without, trees, which we have already seen was characteristic of the regions of Tuscany, Umbria, and the Marche. According to the agricultural cadastre of 1929, the area of simple sowable fields (848,000 hectares) remained much less in comparison both with that occupied by trees, whether in specialized or mixed cultivation (1,292,000 hectares), or by sowable land with trees (1,192,000 hectares). In this sense, the distribution of the open landscape with trees appeared then—and remains today—in Tuscany, Umbria, and the Marche, similar to that of northern Italy, where the landscape with trees occupied 3,094,000 hectares, in contrast with 1,922,000 hectares of sowable land without trees. In southern and insular Italy (including Lazio) instead, this last landscape (4,963,000 hectares) much exceeded that of sowable land with trees (1,163,000), even when one adds to this the area devoted to the specialized cultivation of trees (1,750,000 hectares).

This simple comparison of surface areas, however, is not sufficient by itself to give a concrete idea of the importance of trees and shrubs in the landscape of Tuscany, Umbria, and the Marche, since it does not take account of the density of planting, which has been so much greater here than along the Po. Thus still following data from the agricultural cadastre of 1929, on 2,665,000 hectares of mixed cultivation in northern Italy there were 861 million woody plants, while on 1,192,000 hectares in central Italy (excluding Lazio), there were no less than 604 million, with a mean density of 507 woody plants per hectare in Tuscany, Umbria, and the Marche—which was almost double the 323 plants per hectare on the area in mixed cultivation in northern Italy. And the comparison seems all the more significant when it is limited to zones in the plain, for which the mean density per hectare of cultivation of trees and shrubs was 262 in the North, and 607 in Tuscany, Umbria, and the Marche.

We already know what an important part in the typical *alberata* of Tuscany, Umbria, and the Marche that vines had in the more dense plots of trees and shrubs of central Italy. But what can we say, more precisely, about the extent of the traditional landscape of the *alberata*, and its dynamics over the last decades?

As an index of extent we can take, as a rough approximation, the extent of vineyards in mixed cultivation that present more or less the typical form of the *alberata* in these regions, with its rows of trees festooned with vines. In the three regions just mentioned, the extent of land used in this way passed from 1,197,000 hectares in 1913, to 986,000 in 1929, 988,000 in 1949, and 957,000 in 1957, with a fairly rapid contraction, therefore, during the immediate postwar period of World War II, but a slower and more limited one in recent decades. The relative stasis of this landscape can be compared with the more serious decline of the *piantata* of the Po, and it confirms both the seriousness of the crisis of sharecropping in these regions, and the obstacles that its preservation has posed to adapting forms of the landscape to modern technical and economic needs. These considerations become still more valid when one proceeds to a more minute examination of the landscape of the *alberata*, taking this term in its most precise meaning of a landscape of vines interspersed with live supports. The agricultural cadastre of 1929 permits us to define the extent of the landscape even in this more precise sense, and as an index we will take cultivation of vines with living supports, which still occupied 327,000 hectares in Tuscany, 333,000 in the Marche, and 174,000 in Umbria, a grand total of 834,000 hectares in these regions.

One cannot say, however, that the extent of the *alberata* exhausts, in these regions, the surface devoted to mixed cultivation of vines, and even less the total used for mixed cultivation of other types of herbs and shrubs. In reality, beside the typical landscape of the *alberata*, a notable importance was assumed in these regions by a mixed cultivation of herbs, shrubs, and trees, but that differed from the *alberata* in the disorder and fragmentation of the units cultivated. An index of the extent

Plate 79. *Umbrian Landscape* by A. Salietti.

of this last type of landscape, which we could call disordered mixed cultivation (an image is provided in Plate 79), can usefully be made, with sufficient approximation, from the extent of land devoted to promiscuous cultivation in these regions, by subtracting from the total the extent of land covered with vines on live supports that characterized the landscape of the true and proper *alberata*. According to the agricultural cadastre of 1929, the landscape of disordered mixed cultivation remained clearly in a secondary place in the Marche (55,000 hectares) and in Umbria (75,000) in comparison with the true *alberata*, but in Tuscany, it extended to at least 227,000 hectares and assumed an importance not much less than the *alberata* itself. It does not seem that the situation of this type of landscape has changed substantially since 1929. Even in this regard one can say that in Tuscany above all (but not only in Tuscany), the first successive fragmentation of pieces of land devoted to disordered mixed cultivation was often the prelude in these years to the later abandonment of the pieces of land themselves, and of farms, and a consequent enlargement of the land associated with individual farms, which were, however, difficult to adapt to new and larger dimensions. More and more widely, nonetheless, the landscape of disordered mixed cultivation has invaded and penetrated the *alberata*, specialized plantations of trees, and even sowable land without trees. And this elaborate but disordered landscape, particularly in the hills, is found above all in Tuscany, once the preferred place of specialized systematizations with banks, with plowing around the hill, across the hilltop, or with furrows at angles, and other methods, of which we have followed the history, but of which it is impossible to tell precisely the extent, since no statistical data on these forms exist.

In the agricultural landscape of Tuscany, Umbria, and the Marche, however, with the now almost complete elimination of the agricultural system of fallow and the diffusion of systems of continuous rotation, the absolutely dominant system, as in northern Italy, has become the system of enclosed fields. This was valid, it is understood, for cultivated land, while for pasturage and permanent meadows the system and landscape present, in the central Appennines, notable differences in comparison with the Alpine *malghe*, as well as with the thin and arid pastures of the South. On the whole, the landscape of permanent meadows and mountain pastures of the central Appennine type extended, according to the agricultural cadastre of 1929, over an area of 476,000 hectares, of which 74,000 were in Liguria, 100,000 in Emilia, 74,000 in Tuscany, 135,000 in the Marche, and 93,000 in Umbria. Even with the basic forage of agricultural enterprises, and the breeding of farm animals, however; and still more with regard to the integration of forage crops into the productive cycle of agriculture, the region of Tuscany, Umbria, and the Marche has been far from the dynamism that one encounters in the regions of the Po. It presented instead the symptoms of stasis, which we have already noticed in other matters, but always in relationship with the obstacles that the persistence of sharecropping has placed in the way of a more lively rhythm of agricultural progress. Thus while in the four regions of the Po, between 1948 and 1957, we have seen the area of rotated meadows and grasslands, already quite conspicuous, expand further by another 268,000 hectares, in the four regions of central Italy (including Lazio) instead, in the same period, this area passed from 1,217,000 to 1,372,000 hectares, with an increase of not more than 155,000 hectares. This seems quite modest for regions where large margins undoubtedly existed, in comparison with northern Italy, for the cultivation of forage. But the preoccupying stasis of the agricultural economy of central Italy is still more evident when one considers data relative to forage production. In comparison with what has happened in the Po valley, basic forage crops in these provinces, for the agricultural economy of breeding farm animals, which had been so deficient in the past, has remained at a level not much greater than in 1913.

This is one of the few aspects in which the dynamism of agricultural development has seemed more lively in the South and in the Islands, where, between 1948 and 1957, the area occupied by rotating meadows or grasslands grew from 688,000 to 971,000 hectares, an increase of 283,000 hectares, which is slightly superior even to the Po regions. We find ourselves, as we have already indicated at the beginning of this chapter, confronted with one of the most conspicuous changes that the peasant mass movement succeeded in inducing into the forms of the Italian agricultural landscape, with the liquidation of the traditional system of fallow for pasture that had dominated in a good part of the South. But one should not forget the extent of this increase that was due to the cultivation of forage crops, although this has not brought southern agriculture to the level of the Po valley, or even of central Italy. In fact already in 1948, in the four regions of the Po, the surface of rotated meadows and grasslands (2,008,000 hectares) exceeded by a good 708,000 hectares that of permanent pastures and meadows (1,300,000 hectares); and in 1957, with a further extension of the cultivation of rotated forage crops, which extended over 2,276,000 hectares, the surplus utilized in this way in comparison with meadows and permanent pastures (which had increased to 1,406,000 hectares) appears to have been even greater than in 1948, reaching, in fact, 870,000 hectares. Even in the regions of central Italy, the surface utilized for rotated meadows and grasslands in 1948 (1,217,000 hectares) exceeded by at least 613,000 hectares that devoted to permanent meadows and pastures (604,000 hectares); and the excess had further increased in 1957, reaching 793,000 hectares. In southern and insular Italy, by contrast, in 1948, the surface cultivated in rotated forage crops was at most 688,000 hectares, and remained a good 1,732,000 hectares less than permanent meadows and pastures (2,423,000 hectares). In 1957, the smaller surface area destined to more intensive modern cultivation of forage crops—despite its expansion to more than 971,000 hectares, and a reduction in the surface area in permanent pastures and meadows to 2,304,000 hectares—remained still a good 1,333,000 hectares less that the permanent pastures and meadows were.

These data confirm that, despite the results obtained by the pressure of the peasant movements of the postwar period, the integration of forage crops into the agricultural cycle, which since the second half of the eighteenth century had truly revolutionized agricultural systems, remained still largely to be accomplished in the South and the Islands, where the greater part of the productive surface was still devoted to unrotated forage rather than to forage alternating with other agricultural crops. But still more: in the climatic and environmental conditions of the South and Islands, one should not forget that a simple extension of the surface destined to rotated forage could decisively increase the normal supply of fodder where significant hydraulic works, irrigation, and so on were not also undertaken to increase the productivity of these crops in less favorable years. Precisely in this area—as we have already indicated with regard to the development of irrigation in regions of the North—the contradictory character of the agents operating in Italian agriculture has made all its negative influence felt to the detriment of the South. Here, as we have seen, the relatively rapid development of works of irrigation in the first decade after the war—when the great peasant mass movement to occupy and improve the land occurred—gave way, in the following period to a dangerous slowdown, which has worsened further, in this area as well, the already deep inferiority of the Italian South and Islands.

We find ourselves here, undoubtedly, confronted with the consequences of a deliberate policy of public and private investment by the government and the Italian ruling classes, to impose openly declared criteria of "productivity" (that is, maximum capitalist profit), a policy to which in recent years, it must be admitted, the peasant movement and rebirth of the South has not succeeded in providing an effective and adequate opposition. It is not surprising that in such conditions the tasks of a modern agricultural revolution—at least with regard to basic forage crops in the agricultural economy of the South and Islands—have remained largely still to be resolved. This assertion is all the more valid in that it applies not only to the deficiency of forage crops in rotated meadows, but also to insufficient production in permanent meadows and pastures, however extensive these still are in these parts of Italy. We can thus further reveal that total forage production in the four regions of the Po, tallied in common hay from both rotated and permanent meadows and pastures, passed from 135 million quintals in 1913, to 169 million in 1953, and 178 million in 1957, an increase of 32 percent. In the South and Islands, by contrast, the corresponding total production of forage crops passed from only 49 million quintals in 1913 to 39 million in 1953, and rose to at most 51 million in 1957, an increase of barely 4 percent over 1913.

We find ourselves, in short, confronted with an evident, serious, and worsening deficit, that remains in the basic forage of the southern economy. This, as we have already indicated, should be considered in relation not only to the insufficient extent and production of rotated meadows, but also to the progressive decline of permanent meadows and pastures in the South, whose effective production cannot in any way be compared with that of the Alpine *malga*, or even the permanent meadows and pastures of the central Appennines. And up to the present a large part of the agricultural landscape of the South and Islands has continued to be dominated by imminent devastation of the thin and poor mountain pastures of the central Appennine type. These, according to the data of the agricultural cadastre of 1929, extended over a surface of 734,000 hectares in the continental South (including Lazio, whose pastures were largely of this type), and of 308,000 hectares in the Islands, for a total of 1,042,000 hectares. The problem, in general, as compared with the Alpine region where hay meadows have an important role in supplying fodder, was thin pastures, whose scarce productivity in the South is due not only to less favorable natural conditions, but also to the serious degradation and erosion of the mountainous terrain because of unplanned deforestation, continual overgrazing by livestock, and an insufficient policy of soil conservation. The serious results of this in recent years have struck even in the plain, with frequent, sudden, and deadly floods. In the plains and hills, otherwise—more than is now true of northern and most of central Italy—the South and Islands still have considerable extents of a landscape of plain and hill pastures that at the time of the agricultural cadastre of 1929 covered a surface of 242,000 hectares in the Tuscan and Roman Maremmas (which also largely participated in this type of southern landscape), 607,000 hectares in the continental South, and 1,140,000 hectares in the Islands, for a total of 1,990,000 hectares. While in the North, in fact, out of 913,000 hectares of permanent meadows and pastures that were surveyed in the cadastre of 1929 for zones of the plains and hills, not less than 589,000 hectares consisted of permanent meadows. In the South and the Islands, by contrast, of the 1,990,000 hectares just mentioned, the richest and most productive permanent meadows occupied at most 23,000 hectares, so that precisely the thin and desolate pastures of the southern type are what make the most decisive impression on the landscape.

When one remembers that, in 1929, the agricultural cadastre surveyed, for all of Italy, at least 1,200,000 hectares of sowable land "in repose," at most 16,000 hectares were located in northern Italy, and the remainder exclusively in the southern continent (370,000 hectares) and in the Islands (491,000 hectares), with 323,000 hectares more in central Italy (in the Tuscan and Roman Maremmas). And when one remembers how and to what degree this landscape of sowable land "in repose" was found in the "landscape of the traditional latifundium," it is easy to understand what a great importance this type of naked and desolate landscape retained in a large part of Italy still in 1929 and after. Nor can this significant presence be measured only by adding the territory in permanent pastures of the Mediterranean type to that of sowed land "in repose." The persistent and wide diffusion of this last type, which has remained practically unchanged since the period of the great struggles of southern peasants in the postwar period, indicates, in fact, the persistence on a surface at least twice that of the "reposed" lands themselves, of agricultural systems, like those of bare fallow or even temporary clearings. Thus one is not far from the truth in saying that, almost to our own days, the landscape of the traditional public or private latifundium extended over the plains, hills, and mountains of the southern continent and Islands, and in the Tuscan and Roman Maremmas, for a total extent of not less than 5 million hectares.

Up until nearly 1950, however, if one can speak of a progressive reduction of this territory with respect to 1929, the reduction is to be attributed, it seems, more to clearings and new cultivation of significant areas that had been used for permanent pasture, than to an elimination of land "in repose," that is, by abandoning the system of temporary clearings, or bare fallow in favor of a system of continuous rotation. In the continental South (including Lazio) and the Islands, in effect, the area used for meadows and permanent pasture passed from 2,968,000 hectares in 1929, to 2,825,000 in 1936, and 2,595,000 in 1950. In this last year, when the official statistics still reveal an area of sowable land "in repose" that was substantially unchanged since 1929, the area of permanent pastures appears instead, in comparison with 1929, already reduced by 373,000 hectares. One could say that, for 1950, the territory of sowable land "in repose" was a bit exaggerated by the official statistics, which were probably limited to repeating mechanically the figures provided by the agricultural cadastre. It does not seem, in fact, that the first struggles for land in the postwar years, which increased after the Gullo decrees, were, even here, without positive results. But it is clear that the pace of dismemberment of the old agricultural system quickened, and came to intrude more deeply and organically into the forms of the agricultural landscape only through the further development of pressure for agricultural reform, and then through the, however timid, initiatives of reform legislation. A reduction of the area destined for pasturage thus continued after 1950 in the southern continent (including Lazio) and the Islands. This passed to 2,535,000 hectares in 1953, and to 2,497,000 in 1957, with a decrease, at this second date, of another 98,000 hectares with respect to 1950, and on the whole of 471,000 hectares with respect to 1929. Now, much more than in the preceding years, the traditional landscape of the latifundium, which the clearing of individual plots that were still subject to an alternate cultivation of grain with bare fallow had barely changed, saw its predominance definitively reduced, in the zones affected by reforms and beyond, by the extension of rotated meadows and new plantations of trees and shrubs, by the more exact designation of the network of plots and farm roads, and by the emergence of new human settlements.

We have not failed to give needed emphasis to the severe limits that specific economic, social, and political agents imposed on this progressive development. But the fact remains that it was precisely this development, directly induced in our rural society by the struggle of peasants for land and reforms, that most deeply and broadly affected the structure and forms of the agricultural landscape in the postwar period. It indicates a reorientation—which, in some ways, was the most important change—of direction for further developments, in line with the more modern needs of technology, economics, and society.

But, to continue our summary review of the most characteristic forms of the contemporary Italian agricultural landscape, what forms, precisely, have come to substitute, in the South, for the formless desolation of the latifundium? One can indicate that, as in the past, a mixed cultivation of vines with trees, so typical of the Po valley and the region of Tuscany, Umbria, and the Marche, continued to be much less diffused in the South and Islands. It was practically limited to Lazio and the Abruzzi (where sometimes its form repeats the *alberata* of Tuscany, Umbria, and the Marche), not to mention the Campania of Naples. But this landscape of "southern plantations" differed generally from that of the *piantata* of the Po or the *alberata* of central Italy, because of the lesser importance of hydraulic arrangements, as well as the often irregular form of fields, as is the case, for instance, in the countryside around Frattamaggiore, in the hemp-growing zone of the Campania of Naples, where the irregular fields are traversed with long festoons of vines, held rather high by their supporting poplars.

According to the data from the agricultural cadastre of 1929, this landscape of plantations in the South extended over a surface of 117,000 hectares in Lazio, 29,000 hectares in the Abruzzi, and 77,000 in the Campania of Naples, out of a total of 223,000 hectares, and it does not seem to have increased much in subsequent decades. More intensive, instead—and even often decisive for the transformation of the agricultural landscape in various parts of the continental South and Islands—was the dynamics in these years of zones devoted to the growing of specialized trees and shrubs, which in the southern peninsula (including Lazio) and the Islands increased from 1,137,000 hectares in 1907–10 to 1,750,000 in 1929, 1,908,000 in 1953, and 1,986,000 in 1957. As one sees, the spread of cultivation of specialized trees and shrubs involved in these regions, between 1907 and 1929, an area of more than 600,000 hectares; and, after having undergone a certain slackening in the prewar and war years, it regained a more rapid momentum, leading, in 1957, to an increase of the area devoted to cultivation of specialized trees and shrubs by 78,000 hectares with respect to 1953, and 236,000 with respect to 1929. But how did this cultivation of trees and shrubs express itself in the forms and types of the agricultural landscape that we have seen develop in the South from the Risorgimento onward, and with an accelerated pace after its integration into united Italy?

A first type of landscape, in which the cultivation of specialized trees and shrubs expressed itself particularly after the turn of the century, was the low-growing vineyard, generally without supports or with dry supports, and which thus differed from the landscape of the *piantata* or *alberata*, in its lack of live supports. From this landscape of the vineyard, as it is called simply, where the specialized cultivation of vines predominates, mixed cultivation with other crops is not always excluded (especially in central Italy). According to the data from the agricultural cadastre of 1929, the area with this type of landscape—that is, specialized cultivation of vines, or mixed cultivation of vines without living supports—was distributed over 658,000 hectares in northern Italy, 187,000 in central Italy (excluding Lazio), 78,000 in Lazio, 422,000 in the continental South, and 227,000 in the Islands: a total, for the South in general (including Lazio) of 727,000 hectares, which was located in large zones of the plain and on the slopes of hills. In the North, by contrast, the landscape of vineyards was exclusively limited to hill zones, where it sometimes still predominates, as in Piedmont; while in the plain, generally everywhere in the North, the landscape of the *piantata* has and continues to prevail. The northern hill vineyards (and sometimes even mountainous ones) often took the typical form, as we have already indicated, of irregular fields (*campi a pigola*), arrangements in banks or terraces, or elaborate works of construction, like the famous ones at Carema in the Val d'Aosta. This, however, generally involved relatively modest parcels of land, as to size. And the same can be said, often, for the vineyards of central Italy, which were also almost exclusively limited to hilly and mountainous zones, while in the plain the landscape of the *alberata* has clearly predominated. Tuscany, however— with its 14,000 hectares of specialized cultivation and 101,000 of mixed cultivation without live supports—is among the regions of central Italy (excluding Lazio) where the landscape of vineyards has the most importance. Here there are proper plantations of vineyards, with quite large pieces of land that often have quite elaborate hillside arrangements, like that *a spina* of the Testaferrata Podere at Meleto in the Val d'Elsa; while, even in smaller plots, arrangements in terraces, banks, plowing across the summit of the hill, and so on, are quite frequent, although, as we have already warned, lack of statistical data about such arrangements makes measurement of their precise extent impossible.

In contrast with the regions of central Italy, and above all of the North, the landscape of vineyards has expanded in the South and the Islands, not only in the hills, but also over vast zones of the plain. It is precisely here, in fact, that great plantations of vineyards are to be found, which are managed, as in some parts of Puglia, with the methods of great capitalistic agricultural enterprises employing salaried workers. In other parts of Puglia (for example, around Lecce in the region of Bari), and in many other provinces of the South and Islands, the landscape of the vineyards is fragmented into medium, small, and very small plots, corresponding to enterprises of analogous dimensions, whose forms tend toward the typical landscape of the Mediterranean garden. Developments since 1929, however, do not indicate qualitative changes in the regional distribution of the various types of landscapes with vineyards, in comparison with the situation documented by the agricultural cadastre. Quantitative changes are the most evident. Whereas in the Center and North the area used for the specialized cultivation of vines remained practically unchanged between 1936 and 1939 (185,000 hectares) and 1957 and 1958 (188,000 hectares), in the South and the Islands, by contrast, this passed from 637,000 to 683,000, with an increase, during the same period, of 46,000 hectares. A large part of this area has been developed through the tenacious labor of those to whom the land was assigned from the desolate landscape of former latifundia.

Even, and particularly, in the South, otherwise—where still recently a system of open fields has preserved a certain importance, now unknown in the Center and North—the landscape of vineyards is almost always a landscape of closed fields. It has been, in fact, in the small plots of southern vineyards, and also in the typical landscape of the Mediterranean garden, that the closing off of plots has been elaborated in particularly typical ways. These, as we have already observed, with their frequent houses or rural buildings, often give the landscape an almost suburban appearance, as we see near Anacapri, in the fine painting by Guttuso reproduced in Plate 80.

Even more than in the smaller-sized plots of southern vineyards, however, the typical contours of the Mediterranean garden appear again in citrus groves, or in the enclosed plots used, like the one in Plate 80, for the mixed cultivation of particular types of trees (olives, figs, apricots, almonds, and so on). The statistical information available does not allow us to differentiate exactly the extent of this landscape of Mediterranean gardens from that of "southern plantations," which are characterized by cultivation of the same types of trees, placed however in more regular spatial alignments over a larger area. As an index of the expansion of the Mediterranean garden, however, we can usefully take the surface areas where the cultivation of citrus fruits and also figs—a typical tree in the Mediterranean garden—was practiced in mixed cultivation with other trees. In contrast to other no less typical trees (like olives and

Plate 80. *Anacapri* by Renato Guttuso.

almonds, for instance), the cultivation of figs is, in fact, seldom practiced in great plantations, although figs are never missing from a Mediterranean garden.

According to the information provided by the agricultural cadastre of 1929, the typical area of the landscape of the Mediterranean garden, thus defined, extended apparently over 512,000 hectares, with at most 7,000 hectares in northern Italy, 11,000 in the Center (excluding Lazio), 61,000 in Lazio, 271,000 in the southern peninsula, and 162,000 in the Islands. This has always been, and still is, a landscape almost exclusively typical of the South, where still in 1929 a total of 494,000 hectares can be assigned to it, with 125,000 hectares in citrus fruits and 369,000 in mixed cultivation of trees with figs. The available statistical data do not allow us to tell precisely, after 1929, what direction the extension of the Mediterranean garden as defined in this way has developed. Some indications, however, show that, in the continental South (including Lazio) and the Islands, the area used only for citrus fruits (in specialized or mixed cultivation) passed from 125,000 hectares in 1929 to 143,000 in 1957, with a further increase of 18,000 hectares. When one keeps in mind, as well, the fact that mixed cultivation of figs probably decreased rather than expanding in this period, one seems forced to the conclusion that the traditional landscape of the Mediterranean garden has not showed much dynamism in recent decades. With its contours that were complicated, often, by hillside arrangements of terraces, or works of construction, like those of the coast near Amalfi or along the eastern and northern coasts of Sicily, this type of landscape has continued, however, to stamp its characteristic form on important zones of the South, while almost everywhere in the southern provinces at least some aspect of the Mediterranean garden contributes an integral element, if not more, to the dominant landscape.

Less exclusive, although also widely diffused in the provinces of the South and Islands, is the landscape of the "southern plantation," now dominated by olives, almonds, and hazelnuts: as we have already indicated—since their spread into some northern provinces—we can no longer consider other fruit trees to be exclusively typical of the South. For plantations of olives, however, the South has preserved and accentuated its old preeminence. According to the agricultural cadastre of 1929, out of 817,000 hectares utilized throughout Italy for the specialized cultivation of olive groves, at most 55,000 were in northern Italy and 51,000 in central Italy (excluding Lazio), while Lazio (with 76,000), the peninsular South (with 527,000), and the Islands (with 108,000 hectares) accounted as a whole for at least 711,000 hectares. These passed to 790,000 in 1957, with an increase of 79,000 hectares in these southern regions, as compared with 10,000 in the central ones (excluding Lazio), which increased to 61,000 hectares, and a diminution of 12,000 in the north, which passed to 43,000.

But this is not all. While in the North, and in part central Italy, the mixed cultivation of olives, rather than promoting a landscape of true plantations, has been characterized often instead by the atypical *piantata* or by disordered mixed cultivation, in the south, on the contrary, the area of mixed cultivation of olives generally has taken, precisely, the form of the typical landscape of "southern plantations." A mixed, rather than specialized, cultivation of olives has become more concentrated in the southern provinces in recent decades. This fell, in fact, between 1929 and 1957, from 41,000 to 38,000 hectares in the North and from 401,000 to 176,000 in the regions of central Italy, excluding Lazio, but increased from 392,000 to 654,000 in the continental South and from 172,000 to 279,000 hectares in the Islands, passing—in the southern provinces generally—from 619,000 to 1,010,000 hectares.

This, as one sees, is another fundamental way in which the traditional desolate and naked landscape of the latifundium has changed in recent years. And the preeminence that this particular way of cultivating olives has assumed over, let us say, the more modern plantations of other types of fruit trees—which dominate, as we have seen, in the North—is evidence again of the limits that the contradictory agents, which we have repeatedly indicated, have opposed to decisive progress in the South. One must repeat and further emphasize what we have already said with regard to how these obstacles—for instance, the lack of a policy promoting hydraulic arrangements for irrigation and other factors—have opposed the full effectiveness of rotated meadows, which the struggle and labor of peasants has extended over the lands of former fiefs. Analogous considerations would be needed to explain the relatively little progress, for instance, of another type of landscape: the "southern plantations" of shelled nuts (with the two distinct subtypes of almonds and hazelnuts), which are nonetheless significant in the general picture of the South, where they occupied, in 1957, an area of 207,000 hectares: barely an increase over the 197,000 hectares in 1936. But the limits of space imposed on us, and those of the patience of readers, discourage further expansion, in this concluding chapter of a historical inquiry, beyond the limits of a more than summary survey of the current forms of the Italian agricultural landscape. The present and future of this landscape are entrusted not so much to an inquiry, of whatever depth, that might easily exhaust itself at a theoretical level, but rather to the practical activity of millions of women and men, who are struggling to further the life and civil progress of our countryside, and of the whole national society of Italy.

GLOSSARY

• • • • •

This glossary contains some agricultural terms used in Sereni's text, with their English equivalents or approximate definitions.

alberata. Land in mixed cultivation planted with trees festooned with vines in the denser manner of Tuscany, Umbria, and the Marche.
baulatura. Convex grading of earth in a field to correct for *scolmatura*—heaping of earth along the edges of the field.
bracciante. Agricultural wage day laborer.
campi a pìgola. Arrangement of land in irregularly shaped fields.
campi ed erba. System of temporary clearings typical of primitive agriculture, where the land is allowed to revert to natural vegetation or extended fallow after a few crops when its initial fertility diminishes.
cascina. Central administrative unit of a large capitalist estate, typical of the Po valley.
castra, curtes, massae. Early medieval fortified villas; sometimes small communities.
cavalcapoggio. Method of plowing with furrows across the summit of a hill.
cavaletto. An elongated field with *baulatura*, headed by an embankment and framed with trees, typical of nineteenth-century improvement in the Po valley.
cavedagna. A drainage ditch with its side-long embankment that could serve for drainage, a path, and the boundary of a field. Generally translated as "embankment."
censo. Rent.
centuriatio. Centuriation—the layout of square measurements of land with sides of approximately 710 meters (*centuria*) used by Roman surveyors.
ciglioni; ciglionamento. Banks; hillside systematization of land for planting in banks.
colmata. Landfill executed by sedimentation of flowing water. (*Colmata di piano*: landfill in the plain; *colmata di monte*: landfill in the hills.)
colonie perpetue. Land held by peasants with a traditional permanent right of settlement in the south of Italy.
colono. A peasant, who may be subject to servile, feudal, or communal obligations.
colono parziario. A sharecropper.
contado. Countryside.
debbio. A burned-over area in primitive agriculture, typical of heaths and woodlands, that is cultivated until the initial fertility of the soil diminishes and it is allowed to return to natural vegetation.
difesa. A preserve of land for the exclusive use of the lord in a feudal domain.
enfiteutico. Enfiteutic lease—a long-term lease that was generally free of feudal obligations.

fasce. In terracing, small furrows along the side of a hill.
fattore. The estate agent of a *fattoria*.
fattoria. A traditional estate made up of scattered individual farms (*poderi*), typical of Tuscany.
fideicomissium. An entailed estate.
girapoggio. Method of plowing with horizontal furrows along the slope of a hill.
gradoni. Graded arrangement for hillside cultivation.
jus coloniae. A right to settle land in customary practice.
jus pascendi. A right of pasturage on land in customary practice.
jus serendi. A right to sow land in customary practice.
larga: A large flat improved zone in the open plain, typical of contemporary Emilia.
latifundium. A great landed estate of traditional type with primitive agriculture and a semiservile peasant work force, typical of the Italian South.
livello, livellario. Long-term lease, leaseholder.
lunette. Lunettes—small embankments around trees or shrubs in mountainous cultivation.
maggese. Fallow land that will be sowed in a coming season.
magolato. Landscape arranged annually in *porche*.
malga. An Alpine pasture.
marrasco. Winter wheat.
mezzadria; mezzadro. Sharecropping; sharecropper.
palo secco. Dead-wood stakes for vines, and low-down cultivation of vines by this method.
parziario. A peasant occupying land as a sharecropper.
piantata. Land in mixed cultivation planted with trees festooned with vines in the more open manner of the Po valley.
pigionali. Agricultural wage day laborers and renters.
pìgola [or *spigola*]. Arrangement of land in irregularly shaped fields.
podere. A farm.
porche. Arrangement of fields with narrow elongated beds, or hummocks (*porche*), separated by shallow drainage furrows.
prese, or *prace*. A field or clearing of land for cultivation.
prode; approdata. Arrangement of land in elongated fields separated by drainage ditches, and with a single (or double [*a rivale*]) row of trees with vines along the center or sides of the field.
ripiano. A level plot of land for cultivation on a hillside.
risaia. A rice field.
risone. Unhusked rice.
rittochino. Method of plowing land with furrows running straight down the slope of a hill.
rotazione continua. System of continuous rotation with crop courses.
saltus. Land reserved for pasturage in Roman practice.
seminativo alborato. Sowable land in mixed cultivation with trees.
seminativo nudo. Sowable land without trees.
seminativo vitato. Sowable land in mixed cultivation with vines.

solco acquaio. Small drainage furrow or channel.

spina. Method of plowing on hillsides with furrows at intersecting angles.

starza. A preserve of land for the exclusive use of the lord in a feudal domain.

tagliapoggio. Method of plowing with horizontal furrows following the changing irregular contour of a hillside.

terraticanti. Peasants occupying land with precarious tenure and a land rent in kind, typical of southern Italy.

terrazze. Terraces—stepwise arrangement of plantings along a hillside.

tralcio lungo. Long vine shoot.

transumanza. Transhumance—method of pasturage in which the flocks or herds are kept in pastures in the mountains during the summer and in pastures in the plains during the winter.

traverso. In plowing, crosswise to the slope.

usi civici. Common communal rights of pasturage, sowing, wood gathering, and so on, in customary practice.

vitata. Land planted with vines.

INDEX

• • • • •

Accademia dei Georgofili, 274
agricultural revolution of the 18th–19th centuries, 266–68, 337–39
agronomists, 134–35, 253–56, 267–68
Alamanni, Luigi, 129–30, 134, 141, 158, 165, 173, 277
Alaric, 47
Albani, Francesco, 200
Alberti, Leandro, 136, 138, 158, 174, 176–77
Alberti, Leon Battista, 149
alfalfa, 127
Ambrose, Saint, 55
Andrea di Bartolo, 107
Angelico, Fra, 89, 123
Angiolieri, Cecco, 98
aqueducts, 31
Arabs, 70, 72
Ariosto, Ludovico, 187
artificial pastures, 127, 130, 133–34, 150, 192, 219, 364–65, 370
Athena Polias (temple of), 21–22

barbarian invasions, 47, 58, 62, 65, 74, 236
Barna Senese, 162–63
Bartolo di Fredi, 107
Battista da Vicenza, 163–64
Beaurain, Jean, chevalier de, 217
Bellini, Giovanni, 151
Berti-Pichat, Carlo, 300–301
Bianchini, 284
Blaeu Mortier, 214
Bloch, Marc, 3, 5–7, 10
Boccaccio, Giovanni, 13, 163–64
Bologna, 301, 344
Borgonzoni, Aldo, 359, 364
Bosio, De, 309
Botticelli, Sandro, 140
Brosses, Charles de, 217
Byzantines, 53, 57, 65, 72, 86

cadastres, 296, 306, 315, 319, 367
Calabria, 235, 324
canals, 130, 307, 309
capital investment, 121, 149, 228, 232, 234–35, 238, 243, 266–67, 271, 272, 274, 304, 341, 357
Carducci, Giosue, 276
Carolingians, 74, 83
Carpaccio, Vittore, 127
Cassandro, Giovanni, 4
Cassa per il Mezzogiorno, 353–54
Cassiodorus, 60
Cato the Elder, 38, 40
Cattaneo, Carlo, Count, 306–8
Cavour, Camillo Benso, Count, 308–9
cereals, 60–61; Indian corn, 180–81, 183, 208, 220, 275, 300; wheat, 113–14, 191, 235, 239, 243, 282, 287, 293, 312, 337–38, 355
Champier, 10
Charles the Bald (Emperor), 74
Chevallier, 10
Chiarenti, Francesco, 256
Christian Democratic Party, 356
Cistercians, 81, 84, 92
cities: ancient, 21–22, 49, 55, 65, 67; hill towns, 62–63; medieval communes, 90, 101, 103
Claudio Isidoro, 42
Clemente, Africo, 135, 137, 169
Coleman, Enrico, 285, 325
colonization, 27, 83
Columbus, Christopher, 180
communal rights, 202, 236–37, 273, 280–84, 321; *jus coloniae*, 179, 203, 212, 283; *jus pascendi*, 45, 280, *jus serendi*, 45, 203–4, 280
Cremona, 262–63, 311
Crescenzi, Pietro de', 92, 94–96, 99, 103, 113–14, 121, 134, 157, 256
crop rotation, 35, 181, 208, 220, 266, 274–75, 306, 371
Cuppari, Pietro, 6
Curis, Giovanni, 4

Davanzati, Bernardo, 135, 214
day laborers, 270–71, 274, 287, 311, 333, 340, 350
deforestation, 110, 154, 157–58, 242–46, 324, 373

De Francisci, Piero, 10
D'Elia, 10
Desplanques, Henri, 10
Diodorus Siculus, 21
Dionisotti, 305
Dogana di Puglia, 84, 192, 287
Domenico Veneziano, 145–46
drainage, 81, 84, 92, 248–49, 336
Duccio di Buoninsegna, 93, 94, 160, 162

ecclesiastical properties, 78, 221, 223
Einaudi, Luigi, 334
Emilia, 207, 267, 298, 302, 310, 337, 338–41, 344–45, 359, 364
enclosure, 126, 129–30, 191, 239
Engels, Frederick, 331
erosion, 143, 157, 158, 160, 245, 246, 247
Etruscans, 17–18, 24–27, 29, 206
export, 294

fallow, 18, 35, 44–45, 113, 349, 350–52, 373–74
Fanfani, Ammintore, 356–57
Fascism, 333, 340, 345
Fattori, 286
fencing, 127
Ferdinando d'Aragona (king of Naples), 126
Ferrara, 295–96, 301, 336, 344
fertilizer, 124–27, 306
feudal system, 74–75, 90–91, 94, 280–81; refeudalization, 194–96, 199, 206, 221, 248; suppression of, in the South, 282–84, 321–23
field layout, 21, 136–37, 141, 214–15, 217, 237, 275, 299–302, 319; *campi a pigola*, 6, 119–20, 175, 237, 334–35, 355, 376; closed fields, 5, 22, 35, 70, 202, 237, 240, 323–24; open fields, 6, 202, 236–38, 280, 284; *porche*, 122–24, 137
Fiesole, 90, 164
Florence, 90, 110, 267
Folegno, Teofilo, 137, 150–51, 188
Folgòre da San Gimignano, 89, 93
forage, 37, 114, 125, 127, 129, 213, 236, 266, 275, 282, 330, 339, 361, 371–72
Fossombroni, Vittorio, Count, 249–50
Franks, 74–75

Frederick II (emperor), 84
fruit trees, 261, 338, 366, 377; citrus fruits, 72–73, 178, 211, 294, 366, 377; figs, 377–78

Galilei, Galileo, 187
Gallo, Agostino, 135–36, 138, 141, 169
Gallo, Elio, 42
Gambi, Lucio, 4, 10
garden vegetables, 70–72, 180
Gazzoni, Enzo, 364
Genovesi, Averardo, 276–77
Gentile da Fabriano, 114, 116, 119, 160
Gera, Francesco Agostino, 257, 269, 277
Giotto, 63, 107, 162
Giovannetti, 308
Giulio Romano, 187
Goethe, Johann Wolfgang von, 29, 32, 181, 219, 227
Gonzaga family, 170, 196
Gozzoli, Benozzo, 13, 140, 143, 145, 196
Gramsci, Antonio, 101
Greeks, 17–18, 22, 24–25, 27, 29
Gregory the Great (pope), 55
Grieco, Ruggieri, 350
Guercino, Giovanni Francesco Barberi, called, 207
Guicciardini, Francesco, 177
Gullo, Fausto, 350
Guttoso, Renato, 329, 350

hemp, 207, 220, 300, 338
Hippodamus of Miletus, 21
Horace, 41, 149, 197
hunting, 58–59, 111–12

Imberciadori, Ildobrando, 204
industrialization, 294, 355–56, 361–62
irrigation, 130, 133, 295, 299, 304, 306–7, 309–11, 339, 343, 366–67
Italian mode of development, 230–31, 268, 273–75, 314, 331–32

Jones, Philip, 11
Joseph Bonaparte (king of Naples), 282, 287

La Lande, Joseph Jérôme Le Français de, 218–19, 261
Landeschi, Giovan Battista, 253, 255–58, 277

landfills (*colmate*), 247–50, 276–78
large landholdings: *cascine*, 269, 271; *fattorie*, 229–30, 273–74, 276; *larghe*, 341–45; *latifundia*, 55, 177–79, 349–52, 375. *See also* villas
Lastri, Marco, 253, 255–56
Lautrec (Odet de Foix, seigneur de), 194
Le Lannou, Maurice, 4
Lenin, Nikolaj, 230, 331
Leonardo da Vinci, 131, 140–42, 157, 187, 200, 248, 276
Leopoldo II of Hapsburg-Lorraine (grand duke of Tuscany), 286
Liguria, 165, 168, 228, 254, 297, 365
linen, 207, 220
livestock: dairy farming, 150–51; draft animals, 104, 292; free-range breeding, 58, 60, 84–85, 239, 285–86; goats, 192; herding, 84, 107–8, 126, 151–52; horses, 84–85; sheep, 84–85, 126, 146, 152, 154, 190–92, 287
Lombards, 65, 74–75
Lombardy, 133, 228, 254, 264, 267, 297–98, 305–11, 337, 340, 359, 364–65
Lorenzetti, Pietro, 99, 101, 103–4, 106–8, 119, 160, 162
Lorenzi, Bartolommeo, 241–42, 247, 252, 254
Lucca, 164

Magnasco, Alessandro, 200, 223
Mantegna, Andrea, 171
Mantova, 302
Marche, 72, 136, 213–14, 217, 251, 261–62, 274–75, 280, 314–15, 318, 359, 367–68
Maremma of Siena, 192, 206, 210, 213, 240, 248, 286–87, 349
markets, 190, 292
marshlands, 187–88, 189, 248, 296, 336
Martini, Simone, 75, 77, 162
Marx, Karl, 126, 183, 207, 271
Masaniello revolt, 195
Masi, 4
Medici, Cosimo I de' (grand duke of Tuscany), 195
Medici, Ferdinando I de' (grand duke of Tuscany), 148
Medici, Lorenzo de' (the Magnificent), 120
Mediterranean garden, 6, 22, 36, 63, 73, 176–78, 189, 210–12, 214, 329, 377–79
Melissa, 349
Milan, 267, 291, 306
milling, 104
Mitterpacher, Ludwig, 264
mixed cultivation of vines and trees, 24–26, 63; *alberata*, 72, 136–38, 213–14, 217, 251, 261–62, 275–76, 315–19, 359, 368–69, 375; *piantata*, 72, 97, 136–39, 217–19, 237, 261–64, 295–98, 303, 344, 359–62, 364, 368, 375. *See also* viticulture
Montaigne, Michel de, 136, 138, 164
Montepulciano, 174–75
More, Sir Thomas, 191
mountainous regions, 5, 154, 157, 241–42, 246, 253
mulberry trees, 72, 134, 138, 214, 219, 261, 264, 276, 298, 362

Naples, 242, 291
Nardò, 177–79
Novara, 263–64, 306, 364
Nuccoli, Ser Cecco, 114
nut trees, 211, 327, 377, 380

olive trees, 99, 178, 261, 327, 379
Ortolani, 4

Pajakowski, 10
Palermo, 211
Palizzi, Filippo, 329
Pane, Luigi Dal, 4
Parain, Charles, 10
pasturage, 35, 37, 42, 44, 61, 63, 84, 125–26, 134, 155, 191, 202, 236–37, 239–40, 365, 370–72; *saltus*, 42–45
Piedmont, 188, 219, 228, 235, 264, 267, 297–98, 309–11, 334–35, 337, 359, 365
Pier Francesco Fiorentino, 167
Pietro Leopoldo of Hapsburg-Lorraine (grand duke of Tuscany), 239–40, 242, 248
Pindemonte, Ippolito, 253
Plaisance, Georges, 6
plantation of trees, 36, 40, 63, 78, 213, 251, 261, 316, 325, 367, 379
Pliny the Elder, 25, 42
plowing methods, 103, 119, 121, 173, 252, 256–58, 277, 279, 370

population trends, 80, 92, 251, 332, 355
potatoes, 180, 266
Poussin, Nicolas, 187, 191, 325
Po valley, 24–25, 32, 39, 72, 81, 97, 130–37, 140–41, 150, 181, 187, 189–90, 208, 214, 217–19, 232, 234–35, 261–62, 264–69, 272, 274, 280, 295, 297–306, 311–13, 337, 359–61, 364–65, 368, 371
Prato, Giuseppe, 234–35
price trends, 180, 190
Priuli, Girolamo, 149
productivity of agriculture, 312
Puglia, 287, 377
Pugliese, Salvatore, 234–35, 262

railroads, 291–94
Ramusio, Giambattista, 181
Ravenna, 336
Re, Filippo, 255, 300
Redi, Francesco, 174
regional specialization, 207–8, 292–94
Ricchioni, Vincenzo, 4
rice, 72, 134, 210, 219, 235, 264, 304–6, 364
Ridolfi, Cosimo, Marchese, 6, 124, 276–79
road building, 29, 171
Roger II (king of Sicily), 84
Romagna, 302, 344
Roman campagna, 85, 187, 191–92, 203, 210, 240, 285, 322, 324–25
Romans, 27, 29, 32, 35, 38, 44, 47, 63, 85, 86
Rome, 67
Rosa, Salvator, 200
Rozier, François, 268
rural unrest, 311, 340, 349–53, 356

Sacchetti, Franco, 107
Sardinia, 282, 324
Scelsi, Giacinto, 296
Sforza, Ludovico (duke of Milan), 130
Sforza family, 130–31
sharecropping, 120, 195, 203, 232, 234–35, 266, 268–69, 271, 274, 295, 318
shepherds, 85–86, 104, 107–8
Sicily, 72, 165, 168, 203, 210, 221, 228, 280, 282, 324
Siena, 103, 164
Signorini, Telemaco, 327
silk, 72, 210, 214, 264, 298, 362

Sismondi, Jean Charles Léonard Sismonde de, 255, 275
slash and burn agriculture, 17, 60
Soderini, Piero, 135
Southern Italy, 72, 109, 152, 177, 190, 194–95, 203, 210, 221, 240, 254, 280–84, 321–24, 329, 366, 371
Southern question, 109, 193, 329
Spolverini, Giovanni Battista, Marchese, 246–47
sugar beets, 338
surveying, 21; Roman centuriation, 27, 29, 32
systematization of soil and water, 81, 92, 95, 99, 113, 130–31, 136, 162, 173, 217, 252, 254, 256, 302, 341. *See also* landfills (*colmate*)

Tanaglia, Michelangiolo, 120, 123, 125, 127, 129–30, 134
Tansillo, Luigi, 149
Tarello, Camillo, 135, 181
Targioni-Tozzetti, Ottaviano, 261
Tavola di Alesa, 22
Tavola di Eraclea, 21–22
technology. *See* agronomists; field layout; plowing methods; systematization of soil and water
tenant farmers, 234, 266, 269, 271, 281–82, 333
terracing, 99, 161–68, 252, 254–55, 325, 327
Testaferrata (*fattore*), 253, 276–78
Thurii, 21–22
Tiepolo, Giandomenico, 232, 234
Torelli, Giuseppe, 308–9
Torelli, Pietro, 97
Torricelli, Evangelista, 187
Trajan (emperor), 44
transhumance, 85–86, 152, 287
Treccani, Ernesto, 350
Turati, Filippo, 331
Tuscany, 72, 89–90, 109, 110, 114, 122, 136, 140–43, 147–48, 152, 168, 181, 187, 189, 190, 195, 204, 206, 213–14, 217, 229–30, 232, 240, 242, 251–58, 261–62, 267–68, 273–75, 277, 279–80, 314, 317–19, 359, 367–70

Umbria, 136, 213–14, 217, 251, 261–62, 274–75, 280, 314, 316, 318, 359, 367–68

Varro, 40
Vasari, Giorgio, 158
Vasto, Marchese del, 149
Veneto, 145, 148, 155, 181, 189, 198, 227–28, 243, 253–54, 267, 297, 306, 309–10, 313, 337, 359, 364–65
Venice, 110, 149, 240, 267, 291
Vercelli, 235, 264, 305–6, 311, 364
Vettori, Francesco, 135
Villafranca Veronese, 83, 101
Villari, Rosario, 4
villas: medieval, 55, 57, 58, 78, 89; Renaissance, 147–49, 196–98, 227–28; Roman, 36, 40–41

Virgil, 133, 152–53
viticulture, 96, 121, 145, 178, 216, 345, 377; grape types, 24, 98–99; low-growing vines, 24–25, 39, 63, 376. *See also* mixed cultivation of vines and trees
Viva, Rosina, 327
Volterra, Edoardo, 10

Wittel, Gaspare Van, 228

Young, Arthur, 300, 304

Zangheri, Renato, 4

ABOUT THE AUTHOR AND TRANSLATOR

Emilio Sereni (1907–1971), a Jewish intellectual and an early member of the Italian Communist Party, played an important role in the resistance movement during World War II. After the war, he served as a senator while carrying on his influential work as a historian.

R. Burr Litchfield is Professor of History at Brown University. He is the author of *Emergence of a Bureaucracy: The Florentine Patricians, 1530–1790* and the translator of Franco Venturi's three-volume *The End of the Old Regime in Europe,* both published by Princeton University Press.

Printed by Libri Plureos GmbH in Hamburg, Germany